CAMBRIDGE TRACTS IN MATHEMATICS

General Editors

---

188 Modern Approaches to the Invariant-Subspace Problem

A complete list of books in the series can be found at www.cambridge.org/mathematics.
Recent titles include the following:

150. Harmonic Maps, Conservation Laws and Moving Frames (2nd Edition). By F. HÉLEIN
151. Frobenius Manifolds and Moduli Spaces for Singularities. By C. HERTLING
152. Permutation Group Algorithms. By A. SERESS
153. Abelian Varieties, Theta Functions and the Fourier Transform. By A. POLISHCHUK
154. Finite Packing and Covering. By K. BÖRÖCZKY, JR
155. The Direct Method in Soliton Theory. By R. HIROTA. Edited and translated by A. NAGAI, J. NIMMO, and C. GILSON
156. Harmonic Mappings in the Plane. By P. DUREN
157. Affine Hecke Algebras and Orthogonal Polynomials. By I. G. MACDONALD
158. Quasi-Frobenius Rings. By W. K. NICHOLSON and M. F. YOUSIF
159. The Geometry of Total Curvature on Complete Open Surfaces. By K. SHIOHAMA, T. SHIOYA, and M. TANAKA
160. Approximation by Algebraic Numbers. By Y. BUGEAUD
161. Equivalence and Duality for Module Categories. By R. R. COLBY and K. R. FULLER
162. Lévy Processes in Lie Groups. By M. LIAO
163. Linear and Projective Representations of Symmetric Groups. By A. KLESHCHEV
164. The Covering Property Axiom, CPA. By K. CIESIELSKI and J. PAWLIKOWSKI
165. Projective Differential Geometry Old and New. By V. OVSIENKO and S. TABACHNIKOV
166. The Lévy Laplacian. By M. N. FELLER
167. Poincaré Duality Algebras, Macaulay's Dual Systems, and Steenrod Operations. By D. MEYER and L. SMITH
168. The Cube-A Window to Convex and Discrete Geometry. By C. ZONG
169. Quantum Stochastic Processes and Noncommutative Geometry. By K. B. SINHA and D. GOSWAMI
170. Polynomials and Vanishing Cycles. By M. TIBĂR
171. Orbifolds and Stringy Topology. By A. ADEM, J. LEIDA, and Y. RUAN
172. Rigid Cohomology. By B. LE STUM
173. Enumeration of Finite Groups. By S. R. BLACKBURN, P. M. NEUMANN, and G. VENKATARAMAN
174. Forcing Idealized. By J. ZAPLETAL
175. The Large Sieve and its Applications. By E. KOWALSKI
176. The Monster Group and Majorana Involutions. By A. A. IVANOV
177. A Higher-Dimensional Sieve Method. By H. G. DIAMOND, H. HALBERSTAM, and W. F. GALWAY
178. Analysis in Positive Characteristic. By A. N. KOCHUBEI
179. Dynamics of Linear Operators. By F. BAYART and É. MATHERON
180. Synthetic Geometry of Manifolds. By A. KOCK
181. Totally Positive Matrices. By A. PINKUS
182. Nonlinear Markov Processes and Kinetic Equations. By V. N. KOLOKOLTSOV
183. Period Domains over Finite and p-adic Fields. By J.-F. DAT, S. ORLIK, and M. RAPOPORT
184. Algebraic Theories. By J. ADÁMEK, J. ROSICKÝ, and E. M. VITALE
185. Rigidity in Higher Rank Abelian Group Actions I: Introduction and Cocycle Problem. By A. KATOK and V. NIŢICĂ
186. Dimensions, Embeddings, and Attractors. By J. C. ROBINSON
187. Convexity: An Analytic Viewpoint. By B. SIMON

Modern Approaches to the Invariant-Subspace Problem

ISABELLE CHALENDAR
Université Lyon 1

JONATHAN R. PARTINGTON
University of Leeds

CAMBRIDGE
UNIVERSITY PRESS

CAMBRIDGE UNIVERSITY PRESS
Cambridge, New York, Melbourne, Madrid, Cape Town,
Singapore, São Paulo, Delhi, Mexico City

Cambridge University Press
The Edinburgh Building, Cambridge CB2 8RU, UK

Published in the United States of America by Cambridge University Press, New York

www.cambridge.org
Information on this title: www.cambridge.org/9781107010512

First published 2011

A catalogue record for this publication is available from the British Library

Library of Congress Cataloguing in Publication Data
Chalendar, Isabelle, 1970–
Modern approaches to the invariant subspace problem / Isabelle Chalendar,
Jonathan R. Partington.
p. cm. – (Cambridge tracts in mathematics ; 188)
Includes bibliographical references and index.
ISBN 978-1-107-01051-2 (hardback)
1. Invariant subspaces. 2. Hilbert space. I. Partington, Jonathan R.
(Jonathan Richard), 1955– II. Title. III. Series.
QA322.4.C46 2011
515′.724–dc23
2011019460

ISBN 978-1-107-01051-2 Hardback

Contents

Preface

There is an outstanding problem in operator theory, the so-called 'invariant-subspace problem', which has been open for more than half a century. There have been significant achievements on occasion, sometimes after an interval of more than a decade, but its solution seems nowhere in sight. The invariant-subspace problem for a complex Banach space $\mathcal{X}$ of dimension > 1 concerns whether every bounded linear operator $T : \mathcal{X} \to \mathcal{X}$ has a non-trivial closed T-invariant subspace (a closed linear subspace $\mathcal{M}$ of $\mathcal{X}$ which is different from both $\{0\}$ and $\mathcal{X}$ such that $T(\mathcal{M}) \subset \mathcal{M}$). Throughout this book, when we talk about invariant subspaces, we always assume that they are closed and non-trivial.

For the most important case of Hilbert spaces $\mathcal{H}$ the problem is still open, although Enflo [95, 96] and Read [168, 169] showed that the invariant-subspace problem is false for some Banach spaces.

The general case of the invariant-subspace problem is still open, but there are many positive results in this direction. For example, every finite-rank operator on a non-zero complex space has an eigenvector, and this generates a one-dimensional invariant subspace. Thus the conjecture is easily resolved in the case that the underlying Hilbert space is finite-dimensional. Moreover, every non-zero vector is contained in a smallest invariant subspace, the *cyclic subspace* it generates, which is separable. Thus the question is easily answered for non-separable Hilbert spaces.

Aronszajn and Smith [18] proved that every compact operator on a Banach space of dimension at least 2 has an invariant subspace; the result for Hilbert spaces had already been proved by von Neumann (unpublished work of 1950). It is also known that normal operators have invariant subspaces: this follows easily from the spectral theorem.

In 1966, Bernstein and Robinson [39] proved, using non-standard analysis, that if the operator T on a Hilbert space is polynomially compact (in other words, if $p(T)$ is compact for some non-zero polynomial p) then T has an invariant subspace. Almost immediately, Halmos [117] eliminated the non-standard analysis from their proof and thus obtained a classical proof of this result.

These results were generalized further when Lomonosov [145] gave a very short proof, using the Schauder fixed-point theorem, of a powerful result which implies that if the operator T on a Banach space commutes with a non-zero compact operator then it has a non-trivial invariant subspace (clearly, any operator T commutes with every polynomial in T).

On the other hand, there are now many negative results known in the case of an operator on a Banach space. An example of an operator on a Banach space with no invariant subspaces was found by Enflo [95, 96], and his example was simplified by Beauzamy [28]. Another counterexample was given by Read [168], who later gave the first example on a 'classical' Banach space, namely ℓ^1 [169]. In further work, Read [171] constructed an operator on ℓ^1 without even a non-trivial closed invariant *subset*. We mention also some work of Atzmon [19], who gave an example of an operator without invariant subspaces on a nuclear Fréchet space.

Another very recent example is due to Argyros and Haydon [17], who constructed an infinite-dimensional Banach space such that every bounded operator is the sum of a compact operator and a scalar operator (i.e, a multiple of the identity). Therefore, in particular, every operator on this space has an invariant subspace. Gowers and Maurey [114] had earlier found a space where every bounded operator is the sum of a strictly singular operator and a scalar operator, but since Read [172] had already given an example of a strictly singular operator without invariant subspaces, the Gowers–Maurey example had no direct consequences for the invariant-subspace problem.

Some classic texts which treat the invariant-subspace problem are the books of Radjavi and Rosenthal [167] and Beauzamy [29]. Many other significant results can be found in the books of Herrero [122, 16]. A recent book closely linked to the theme of this book is that of Bayart and Matheron [26].

This book is intended to be suitable for researchers in many areas of operator theory and function theory. It should be accessible to postgraduate students; indeed, we begin with a section establishing the basic background results on Hardy spaces, operator theory, geometry of Banach spaces etc. We omit the proofs, which can be found in many places, for example Rudin [180, 179], Conway [84] and Nikolski [154].

There have been many significant developments in this branch of operator theory. Therefore, it was necessary to be selective in our choice of material. Some themes to be discussed in the book, which are not, as far as we know, to be found elsewhere in books, include the following:

- Use of the operator-valued Poisson kernel, factorizations, H^∞ functional calculus (initiated by S. Brown in a Hilbert space context, as the basis of dual-algebra theory). Very recent Banach space developments due to Ambrozie and Müller.
- Functional calculus for Beurling algebras, the Atzmon–Wermer results. A more transparent presentation of Davie's work on Bishop operators (and his own functional calculus), and subsequent developments.
- The work of Lomonosov and Simonič, applying fixed-point theorems to the study of compact and essentially self-adjoint operators.
- The technique of minimal vectors pioneered by Ansari and Enflo in a Hilbert space context; further developments in the general context of Banach spaces, showing the existence of non-trivial hyperinvariant subspaces in various contexts.
- Universal operators. The study of composition operators and their invariant subspaces as key examples.
- Moment sequences, and the geometrical approach of Atzmon and Godefroy, with striking applications to tridiagonal matrices. Binomial sums and their applications in various areas of functional analysis.
- Particular results applying to special kinds of operators, such as positive (order-preserving) and strictly singular operators.

Chapter 1 contains the basic background material for the book. After this, Chapters 2, 3 and 4 follow a logical sequence and should be read in this order. The other chapters are largely independent and self-contained, with occasional references to one another. For example, compact operators appear in Chapters 6 and 7. Some supplementary results are contained in the Exercises at the end of each chapter.

We are grateful to many people for comments on the manuscript of this book. In particular we mention Nicolas Chevrot, Sam Elliott, Antoine Flattot, Emmanuel Fricain, Eva Gallardo-Gutiérrez, Pamela Gorkin, Nazar Miheisi and Elodie Pozzi. We also appreciate the help given to us by Cambridge University Press, especially Silvia Barbina.

Certainly, this work would not have been possible without the patient support of Patrick and Hélène.

Finally, this book is dedicated to our parents.

1
Background

In this chapter we begin by reviewing the main definitions and theorems from the basic theory of functional analysis, linear operators and geometry of Banach spaces. It is not our intention to summarize the whole of analysis within a few pages, but we do supply the necessary background to the results used later in the book. This material is very standard and likely to be met in any basic course on functional analysis, and so we give just the essentials of the subject, without proofs.

In the last sections of this chapter, we also recall some basic facts of function theory. In particular we discuss the fundamental properties of Hardy spaces, which are Banach spaces of holomorphic functions defined in the unit disc and extended to the unit circle $\mathbb{T}$. We also briefly review the definitions of the disc algebra, functions of bounded mean oscillation, and the Hilbert transform of real functions defined on the unit circle.

1.1 Functional analysis

1.1.1 Weak topology

The term *weak topology* is most commonly used for the topology of a normed vector space or topological vector space induced by its (continuous) dual.

One may call subsets of a topological vector space *weakly closed* (respectively, *compact* etc.) if they are closed (respectively, compact etc.) in the weak topology. Likewise, functions are sometimes called *weakly continuous* (respectively, *differentiable, analytic* etc.) if they are continuous (respectively, differentiable, analytic etc.) in the weak topology.

The strong and weak topologies

Let $\mathcal{X}$ be a topological vector space; then in particular $\mathcal{X}$ is a topological space carrying a topology as part of its definition. (For example, a normed vector space $\mathcal{X}$ is, by using the norm to measure distances, a metric space, and hence also a topological vector space.) This topology is also called the *strong topology* on $\mathcal{X}$.

The weak topology on $\mathcal{X}$ is defined using the continuous dual space $\mathcal{X}^*$. This dual space consists of all linear functions from $\mathcal{X}$ into the base field $\mathbb{R}$ or $\mathbb{C}$ which are continuous with respect to the strong topology. The *weak topology* on $\mathcal{X}$ is the weakest topology (the topology with the fewest open sets) such that all elements of $\mathcal{X}^*$ remain continuous. Explicitly, a sub-base for the weak topology is the collection of sets of the form $\phi^{-1}(\mathcal{U})$ where $\phi \in \mathcal{X}^*$ and $\mathcal{U}$ is an open subset of the base field $\mathbb{R}$ or $\mathbb{C}$. In other words, a subset of $\mathcal{X}$ is open in the weak topology if and only if it can be written as a union of (possibly infinitely many) sets, each of which is an intersection of finitely many sets of the form $\phi^{-1}(\mathcal{U})$.

The weak topology is characterized by the following condition: a net $(x_\lambda)_\lambda$ in $\mathcal{X}$ converges in the weak topology to the element x of $\mathcal{X}$ if and only if $(\phi(x_\lambda))_\lambda$ converges to $\phi(x)$ in $\mathbb{R}$ or $\mathbb{C}$ for all ϕ in $\mathcal{X}^*$.

In particular, if $(x_n)_n$ is a sequence in $\mathcal{X}$, then $(x_n)_n$ converges weakly to x if

$$\phi(x_n) \to \phi(x) \qquad \text{as} \quad n \to \infty,$$

for all ϕ in $\mathcal{X}^*$. In this case, it is customary to write

$$x_n \xrightarrow{w} x$$

or, sometimes,

$$x_n \rightharpoonup x.$$

If $\mathcal{X}$ is equipped with the weak topology, then addition and scalar multiplication remain continuous operations, and $\mathcal{X}$ is a locally convex topological vector space.

The weak* topology

A normed space $\mathcal{X}$ can be embedded into $\mathcal{X}^{**}$ by

$$x \mapsto Tx,$$

where

$$Tx(\phi) = \phi(x).$$

In fact, $T : \mathcal{X} \to \mathcal{X}^{**}$ is an injective linear mapping. In the particular case where T is surjective, one says that $\mathcal{X}$ is *reflexive*.

The *weak* topology* on $\mathcal{X}^*$ is the weak topology induced by the image of $T : \mathcal{X} \to \mathcal{X}^{**}$. A net $(\phi_\lambda)_\lambda$ in $\mathcal{X}^*$ is convergent to ϕ in the weak* topology if it converges pointwise:

$$\phi_\lambda(x) \to \phi(x),$$

for all x in $\mathcal{X}$. In particular, a sequence $(\phi_n)_n \in \mathcal{X}^*$ converges to ϕ provided that

$$\phi_n(x) \to \phi(x),$$

for all x in X. In this case, one writes

$$\phi_n \stackrel{w^*}{\to} \phi$$

as $n \to \infty$.

This weak* topology is sometimes called the topology of *simple convergence* or the topology of *pointwise convergence*. Indeed, it coincides with the topology of pointwise convergence of linear functionals.

By definition, the weak* topology is weaker than the weak topology on $\mathcal{X}^*$. An important fact about the weak* topology is the *Banach–Alaoglu theorem*: if $\mathcal{X}$ is normed, then the unit ball in $\mathcal{X}^*$ is weak*-compact. Moreover, the unit ball in a normed space $\mathcal{X}$ is compact in the weak topology if and only if $\mathcal{X}$ is reflexive.

If a normed space $\mathcal{X}$ is separable, then the weak* topology is metrizable on (norm-)bounded subsets of $\mathcal{X}^*$.

1.1.2 Hahn–Banach theorem

Suppose that $\mathcal{X}$ is a normed space (real or complex). Then the *norm* of a continuous linear map $f : \mathcal{X} \to \mathbb{C}$ (a continuous linear *functional*) is given by

$$\|f\| = \sup\{|f(x)| : \|x\| \le 1\}.$$

Sometimes, for convenience, we shall use the alternative notation $\langle x, f\rangle$ instead of $f(x)$.

The *dual space*, $\mathcal{X}^*$, is the space of linear functionals equipped with the above norm.

Suppose now that $\mathcal{X}$ and $\mathcal{Y}$ are two normed spaces with $\mathcal{X} \subset \mathcal{Y}$. Then an element $g \in \mathcal{Y}^*$ clearly determines a unique element $g_{|\mathcal{X}}$ of $\mathcal{X}^*$ by restricting its action to $\mathcal{X}$. Moreover, $\|g_{|\mathcal{X}}\| \le \|g\|$. The Hahn–Banach theorem is concerned with the converse situation: the extension of a linear functional to a larger normed space. In its most common form it is stated as follows.

Theorem 1.1.1 (Hahn–Banach) *If $\mathcal{X}$ and $\mathcal{Y}$ are normed spaces with $\mathcal{X} \subset \mathcal{Y}$ and $f \in \mathcal{X}^*$, then there exists a functional $\tilde{f} \in \mathcal{Y}^*$ such that $\tilde{f}(x) = f(x)$ for all $x \in \mathcal{X}$, and such that $\|\tilde{f}\|_{\mathcal{Y}^*} = \|f\|_{\mathcal{X}^*}$.*

Another form of the Hahn–Banach theorem is more geometrical.

Definition 1.1.2 A set $\mathcal{S}$ in a normed space is *convex* if, for all $s, t \in \mathcal{S}$, the line segment joining s and t is contained in $\mathcal{S}$, i.e.,

$$\lambda s + (1 - \lambda)t \in \mathcal{S} \quad \text{for all} \quad 0 \le \lambda \le 1.$$

A set $\mathcal{S}$ in a normed space is *absolutely convex* if it is convex and if, in addition,

$$\lambda s \in \mathcal{S} \quad \text{for all} \quad s \in \mathcal{S}, \quad |\lambda| \le 1,$$

λ being real or complex, as appropriate.

The theorem of the separating hyperplane can now be stated.

Theorem 1.1.3 (separating hyperplane theorem) *Let $\mathcal{X}$ be a normed space, $\mathcal{S}$ a closed absolutely convex subset of $\mathcal{X}$, and x a point of $\mathcal{X}$ which is not in $\mathcal{S}$. Then there exists a functional $f \in \mathcal{X}^*$ such that $|f(s)| \le 1$ for all $s \in \mathcal{S}$, and $|f(x)| > 1$.*

A further application of the Hahn–Banach theorem is the following (see Rudin [180, Theorem 3.4]).

Theorem 1.1.4 *Let A and B be disjoint non-empty convex open subsets of a normed space $\mathcal{X}$. Then there are a functional $\Lambda \in \mathcal{X}^*$ and a real number γ such that*

$$\operatorname{Re} \Lambda(x) < \gamma < \operatorname{Re} \Lambda(y)$$

for all $x \in A$ and $y \in B$.

1.1.3 Stone–Weierstrass theorem

Let K be a compact metric space, for example $[0, 1]$ or the unit circle $\mathbb{T}$. We write $C(K, \mathbb{R})$ for the space of continuous real-valued functions on K, and $C(K)$ for the space of continuous complex-valued functions. Each is a normed space over the appropriate field, with the (supremum) norm

$$\|f\| = \sup\{|f(x)| : x \in K\}.$$

The classical *Weierstrass approximation theorem* states that the polynomials are dense in $C([0,1],\mathbb{R})$, that is, that a real continuous function can be uniformly approximated by polynomials on the interval $[0, 1]$.

The Stone–Weierstrass theorem is a generalization of this, and requires us to consider an *algebra* of functions, that is, a set of functions that forms a vector space and is also closed under multiplication. So, for example, the polynomials form an algebra, as do the trigonometric polynomials (polynomials in e^{it} and e^{-it}).

An algebra $\mathcal{A}$ of functions is said to *separate points* if, given any two distinct points $x, y \in K$, there is a function $f \in \mathcal{A}$ such that $f(x) \neq f(y)$. Over the reals we then have the simplest form of the Stone–Weierstrass theorem, as follows.

Theorem 1.1.5 (Stone–Weierstrass over $\mathbb{R}$) *If $\mathcal{A}$ is a real algebra of continuous functions on a compact metric space K, which separates points and contains the constant functions, then $\mathcal{A}$ is dense in $C(K,\mathbb{R})$. In other words, every function in $C(K,\mathbb{R})$ can be approximated arbitrarily closely (in the uniform norm) by functions in $\mathcal{A}$.*

Over the complex numbers the above form of the Stone–Weierstrass theorem does not hold, since, for example, the function $f(z) = \overline{z}$ cannot be approximated arbitrarily closely on $\mathbb{T}$ by polynomials in z, since it is not analytic. However, taking into account this special case, we obtain a complex form of the theorem: it can be deduced from the real form by taking real and imaginary parts.

Theorem 1.1.6 *If $\mathcal{A}$ is a complex algebra of continuous functions on a compact metric space K, which separates points, contains the constant functions, and is closed under complex conjugation, then $\mathcal{A}$ is dense in $C(K)$.*

1.1.4 Banach–Steinhaus theorem

Given two normed spaces $\mathcal{X}$ and $\mathcal{Y}$, we write $\mathcal{L}(\mathcal{X},\mathcal{Y})$ for the space of all bounded linear operators from $\mathcal{X}$ to $\mathcal{Y}$. Moreover, we write $\mathcal{L}(\mathcal{X})$ for $\mathcal{L}(\mathcal{X},\mathcal{X})$.

The Banach–Steinhaus theorem (or *uniform boundedness theorem*) may be stated as follows.

Theorem 1.1.7 *Suppose that $\mathcal{X}$ is a Banach space and $\mathcal{Y}$ a normed space. Then a collection of operators $\mathcal{S} \subset \mathcal{L}(\mathcal{X},\mathcal{Y})$ is* uniformly bounded in norm, *i.e.,*

$$\sup\{\|T\| : T \in \mathcal{S}\} < \infty,$$

if and only if it is pointwise bounded, *that is,*

$$\sup\{\|Tx\| : T \in \mathcal{S}\} < \infty \qquad \textit{for each} \quad x \in \mathcal{X}.$$

We may deduce from Theorem 1.1.7 that, when $\mathcal{X}$ is a Banach space, a sufficient condition for a family $\mathcal{S} \subset \mathcal{L}(\mathcal{X})$ to be uniformly bounded in norm is that

$$\sup\{|\langle Tx, x^*\rangle| : T \in \mathcal{S}\} < \infty, \qquad \text{for each} \quad x \in \mathcal{X} \quad \text{and} \quad x^* \in \mathcal{X}^*.$$

1.1.5 Complex measures

Definition

Definition 1.1.8 We define a *σ-ring* of subsets of a set X to be a non-empty collection Σ of sets closed under taking countable unions $\bigcup_{n=1}^{\infty} A_n$ and complements $A \setminus B$.

A *real positive measure* μ on a *measure space* (X, Σ) is a function

$$\mu : \Sigma \to [0, \infty],$$

defined on a σ-ring Σ, which is *σ-additive*; that is, for any sequence $(A_n)_n$ of disjoint sets in Σ one has

$$\mu\left(\bigcup_{n=1}^{\infty} A_n\right) = \sum_{n=1}^{\infty} \mu(A_n),$$

where the sum on the right may be finite or may diverge to ∞.

We shall also require more general (complex) measures, but here we shall always suppose that they are finite.

Definition 1.1.9 A *finite complex measure* μ on a *measure space* (X, Σ) is a function

$$\mu : \Sigma \to \mathbb{C},$$

defined on a σ-ring Σ, which is σ-additive in the sense that for any sequence $(A_n)_n$ of disjoint sets in Σ one has

$$\mu\left(\bigcup_{n=1}^{\infty} A_n\right) = \sum_{n=1}^{\infty} \mu(A_n),$$

where the sum on the right converges absolutely.

The set of complex measures defined on the unit circle $\mathbb{T}$ will be denoted by $\mathcal{M}(\mathbb{T})$.

Integration with respect to a complex measure

One can define the integral of a complex-valued measurable function with respect to a complex measure in the same way as the Lebesgue integral of a real-valued measurable function with respect to a non-negative measure, by approximating a measurable function with simple functions, beginning with the formula

$$\int_X \left(\sum_{j=1}^{N} a_j \chi_{A_j} \right) d\mu = \sum_{j=1}^{N} a_j \mu(A_j) \quad (a_1, \ldots, a_N \in \mathbb{C}, \quad A_1, \ldots, A_N \in \Sigma),$$

where χ_A is the *characteristic function* or *indicator function* of A, defined by

$$\chi_A(x) = \begin{cases} 1 & \text{if } x \in A, \\ 0 & \text{if } x \in X \setminus A. \end{cases}$$

Just as in the case of ordinary integration, this more general complex-valued integral may fail to exist, or its value may be infinite.

Another approach is to start with the concept of the integral of a real-valued function with respect to a non-negative measure. To that end, it is easy to verify that the real and imaginary parts μ_1 and μ_2 of a complex measure μ are finite-valued signed measures. One can apply the *Hahn–Jordan decomposition* to these measures to split them as

$$\mu_1 = \mu_1^+ - \mu_1^-$$

and

$$\mu_2 = \mu_2^+ - \mu_2^-,$$

where μ_1^+, μ_1^-, μ_2^+, μ_2^- are finite-valued non-negative measures (unique in some sense). Then, for a real-valued measurable function f, one can define

$$\int_X f \, d\mu = \left(\int_X f \, d\mu_1^+ - \int_X f \, d\mu_1^- \right) + i \left(\int_X f \, d\mu_2^+ - \int_X f \, d\mu_2^- \right)$$

as long as the expression on the right-hand side is defined, that is, all four integrals exist and when adding them up one does not encounter the indeterminate $\infty - \infty$.

Now, given a complex-valued measurable function, one can integrate its real and imaginary components separately as above and define, as expected,

$$\int_X f \, d\mu = \int_X \mathrm{Re}(f) \, d\mu + i \int_X \mathrm{Im}(f) \, d\mu.$$

We shall have recourse several times to *Fubini's theorem*, which deals with integration of a function $f : X \times Y \to \mathbb{C}$, supposed measurable with respect to a product measure $\mu \times \nu$. It is necessary to restrict ourselves to the case when μ and ν are σ*-finite* measures: this means that X is the countable union of a sequence of sets with finite μ-measure, and similarly for Y. Now, provided that we have the absolute convergence condition

$$\int_X \left(\int_Y |f(x, y)| \, d\nu(y) \right) d\mu(x) < \infty$$

(or, indeed, the same condition with the roles of X and Y interchanged), we may conclude that

$$\int_X \left(\int_Y f(x, y) \, d\nu(y) \right) d\mu(x) = \int_Y \left(\int_X f(x, y) \, d\mu(x) \right) d\nu(y). \quad (1.1)$$

We recall that for a measure space (X, μ), where μ is a positive measure, the spaces $L^p(X, \mu)$, for p satisfying $1 \leq p < \infty$, consist of the measurable functions f such that

$$\|f\|_p := \left(\int_X |f|^p \, d\mu \right)^{1/p} < \infty,$$

two functions being identified if they are equal *almost everywhere* (*a.e.*) (i.e., except on a set of measure 0). The space $L^\infty(X, \mu)$ consists of all functions such that

$$\|f\|_\infty := \inf\{K \geq 0 : |f| \leq K \quad \text{a.e.}\} < \infty,$$

with the same identification of functions equal a.e. There is a duality formula, namely $(L^p)^* = L^q$, where $1 \leq p < \infty$, $1 < q \leq \infty$ and $1/p + 1/q = 1$, which arises from *Hölder's inequality*,

$$\|fg\|_1 \leq \|f\|_p \|g\|_q. \quad (1.2)$$

If neither f nor g is identically zero, then the equality in (1.2) occurs if and only if there exists $c > 0$ such that

$$|f|^p = c|g|^q \quad \text{a.e.} \quad (1.3)$$

Let μ be a positive measure. We recall that a sequence (f_n) of measurable functions *converges in measure* to a function f if

$$\mu\{x \in X : |f_n(x) - f(x)| > \varepsilon\} \to 0$$

for all $\varepsilon > 0$. This condition is implied by convergence in the L^p norm.

We shall also require *Lusin's theorem*, in the following form.

Theorem 1.1.10 *Let f be a measurable function on a closed bounded interval $[a,b]$. Then, for every $\varepsilon > 0$ there is a closed subset $K_\varepsilon \subset [a,b]$ with $\mu([a,b] \setminus K_\varepsilon) < \varepsilon$, such that the restriction of f to K_ε is continuous.*

The idea underlying the proof of this result is that every bounded measurable function is the pointwise limit of a sequence of continuous functions, and indeed in this case the sequence converges uniformly on the complement of some set of measure less than ε.

Variation of a complex measure and polar decomposition

For a complex measure μ, one defines its *variation* or *absolute value* $|\mu|$ by the formula

$$|\mu|(A) = \sup \sum_{n=1}^{\infty} |\mu(A_n)|,$$

where A is in Σ and the supremum runs over all sequences of disjoint sets $(A_n)_n$ whose union is A. Taking only finite partitions of the set A into measurable subsets, one obtains an equivalent definition.

It turns out that $|\mu|$ is a non-negative finite measure. In the same way that a complex number can be represented in a polar form, one has a *polar decomposition* for a complex measure: there exists a measurable function θ with real values such that

$$d\mu = e^{i\theta} d|\mu|,$$

meaning that

$$\int_X f \, d\mu = \int_X f e^{i\theta} \, d|\mu|$$

for any absolutely integrable measurable function f, i.e., f satisfying

$$\int_X |f| \, d|\mu| < \infty.$$

One can use the Radon–Nikodým theorem (see below) to prove that the variation is a measure and the existence of the polar decomposition.

The space of complex measures

The sum of two complex measures is a complex measure, as is the product of a complex measure by a complex number. That is to say, the set of all complex measures on a measure space (X, Σ) forms a vector space. Moreover, the *total variation* $\|\cdot\|$, defined by

$$\|\mu\| = |\mu|(X),$$

is a norm, with respect to which the space of complex measures of finite variation is a Banach space.

Absolute continuity

Definition 1.1.11 Let ν and μ be positive, σ-finite measures on (X, Σ).

- The measure μ is *absolutely continuous* with respect to ν if, for all $A \in \Sigma$ such that $\nu(A) = 0$, we have also $\mu(A) = 0$. We denote this by
$$\mu \ll \nu.$$
- The measure μ is *supported* by $E \in \Sigma$ if, for all $A \in \Sigma$, we have
$$\mu(A) = \mu(A \cap E) \quad \text{or, equivalently,} \quad \mu(A \setminus E) = 0.$$
- The measures μ and ν are *mutually singular* if there exists $E \in \Sigma$ such that μ is supported by E and ν is supported by its complement E^c. We denote this by
$$\mu \perp \nu.$$

The Radon–Nikodým theorem and its consequences

Theorem 1.1.12 (Radon–Nikodým theorem) *Let ν and μ be positive σ-finite measures on (X, Σ). Then we have*

1. *There exists a unique pair of measures μ_1 and μ_2 such that*
   - $\mu = \mu_1 + \mu_2$;
   - $\mu_1 \ll \nu$;
   - $\mu_2 \perp \nu$.

   The measures μ_1 and μ_2 are positive and σ-finite.
2. *There exists a unique (ν almost everywhere) positive ν-integrable function h, such that, for all $A \in \Sigma$, we have*
$$\mu_1(A) = \int_A h \, d\nu = \int_X \chi_A \, h \, d\nu.$$

We now mention various consequences of the Radon–Nikodým theorem.

Definition 1.1.13 Let ν be a positive σ-finite measure on (X, Σ) and let μ be a positive σ-finite measure on (X, Σ). One says that μ has a *density* h with respect to ν if h is a positive ν-integrable function, such that for all $A \in \Sigma$ we have
$$\mu(A) = \int_A h \, d\nu = \int_X \chi_A \, h \, d\nu.$$

We write

$$h = \frac{d\mu}{d\nu},$$

and sometimes call it the *Radon–Nikodým derivative* of μ with respect to ν.

As a consequence of the Radon–Nikodým theorem, we have the following property.

Proposition 1.1.14 *Let ν and μ be positive σ-finite measures on (X, Σ). Then the following assertions are equivalent:*

- $\mu \ll \nu$;
- *μ has a density with respect to ν.*

The Radon–Nikodým theorem asserts that any complex measure μ absolutely continuous with respect to some positive σ-finite measure ν (for example, Lebesgue measure) is given by the integral of some $L^1(\nu)$-function f,

$$\mu(E) = \int_E f\, d\nu.$$

The function f is again a density function for the measure.

A closely related theorem says that any complex measure μ decomposes into an absolutely continuous measure μ_a and a singular measure μ_c. This is the *Lebesgue decomposition*,

$$\mu = \mu_a + \mu_c.$$

1.1.6 Riesz representation theorem

There are several well-known theorems in functional analysis known as the Riesz representation theorem. They are named in honour of Frigyes Riesz.

The Hilbert space representation theorem

This theorem establishes an important connection between a Hilbert space and its (continuous) dual space. If the underlying field is the real numbers, the two are isometrically isomorphic; if the field is the complex numbers, the two are isometrically anti-isomorphic. The (anti-) isomorphism is a particular natural one as will be described next.

Let $\mathcal{H}$ be a Hilbert space, and let $\mathcal{H}^*$ denote its dual space, consisting of all continuous linear functionals from $\mathcal{H}$ into the field $\mathbb{R}$ or $\mathbb{C}$. If x is an element of $\mathcal{H}$, then the function ϕ_x defined by

$$\phi_x(y) = \langle y, x \rangle \quad \text{for all } y \in \mathcal{H},$$

where $\langle \cdot, \cdot \rangle$ denotes the inner product of the Hilbert space, is an element of $\mathcal{H}^*$ (note that a similar notation is used for Banach space duality, but no confusion is likely). The Riesz representation theorem states that every element of $\mathcal{H}^*$ can be written uniquely in this form.

Theorem 1.1.15 *The mapping*

$$\Phi : \mathcal{H} \to \mathcal{H}^*, \quad \Phi(x) = \phi_x$$

is an isometric (anti-) isomorphism, meaning that

- *Φ is bijective.*
- *The norms of x and $\Phi(x)$ agree: $\|x\| = \|\Phi(x)\|$.*
- *Φ is additive: $\Phi(x_1 + x_2) = \Phi(x_1) + \Phi(x_2)$.*
- *If the base field is $\mathbb{R}$, then $\Phi(\lambda x) = \lambda\Phi(x)$ for all real numbers λ.*
- *If the base field is $\mathbb{C}$, then $\Phi(\lambda x) = \bar{\lambda}\Phi(x)$ for all complex numbers λ, where $\bar{\lambda}$ denotes the complex conjugate of λ.*

The inverse map of Φ can be described as follows. Given an element φ of $\mathcal{H}^*$, the orthogonal complement of the kernel of φ is a one-dimensional subspace of $\mathcal{H}$. Take a non-zero element z in that subspace, and let $x \in \mathcal{H}$ be given by $x = \overline{\varphi(z)} \cdot z/\|z\|^2$. Then $\Phi(x) = \varphi$.

The representation theorem for linear functionals on $C_0(X)$ and $C_c(X)$ Let $C_0(X)$ be the space of continuous complex-valued functions f which vanish at infinity, that is, such that for every $\varepsilon > 0$ there is a compact set $K_\varepsilon \subset X$ with $|f(x)| \leq \varepsilon$ for $x \in X \setminus K_\varepsilon$. The *Borel sets* in the following statement refer to the σ-algebra generated by the open sets.

Definition 1.1.16 1. A non-negative countably additive Borel measure μ on a locally compact Hausdorff space X is *regular* if and only if

- $\mu(K) < \infty$ for every compact K;
- For every Borel set E, $\mu(E) = \inf\{\mu(U) : E \subset U, U \text{ open}\}$;
- The relation $\mu(E) = \sup\{\mu(K) : K \subset E, K \text{ compact}\}$ holds whenever E is open or when E is Borel and $\mu(E) < \infty$.

2. If μ is a complex-valued countably additive Borel measure, then μ is regular if and only if the non-negative countably additive measure $|\mu|$ is regular.

The following theorem gives a concrete realization of the dual space of $C_0(X)$.

Theorem 1.1.17 (Riesz–Markov theorem) *Let X be a locally compact Hausdorff space. For any continuous linear functional Ψ on $C_0(X)$, there is a unique regular countably additive complex Borel measure μ on X such that*

$$\Psi(f) = \int_X f(x)\,d\mu(x)$$

for all $f \in C_0(X)$. The norm of Ψ as a linear functional is the total variation of μ; that is,

$$\|\Psi\| = |\mu|(X).$$

Finally, Ψ is positive if and only if the measure μ is non-negative.

Denote by $C_c(X)$ the space of continuous complex-valued functions of compact support. The following theorem represents positive linear functionals on $C_c(X)$.

Theorem 1.1.18 *Let X be a locally compact Hausdorff space. For any positive linear functional Ψ on $C_c(X)$, there is a unique regular Borel measure μ on X such that*

$$\Psi(f) = \int_X f(x)\,d\mu(x)$$

for all $f \in C_c(X)$.

Remark 1.1.19 A positive linear functional on $C_c(X)$ need not extend to a bounded linear functional on $C_0(X)$. For this reason the above results apply to slightly different situations.

1.1.7 Geometry of Banach spaces

Definition 1.1.20 A Banach space $\mathcal{X}$ is said to be *smooth* if for every vector $x \in \mathcal{X} \setminus \{0\}$ there is a unique linear functional $\nu_{\mathcal{X}}(x)$ in $\mathcal{X}^*$ such that $\|\nu_{\mathcal{X}}(x)\| = 1$ and $\nu_{\mathcal{X}}(x)x = \|x\|$. When there is no ambiguity, we shall write $\nu(x)$ for $\nu_{\mathcal{X}}(x)$.

Furthermore, $\mathcal{X}$ is said to be *strictly convex* if, whenever x and y are linearly independent vectors in $\mathcal{X}$, we have $\|x + y\| < \|x\| + \|y\|$.

We shall later require the following result (see Beauzamy [27, p. 183]).

Proposition 1.1.21 *Let $\mathcal{X}$ be a smooth Banach space, and let $x \in \mathcal{X}$ satisfy $\|x\| = 1$. Suppose that* $\operatorname{Re}\nu(x)y < 1$ *for some $y \in \mathcal{X}$. Then there exists a $\lambda \in (0, 1)$ such that $\|\lambda x + (1 - \lambda)y\| < 1$.*

We shall also require the following well-known fixed-point theorem.

Theorem 1.1.22 (Schauder–Tychonoff) *Let $\mathcal{X}$ be a normed space. Suppose that $B \subset \mathcal{X}$ is convex and that $C \subset B$ is compact. Then every continuous map $f : B \to C$ has a fixed point.*

1.2 Operator theory

1.2.1 Basic definitions and spectral properties

Let $\mathcal{X}$ and $\mathcal{Y}$ be normed spaces, and T a function on $\mathcal{X}$ with values in $\mathcal{Y}$. Then T is a *linear operator* if

$$T(\lambda x + \mu y) = \lambda Tx + \mu Ty \quad \text{for all} \quad x, y \in \mathcal{X} \quad \text{and} \quad \lambda, \mu \in \mathbb{C}.$$

The operator T is *continuous* (or *bounded*) if there exists a positive constant C with $\|Tx\| \leq C\|x\|$ for all $x \in \mathcal{X}$. The least such C for which the inequality holds is called the *norm* of T and is denoted by $\|T\|$. Equivalently, the norm of T can be defined as follows:

$$\|T\| = \sup\{\|Tx\| : x \in \mathcal{X}, \|x\| \leq 1\}.$$

The space of all bounded linear operators from $\mathcal{X}$ to $\mathcal{Y}$ is denoted by $\mathcal{L}(\mathcal{X}, \mathcal{Y})$. We write $\mathcal{L}(\mathcal{X})$ for $\mathcal{L}(\mathcal{X}, \mathcal{X})$. The product of two operators S and T in $\mathcal{L}(\mathcal{X})$ is given by composition

$$(ST)(x) = (S \circ T)(x) = S(Tx), \quad x \in \mathcal{X}.$$

Then, trivially, $\|ST\| \leq \|S\|\|T\|$. We write $\mathrm{I}_{\mathcal{X}}$ (or simply I when there is no ambiguity) for the identity mapping on $\mathcal{X}$.

Definition 1.2.1 Let $\mathcal{A}$ be an algebra on $\mathbb{C}$ endowed with a norm denoted by $\|\cdot\|$. The algebra $\mathcal{A}$ is a *Banach algebra* if the following two conditions are satisfied:

(*i*) The algebra $\mathcal{A}$ is a complex Banach space (i.e., a complete and normed complex vector space).
(*ii*) For all $x, y \in \mathcal{A}$, we have

$$\|xy\| \leq \|x\|\|y\|.$$

If, moreover, there exists an element $e \in \mathcal{A}$ such that

$$xe = ex = x \quad (x \in \mathcal{A}) \quad \text{and} \quad \|e\| = 1,$$

then we say that $\mathcal{A}$ is a *unital Banach algebra*. The element e is called the *unit* or *identity* of the algebra $\mathcal{A}$.

Finally, an algebra $\mathcal{A}$ is *commutative* if $xy = yx$ for all $x, y \in \mathcal{A}$.

The set $\mathcal{L}(\mathcal{X})$ is thus an example of non-commutative unital Banach algebra. Now let $T \in \mathcal{L}(\mathcal{X})$.

- The *spectrum* of T, $\sigma(T)$, is the set of those $\lambda \in \mathbb{C}$ for which $T - \lambda I$ fails to have a continuous inverse.
- The *point spectrum* of T, $\sigma_p(T)$, is the set of those $\lambda \in \mathbb{C}$ for which $T - \lambda I$ is not injective. The elements of $\sigma_p(T)$ are called the *eigenvalues* of T, and $\operatorname{Ker}(T - \lambda I)$ is called the *eigenspace* associated with the eigenvalue λ. The non-zero elements of $\operatorname{Ker}(T - \lambda I)$ are the *eigenvectors* of T associated with λ.
- The *approximate point spectrum* is the set of $\lambda \in \mathbb{C}$ for which there exists a sequence of unit vectors $(x_n)_n$ in $\mathcal{X}$ satisfying

$$\lim_{n\to\infty} \|(T - \lambda I)x_n\| = 0.$$

  The elements of the approximate point spectrum are called the *approximate eigenvalues*.

Note that the point spectrum is included in the approximate point spectrum. Moreover, the point spectrum may be empty, unlike the approximate point spectrum.

Proposition 1.2.2 *Let $T \in \mathcal{L}(\mathcal{X})$, where $\mathcal{X}$ is a complex Banach space, and let $\lambda \in \mathbb{C}$. Then the following statements are equivalent:*

(*i*) *The scalar λ is not an approximate eigenvalue of T.*
(*ii*) *The kernel of $T - \lambda I$ is reduced to $\{0\}$ and its range is closed.*
(*iii*) *There exists a constant $c > 0$ such that $\|(T - \lambda I)x\| \geq c\|x\|$.*

Definition 1.2.3 Let $T \in \mathcal{L}(\mathcal{X}, \mathcal{Y})$, where $\mathcal{X}$ and $\mathcal{Y}$ are complex Banach spaces. The operator T is *compact* if the closure of $T(\{x \in \mathcal{X} : \|x\| \leq 1\})$ is a compact subset of Y.

The set of compact operators in $\mathcal{L}(\mathcal{X}, \mathcal{Y})$ is denoted $\mathcal{K}(\mathcal{X}, \mathcal{Y})$, and we write $\mathcal{K}(\mathcal{X})$ for $\mathcal{K}(\mathcal{X}, \mathcal{X})$.

A *polynomially compact* operator $T \in \mathcal{L}(\mathcal{X})$ is an operator such that there exists a non-trivial polynomial p for which $p(T)$ is compact.

The compact operators $\mathcal{K}(\mathcal{X})$ form a closed ideal in the algebra $\mathcal{L}(\mathcal{X})$ of all operators on $\mathcal{X}$; the quotient $\mathcal{L}(\mathcal{X})/\mathcal{K}(\mathcal{X})$ is called the *Calkin algebra* on $\mathcal{X}$. This motivates some further definitions.

Definition 1.2.4 The *essential norm* of an operator $T \in \mathcal{L}(\mathcal{X})$ is given by

$$\|T\|_e = \inf\{\|T - K\| : K \in \mathcal{K}(\mathcal{X})\},$$

and the *essential spectrum* $\sigma_e(T)$ of T is the set of all $\lambda \in \mathbb{C}$ such that the coset $[T - \lambda I]$ is non-invertible in the Calkin algebra.

Definition 1.2.5 Let $T \in \mathcal{L}(\mathcal{X})$, where $\mathcal{X}$ is a Banach space. The *singular values* $(\sigma_k(T))_{k=1}^{\infty}$ are defined by

$$\sigma_k(T) = \inf\{\|T - S\| : S \in \mathcal{L}(\mathcal{X}), \dim \operatorname{Im}(S) < k\}.$$

Thus, in particular, $\sigma_1(T) = \|T\|$.

Definition 1.2.6 An operator $T \in \mathcal{L}(\mathcal{H})$ is said to be of *trace class* if

$$\|T\|_{\mathcal{S}_1} := \sum_{k=1}^{\infty} \sigma_k(T) < \infty.$$

The trace-class operators form a Banach space $\mathcal{S}_1(\mathcal{H})$ under this norm. Moreover, we have $\mathcal{K}(\mathcal{H})^* = \mathcal{S}_1(\mathcal{H})$ and $\mathcal{S}_1(\mathcal{H})^* = \mathcal{L}(\mathcal{H})$, under the duality pairing

$$[S, T] = \sum_{k=1}^{\infty} \langle STe_k, e_k \rangle, \tag{1.4}$$

where $(e_k)_k$ is any orthonormal basis of $\mathcal{H}$. This duality enables us to talk about a weak* topology on $\mathcal{L}(\mathcal{H})$.

For $T \in \mathcal{L}(\mathcal{H})$, where $\mathcal{H}$ is a Hilbert space, recall that the *adjoint* T^* of T is defined as follows:

$$\langle Tx, y \rangle = \langle x, T^* y \rangle,$$

for all $x, y \in \mathcal{H}$. Then $T^* \in \mathcal{L}(\mathcal{H})$, $\|T^*\| = \|T\|$ and $T^{**} = T$. Moreover, using the adjoint, there are useful orthogonal decompositions of $\mathcal{H}$:

$$\mathcal{H} = \operatorname{Ker} T \oplus \overline{\operatorname{Im}(T^*)} = \operatorname{Ker} T^* \oplus \overline{\operatorname{Im}(T)}.$$

We may now state the following proposition, which will be particularly useful in Chapter 8.

Proposition 1.2.7 *An operator $T \in \mathcal{L}(\mathcal{H})$ is surjective if and only if 0 is not an approximate eigenvalue of T^*.*

We write $\rho(T)$ for the *spectral radius* of T, which is defined by

$$\rho(T) = \sup\{|\lambda| : \lambda \in \sigma(T)\}.$$

Theorem 1.2.8 *The spectrum of $T \in \mathcal{L}(\mathcal{X})$ is a non-empty compact subset of $\mathbb{C}$ and, moreover, $\rho(T) = \lim_{n\to\infty} \|T^n\|^{1/n}$.*

Therefore, in particular, we have $\rho(T) = \max\{|\lambda| : \lambda \in \sigma(T)\}$ and, clearly, $\rho(T) \leq \|T\|$.

Definition 1.2.9 An operator is said to be *quasinilpotent* if $\sigma(T) = \{0\}$; clearly this is equivalent to the condition $\rho(T) = 0$.

We shall require some elements of the Gelfand theory of commutative (complex) Banach algebras. Again the *spectrum* $\sigma(a)$ of an element a in a unital Banach algebra $\mathcal{A}$ consists of the set of $\lambda \in \mathbb{C}$ such that $a - \lambda e$ has no multiplicative inverse in $\mathcal{A}$. In the commutative case, this can be linked to the *character space* or *maximal ideal space* $\Phi_{\mathcal{A}}$ of continuous homomorphisms $\phi : \mathcal{A} \to \mathbb{C}$, i.e., the linear mappings ϕ such that $\phi(e) = 1$ and $\phi(ab) = \phi(a)\phi(b)$ for all $a, b \in \mathcal{A}$. These mappings are called *characters* and their kernels are the *maximal ideals* of $\mathcal{A}$. The link is as follows.

Theorem 1.2.10 *For an element a in a commutative unital Banach algebra $\mathcal{A}$, we have*

$$\sigma(a) = \{\phi(a) : \phi \in \Phi_{\mathcal{A}}\}.$$

In particular, a is invertible if and only if $\phi(a) \neq 0$ for all characters $\phi \in \Phi_{\mathcal{A}}$.

We note that for $\mathcal{A} = C(K)$, where K is a compact Hausdorff space, the maximal ideal space can be identified with K, as the characters are the point evaluations $\phi_x : f \mapsto f(x)$, for $x \in K$.

Theorem 1.2.11 *Let $\mathcal{A} = \ell^1(\mathbb{Z})$. Then $\mathcal{A}$ is a Banach algebra if the multiplication is given by convolution:*

$$(f * g)_k = \sum_{j=-\infty}^{\infty} f_j g_{k-j} \qquad (f, g \in \mathcal{A}).$$

Its maximal ideal space may be identified with $\mathbb{T}$ by the evaluation mapping

$$\phi_{e^{it}}(f) = \sum_{j=-\infty}^{\infty} f_k e^{ikt} \qquad (f \in \mathcal{A}, \quad e^{it} \in \mathbb{T}).$$

From this it is easy to deduce the following result.

Theorem 1.2.12 (Wiener's theorem) *Suppose that $f \in C(\mathbb{T})$ has an absolutely convergent Fourier series*

$$f(e^{it}) = \sum_{j=-\infty}^{\infty} a_k e^{ikt}, \qquad with \qquad \sum_{j=-\infty}^{\infty} |a_k| < \infty.$$

Then, if $f \neq 0$ *on* $\mathbb{T}$*, the function* $1/f$ *also has an absolutely convergent Fourier series.*

We mention also some properties of the spectrum of a compact operator.

Theorem 1.2.13 *Let* $\mathcal{X}$ *be an infinite-dimensional Banach space, and suppose that* $T \in \mathcal{K}(\mathcal{X})$*. Then,*

(*i*) $\sigma(T)$ *is a countable set of the form* $\{z_n : n \in \mathbb{N}\} \cup \{0\}$*, where the set* $\{z_n : n \in \mathbb{N}\}$ *is either finite or a sequence which converges to* 0*.*

(*ii*) *Each* $z \in \sigma(T) \setminus \{0\}$ *is an eigenvalue of* T*, and the corresponding eigenspace* $\{x \in \mathcal{X} : Tx = zx\}$ *is finite-dimensional.*

(*iii*) *For each* $z \in \mathbb{C} \setminus \{0\}$*, the subspace* $(T - z\mathrm{I})(\mathcal{X})$ *is closed in* $\mathcal{X}$ *and*

$$\dim \operatorname{Ker}(T - z\mathrm{I}_{\mathcal{X}}) = \dim(\mathcal{X}/(T - z\mathrm{I}_{\mathcal{X}})(\mathcal{X})),$$

so that, in particular, $T - z\mathrm{I}_{\mathcal{X}}$ *is surjective if and only if it is injective.*

We are now ready for the definitions which motivate the whole of this book.

Definition 1.2.14 Let $T \in \mathcal{L}(\mathcal{X})$, where $\mathcal{X}$ is a Banach space.

1. An *invariant subspace* $\mathcal{M}$ for T is a closed subspace of $\mathcal{X}$ such that $T\mathcal{M} \subset \mathcal{M}$.
2. A subspace $\mathcal{M} \subset \mathcal{X}$ is *non-trivial* if $\mathcal{M} \neq \{0\}$ and $\mathcal{M} \neq \mathcal{X}$.
3. A subspace $\mathcal{M}$ is said to be *hyperinvariant* for T if $A\mathcal{M} \subset \mathcal{M}$ for all A in the *commutant* of T, which is given by

$$\{T\}' = \{A \in \mathcal{L}(\mathcal{X}) : AT = TA\}.$$

4. Suppose now that $\mathcal{X} = \mathcal{H}$, where $\mathcal{H}$ is a complex Hilbert space. A *reducing subspace* $\mathcal{K}$ for $T \in \mathcal{L}(\mathcal{H})$ is a subspace invariant for both T and T^*. Thus we may write

$$\mathcal{H} = \mathcal{K} \oplus \mathcal{K}^{\perp},$$

and each of $\mathcal{K}$ and $\mathcal{K}^{\perp}$ is invariant under both T and T^*.

We shall use the notation $\operatorname{Lat}(T)$ to denote the lattice of all closed invariant subspaces of an operator T.

The most important version of the *invariant-subspace problem* is whether every operator on a complex Hilbert space has a non-trivial invariant subspace. This is unsolved, as is the analogous *hyperinvariant-subspace problem*.

For Banach spaces there are counterexamples due to Enflo [95, 96], Read [168, 169, 170, 171] and Beauzamy [28], even on spaces as simple as ℓ^1.

At present it is unknown whether there is any operator $T \in \mathcal{L}(\mathcal{X})$ for which the adjoint $T^* \in \mathcal{L}(\mathcal{X}^*)$ has no non-trivial closed invariant subspaces. In particular, there is no known counterexample to the invariant-subspace problem on a reflexive Banach space.

A vector $x \in \mathcal{X}$ is *cyclic* for an operator $T \in \mathcal{L}(\mathcal{X})$ if the smallest invariant subspace containing x is $\mathcal{X}$ itself; equivalently, if $\operatorname{span}\{x, Tx, T^2x, \ldots\}$, the *closed linear span* of x and its iterates under T, is the whole of $\mathcal{X}$. Clearly the invariant-subspace problem for T is equivalent to asking whether every non-zero vector $x \in \mathcal{X}$ is cyclic for T.

Certain particular classes of operator are of particular interest. Let $\mathcal{H}$ be a complex Hilbert space.

Definition 1.2.15 Let $T \in \mathcal{L}(\mathcal{H})$. We say that

1. The operator T is *self-adjoint* if $T^* = T$.
2. The operator T is *normal* if $T^*T = TT^*$.
3. The operator T is *subnormal* if there exists a Hilbert space $\mathcal{K}$ containing $\mathcal{H}$ and a normal operator $N \in \mathcal{L}(\mathcal{K})$ such that $T = N_{|\mathcal{H}}$. Thus $\mathcal{H}$ is an invariant subspace for N.
4. The operator T is an *isometry* (or is *isometric*) if $T^*T = \mathrm{I}$ or, equivalently, if $\|Tx\| = \|x\|$ for all $x \in \mathcal{H}$.
5. The operator T is *unitary* if $T^*T = TT^* = \mathrm{I}$.
6. The operator T is *positive* if $\langle Tx, x\rangle \geq 0$ for all $x \in \mathcal{H}$.

A convenient representation of normal operators is as follows.

Theorem 1.2.16 *An operator $T \in \mathcal{L}(\mathcal{H})$ is normal if and only if there is a Borel set $\Omega \subset \mathbb{C}$, a positive measure μ defined on the Borel subsets of Ω, and a function $\Phi \in L^\infty(\Omega, \mu)$ such that T is unitarily equivalent to the operator M_Φ of multiplication by Φ on $L^2(\Omega, \mu)$. Moreover, if T has a cyclic vector, then we may take $\Phi(z) = z$.*

From this result it is easy to deduce the following.

Corollary 1.2.17 *Let $T \in \mathcal{L}(\mathcal{H})$ be a positive operator. Then there exists a unique positive operator A such that $A^2 = T$.*

The operator A is called the *square root* of T, and is denoted by $\sqrt{T}$ or $T^{1/2}$.

Normal operators also satisfy the following useful properties.

Proposition 1.2.18 *Let $N \in \mathcal{L}(\mathcal{H})$ be a normal operator. Then*

1. *For all $p \in \mathbb{C}[X, Y]$, the operator $p(N, N^*)$ is normal and, moreover,*

$$\{p(\lambda, \overline{\lambda}) : \lambda \in \sigma(N)\} = \sigma(p(N, N^*)).$$

2. *The spectral radius of N coincides with its norm.*

The best-known classes of operators for which the existence of invariant subspaces are known are the normal operators on a Hilbert space (see Theorem 3.4.1) and the compact operators (on any Banach space) (see Theorem 6.1.2). It is not too hard to see that *algebraic operators*, those T for which there is a non-trivial polynomial T such that $p(T) = 0$, also have invariant subspaces. A generalization of this result appears later (Theorem 2.2.7).

1.2.2 Wold decomposition of an isometry

Definition 1.2.19 Let $\mathcal{H}$ be a complex Hilbert space.

1. An operator $T \in \mathcal{L}(\mathcal{H})$ is *completely non-unitary* if there is no reducing subspace $\mathcal{M}$ for T such that the restriction of T to $\mathcal{M}$ is unitary.
2. An isometry $T \in \mathcal{L}(\mathcal{H})$ is a *shift* (or is *pure*) if T is completely non-unitary.

The next result provides an explicit expression of the unitary and the shift parts of an isometry.

Theorem 1.2.20 (Wold decomposition) *Suppose that $\mathcal{H}$ is a complex Hilbert space and let $V \in \mathcal{L}(\mathcal{H})$ be an isometry. Then there exist two reducing subspaces $\mathcal{H}_\infty$ and $\tilde{\mathcal{H}}$, such that $\mathcal{H} = \mathcal{H}_\infty \oplus \tilde{\mathcal{H}}$, as an orthogonal direct sum, and such that*

$$V_{|\mathcal{H}_\infty} \quad \textit{is unitary} \quad \textit{and} \quad V_{|\tilde{\mathcal{H}}} \quad \textit{is a shift.}$$

Moreover, we have $\mathcal{H}_\infty = \bigcap_{n \geq 0} V^n \mathcal{H}$.

Corollary 1.2.21 *Let $\mathcal{H}$ be a complex Hilbert space and let $V \in \mathcal{L}(\mathcal{H})$ be an isometry. The following assertions are equivalent:*

1. *The operator V is a shift.*
2. $\lim_{n \to \infty} \|V^{*n} x\| = 0$ *for all $x \in \mathcal{H}$.*
3. $\bigcap_{n \geq 0} V^n \mathcal{H} = \{0\}$.

The Wold decomposition can be taken further for a shift S, by writing $\mathcal{K} = \mathcal{H} \ominus \operatorname{Im} S$. It can then be verified that

$$\mathcal{H} = \mathcal{K} \oplus S\mathcal{K} \oplus S^2\mathcal{K} \oplus \ldots,$$

as an orthogonal direct sum of closed subspaces.

1.2.3 Riesz–Dunford functional calculus

Given a locally holomorphic function f of a complex argument z and an operator $T \in \mathcal{L}(\mathcal{X})$, where $\mathcal{X}$ is a Banach space, the aim is to construct an operator $f(T)$, which in a sense extends the function f from a complex argument to an operator argument.

Definition

Let us start with a formula that justifies the definition of the holomorphic functional calculus.

Lemma 1.2.22 *Let $\mathcal{X}$ be a Banach space and $T \in \mathcal{L}(\mathcal{X})$. Then, for all $n \in \mathbb{N}$, we have*

$$T^n = \frac{1}{2\pi i} \int_{\Gamma_r} z^n (z\mathrm{I} - T)^{-1} \, dz, \tag{1.5}$$

where Γ_r is the circle centred at 0 *and of radius r, positively oriented, with $r > \rho(T)$.*

Since the resolvent mapping $z \mapsto (z\mathrm{I} - T)^{-1}$ is holomorphic on the resolvent set, it follows that the integral

$$\frac{1}{2\pi i} \int_{\Gamma} f(z)(z\mathrm{I} - T)^{-1} \, dz$$

makes sense when f is holomorphic.

We write $\operatorname{Hol}(\Omega)$ for the set of holomorphic functions on Ω. Cauchy's integral formula suggests the following definition of the holomorphic functional calculus.

Definition 1.2.23 Let $\mathcal{X}$ be a Banach space, $T \in \mathcal{L}(\mathcal{X})$ and Ω be an open set of $\mathbb{C}$ containing the spectrum of T. For $f \in \operatorname{Hol}(\Omega)$ and

$$\Gamma = \{\gamma_i\}_{i=1}^m,$$

a collection of Jordan curves in Ω such that $\sigma(T)$ lies in the inside of Γ (that is, the total winding number of Γ about each point of $\sigma(T)$ is 1), we set

$$f(T) = \frac{1}{2\pi i} \int_\Gamma f(z)(zI - T)^{-1}\, dz. \tag{1.6}$$

Remark 1.2.24 The formula (1.6) defines an element in $\mathcal{L}(\mathcal{X})$ that is independent of the choice of Γ. Moreover, the open set Ω may depend on f, and need not be connected.

Properties

The next theorem summarizes the essential properties of the holomorphic functional calculus.

Theorem 1.2.25 *Let $\mathcal{X}$ be a Banach space, $T \in \mathcal{L}(\mathcal{X})$ and Ω be an open set of $\mathbb{C}$ containing the spectrum of T. Consider then*

$$\begin{array}{rccc} \psi : & \mathrm{Hol}(\Omega) & \longrightarrow & \mathcal{L}(\mathcal{X}) \\ & f & \longmapsto & f(T). \end{array}$$

Then

(a) ψ is a homomorphism of unital algebras.
(b) ψ is continuous if $\mathrm{Hol}(\Omega)$ is endowed with the topology of uniform convergence on compact sets. In other words, if $f_n, f \in \mathrm{Hol}(\Omega)$ and (f_n) converges to f uniformly on every compact subset of Ω, then $f_n(T)$ tends to $f(T)$ as $n \to +\infty$.
(c) For each function $f \in \mathrm{Hol}(\Omega)$, we have

$$\sigma(f(T)) = f(\sigma(T)).$$

(d) Let $f \in \mathrm{Hol}(\Omega)$, $\mathcal{U}$ be an open set containing $f(\sigma(T))$ and $g \in \mathrm{Hol}(\mathcal{U})$. Then

$$(g \circ f)(T) = g(f(T)).$$

1.3 The Poisson kernel

Let $\mathbb{D}$ denote the unit disc $\{z \in \mathbb{C} : |z| < 1\}$. For $\lambda = re^{it} \in \mathbb{D}$ and $e^{i\theta} \in \mathbb{T}$, the (scalar) Poisson kernel is defined by

$$\begin{aligned} P_\lambda(\theta) = P_{r,t}(e^{i\theta}) &= \frac{1}{1 - re^{it}e^{-i\theta}} + \frac{1}{1 - re^{-it}e^{i\theta}} - 1 \\ &= \frac{1 - r^2}{|1 - re^{it}e^{-i\theta}|^2} \\ &= \sum_{n \ge 0} r^n e^{int} e^{-in\theta} + \sum_{n \ge 0} r^n e^{-int} e^{in\theta} - 1. \end{aligned}$$

We also have

$$P_{r,t}(e^{i\theta}) = \frac{1-r^2}{1+r^2-2r\cos(\theta-t)} = \operatorname{Re}\left(\frac{e^{i\theta}+re^{it}}{e^{i\theta}-re^{it}}\right).$$

By a minor abuse of notation, we also regard P_λ as a function in $L^1(\mathbb{T})$.

Note that $P_{r,t}(e^{i\theta}) \geq 0$. Recall also that each function $f \in L^1(\mathbb{T})$ has a natural *harmonic extension* into $\mathbb{D}$, which will also be denoted by f. The extension is defined by Poisson's formula,

$$f(re^{it}) = \frac{1}{2\pi}\int_0^{2\pi} f(e^{i\theta})P_{r,t}(e^{i\theta})d\theta.$$

In later sections we will need the following theorem ([179], chapter 11).

Theorem 1.3.1 *Let $\mathcal{E}$ be the set of harmonic functions f on $\mathbb{D}$ such that*

$$\nu(f) := \sup_{0\leq s<1} \frac{1}{2\pi}\int_0^{2\pi} |f(se^{it})|dt < \infty.$$

For $\mu \in \mathcal{M}(\mathbb{T})$, denote by $P(\mu)$ the harmonic function on $\mathbb{D}$ defined by

$$P(\mu)(re^{it}) = \int_0^{2\pi} P_{r,t}(e^{i\theta})\,d\mu(e^{i\theta}).$$

The mapping

$$\begin{array}{ccc} \mathcal{M}(\mathbb{T}) & \to & \mathcal{E} \\ \mu & \longmapsto & P(\mu) \end{array}$$

is bijective with $\nu(P(\mu)) = \|\mu\|$. Moreover, $\lim_{r\to 1^-} P(\mu)(re^{it})$ exists a.e. and is equal to the $L^1(\mathbb{T})$-function which is the Radon–Nikodým derivative of μ, that is, the density of the absolutely continuous part of μ.

1.4 Hardy spaces

For $1 \leq p < \infty$ the Hardy space H^p or $H^p(\mathbb{D})$ is defined as the space of all analytic functions f in the unit disc $\mathbb{D}$ for which the norm

$$\|f\|_p = \sup_{0\leq r<1}\left(\frac{1}{2\pi}\int_0^{2\pi} |f(re^{it})|^p\,dt\right)^{1/p}$$

is finite. The space H^∞ consists of all bounded analytic functions f in $\mathbb{D}$, and the norm is now

$$\|f\|_\infty = \sup_{|z|<1} |f(z)|.$$

For functions in $H^p(\mathbb{D})$, $1 \le p \le \infty$, the *radial limit*

$$\tilde{f}(e^{it}) = \lim_{r \to 1} f(re^{it})$$

exists almost everywhere in t (*Fatou's theorem*), and indeed $\tilde{f} \in L^p(\mathbb{T})$. Moreover,

$$\|f\|_p = \left(\frac{1}{2\pi} \int_0^{2\pi} |\tilde{f}(e^{it})|^p dt \right)^{1/p} = \|\tilde{f}\|_{L^p(\mathbb{T})}.$$

We normally identify f with $\tilde{f}$, and can just regard H^p as the subspace of those $L^p(\mathbb{T})$ functions for which the negative Fourier coefficients vanish; that is

$$\frac{1}{2\pi} \int_0^{2\pi} \tilde{f}(e^{it}) e^{-int} dt = 0$$

for all $n < 0$. Then a function $\tilde{f}$ with $\tilde{f}(e^{it}) \sim \sum_{n=0}^{\infty} a_n e^{int}$ can be naturally identified with the power series $f(z) = \sum_{n \ge 0} a_n z^n$, defining an analytic function f in $\mathbb{D}$. Thus it will be natural to use the notation $H^p(\mathbb{T})$ for $H^p(\mathbb{D})$ when we wish to regard it as a closed subspace of $L^p(\mathbb{T})$.

One can also obtain f from $\tilde{f}$ by convolving with the Poisson kernel $P_{r,t}$:

$$f(re^{it}) = \frac{1}{2\pi} \int_0^{2\pi} P_{r,t}(e^{i\theta}) \tilde{f}(e^{i\theta}) d\theta.$$

Calculating the norm in the case $p = 2$ is simpler, since for a function $f : z \mapsto \sum_{n=0}^{\infty} a_n z^n$ we have

$$\|f\|_2 = \left(\sum_{n=0}^{\infty} |a_n|^2 \right)^{1/2}.$$

We use P_+ to denote the orthogonal projection from $L^2(\mathbb{T})$ onto H^2, so that

$$P_+ : \sum_{n=-\infty}^{\infty} a_n e^{int} \mapsto \sum_{n=0}^{\infty} a_n e^{int}.$$

We have the following inclusions:

$$H^\infty \subset H^p \subset H^q \subset H^1,$$

for $1 \le q \le p \le \infty$.

We shall also require some of the theory of Hardy spaces on the right-hand half-plane $\mathbb{C}_+ = \{z \in \mathbb{C} : \operatorname{Re} z > 0\}$, which is given in more detail, for example, by Partington [161]. Briefly, then, for $1 \le p < \infty$ the Hardy space $H^p(\mathbb{C}_+)$ consists of all analytic functions $f : \mathbb{C}_+ \to \mathbb{C}$ such that the norm

$$\|f\|_p = \left(\sup_{x>0} \int_{-\infty}^{\infty} |f(x+iy)|^p \, dy \right)^{1/p}$$

is finite. Such functions have well-defined boundary values in $L^p(i\mathbb{R})$, and in this way $H^p(\mathbb{C}_+)$ embeds naturally as a closed subspace of $L^p(i\mathbb{R})$. A similar story can be told in the case $p = \infty$, where we use the norm

$$\|f\|_\infty = \sup\{|f(z)| : z \in \mathbb{C}_+\}.$$

There is a link with the *Laplace transform*, defined for $z \in \mathbb{C}_+$ by

$$\hat{h}(z) = \int_0^\infty e^{-zt} h(t) \, dt,$$

namely the following result.

Theorem 1.4.1 (Paley–Wiener) *The Laplace transform gives a linear isomorphism from $L^2(0, \infty)$ onto $H^2(\mathbb{C}_+)$, such that*

$$\|\hat{h}\|_{H^2(\mathbb{C}_+)} = \sqrt{2\pi} \|h\|_2 \qquad \text{for} \quad h \in L^2(0, \infty).$$

It is important also to consider the *bilateral Laplace transform* or *Fourier–Laplace transform*. To define this we proceed first by supposing that $f \in L^1(\mathbb{R}) \cap L^2(\mathbb{R})$, and define

$$\hat{f}(i\omega) = \int_{-\infty}^{\infty} e^{-i\omega t} f(t) \, dt \qquad (\omega \in \mathbb{R}). \tag{1.7}$$

Then $\hat{f} \in L^2(i\mathbb{R})$ and $\|\hat{f}\|_2 = \sqrt{2\pi} \|f\|$. We may thus extend the definition of $\hat{f}$ continuously and linearly to $L^2(\mathbb{R})$, even though the formal integral in (1.7) does not converge for all $f \in L^2(\mathbb{R})$. The full result is the following.

Theorem 1.4.2 (Plancherel) *The bilateral Laplace transform defines a linear isomorphism from $L^2(\mathbb{R})$ onto $L^2(i\mathbb{R})$ satisfying*

$$\|\hat{f}\|_{L^2(i\mathbb{R})} = \sqrt{2\pi} \|f\|_{L^2(\mathbb{R})} \qquad \text{for} \quad f \in L^2(\mathbb{R}).$$

1.4.1 Inner and outer functions

In this section we are concerned with the multiplicative structure of the Hardy spaces, in that we want to factorize a general Hardy class function as the product of two somewhat simpler functions, an inner factor and an outer factor. Here are their definitions.

Definition 1.4.3 An *inner* function is an H^∞ function that has unit modulus almost everywhere on $\mathbb{T}$. An *outer* function is a function $f \in H^1$ which can be written in the form

$$f(re^{i\theta}) = \alpha \exp\left(\frac{1}{2\pi}\int_0^{2\pi} \frac{e^{it}+re^{i\theta}}{e^{it}-re^{i\theta}} k(e^{it})\, dt\right), \tag{1.8}$$

for $re^{i\theta} \in \mathbb{D}$, where k is a real-valued integrable function and $|\alpha| = 1$.

Proposition 1.4.4 *Let f be an outer function, satisfying (1.8). Then*

$$\log|f(e^{i\theta})| = k(e^{i\theta})$$

almost everywhere.

Clearly, an outer function cannot have any zero in the disc, since it is the exponential of a complex function. Any function that is invertible in H^∞ is outer (e.g., $z - a$ where $|a| > 1$); in fact $z - a$ is also outer when $|a| = 1$.

Examples of inner functions include Blaschke products (see below), which have zeroes, but also some functions without zeroes, such as the function $z \mapsto \exp((z-1)/(z+1))$.

Example 1.4.5 A *finite Blaschke product* is a function of the form

$$B(z) = e^{i\varphi} \prod_{j=1}^{n} \frac{z - z_j}{1 - \overline{z_j} z},$$

where $\varphi \in \mathbb{R}$ and $|z_j| < 1$ for $j = 1, \dots, n$. It is easy to verify that B has the following properties:

1. B is analytic in $\mathbb{D}$ and continuous in $\overline{\mathbb{D}}$.
2. B is inner.
3. B has zeroes at $z_1, \dots, z_n$ only, and poles at $1/\overline{z_1}, \dots, 1/\overline{z_n}$ only.

Theorem 1.4.6 (Inner–outer factorization theorem) *Let f be a non-zero function in H^1. Then f has a factorization $f = \theta \cdot u$, where θ is inner and u is outer. This factorization is unique up to a constant of modulus 1.*

We can break the inner part into two factors, an inner function with zeroes (which will be an infinite Blaschke product) and an inner function without zeroes (a so-called *singular inner function*). To do this we need to understand the properties of the zero set of a function in H^p.

Theorem 1.4.7 (G. Szegő) *Let $f \in H^1$ be such that f is not identically zero. Then the zeroes (z_n) of f are countable in number and satisfy the Blaschke condition,*

$$\sum_1^\infty (1 - |z_n|) < \infty.$$

Theorem 1.4.8 *Let $f \in H^1$. Then the* infinite Blaschke product

$$B(z) = z^m \prod_{|z_n| \neq 0} \frac{\overline{z}_n}{|z_n|} \frac{z_n - z}{1 - \overline{z}_n z},$$

where (z_n) are the zeroes of f, m of them being at 0, converges uniformly on compact sets to an H^∞ function the only zeroes of which are the (z_n), with the correct multiplicities. Moreover, $|B(z)| \leq 1$ and $|B(e^{i\theta})| = 1$ almost everywhere.

For any H^p function we can remove a Blaschke factor which accounts for the zeroes, as follows.

Lemma 1.4.9 (F. Riesz) *Let $f \in H^p$, $f \not\equiv 0$, and let $B(z)$ be the (possibly infinite) Blaschke product formed using the zeroes (z_n) of f. Then $f(z) = g(z)B(z)$ for some $g \in H^p$ with $\|f\|_p = \|g\|_p$.*

Corollary 1.4.10 *Any non-zero function $f \in H^1$ can be written as*

$$f = B \cdot S \cdot u,$$

where B is a Blaschke product, S a singular inner function and u an outer function. This factorization is unique up to constants of modulus 1.

In order to study singular inner functions, we recall the following result. It has two equivalent formulations, since any harmonic function in the disc is the real part of an analytic function.

Theorem 1.4.11 (G. Herglotz) *A complex-valued harmonic function u in the disc is the Poisson integral of a finite positive measure μ on $\mathbb{T}$; that is,*

$$u(re^{i\theta}) = \int_0^{2\pi} P_{r,\theta}(e^{it})\, d\mu(e^{it}),$$

if and only if it is non-negative. If h is an analytic function in the unit disc with values in the right-hand half-plane, such that $h(0) > 0$, then

$$h(re^{i\theta}) = \int_0^{2\pi} \frac{e^{it} + re^{i\theta}}{e^{it} - re^{i\theta}}\, d\mu(e^{it})$$

for some positive measure μ defined on $\mathbb{T}$.

The next result explains why an inner function without zeroes is called a singular inner function.

Corollary 1.4.12 *Let g be an inner function without zeroes. Then there is a unique positive measure μ, singular with respect to Lebesgue measure, and a constant α of modulus 1, such that*

$$g(re^{i\theta}) = \alpha \exp\left(-\int_0^{2\pi} \frac{e^{it} + re^{i\theta}}{e^{it} - re^{i\theta}}\, d\mu(e^{it})\right). \tag{1.9}$$

Formula (1.9) is very similar to (1.8), except that the integral is now taken with respect to a singular measure, rather than $k(t)\,dt$.

Example 1.4.13 Let $a \in [0, 2\pi)$ and consider the singular measure $\mu_a := \delta_a$, where δ_a is the Dirac measure at a. Then the singular inner function associated with μ_a is

$$z \mapsto \exp\left(-\frac{e^{ia} + z}{e^{ia} - z}\right). \tag{1.10}$$

In particular, $z \mapsto \exp\left(\frac{z+1}{z-1}\right)$ and $z \mapsto \exp\left(\frac{z-1}{z+1}\right)$ are singular inner functions (take $a = 0$ and $a = \pi$), and a finite product of functions of the form (1.10) is also a singular inner function.

This second version of the Herglotz theorem is also useful in applications.

Theorem 1.4.14 *Let S denote the set of harmonic functions f on $\mathbb{D}$ such that*

$$\sup_{0 \le s < 1} \int_0^{2\pi} |f(se^{it})| dt < \infty.$$

Then the transformation $\mu \mapsto P(\mu)$, where $P(\mu)$ is the Poisson integral of μ, i.e.,

$$P(\mu)(re^{i\theta}) = \int_0^{2\pi} P_{r,\theta}(e^{it})\, d\mu(e^{it}),$$

is a bijection between the set of complex measures on $\mathbb{T}$ and S.

1.4.2 Consequences of the inner–outer factorization

It is easy to check (using the Cauchy–Schwarz inequality) that the product of two H^2 functions is always in H^1. The converse, which is harder, is also true: namely, that any H^1 function can be written as the product of two H^2 functions.

Theorem 1.4.15 (Riesz factorization theorem) *A function f is in H^1 if and only if there exist $g, h \in H^2$ with $f = g \cdot h$ and $\|f\|_1 = \|g\|_2 \|h\|_2$.*

We are now ready to look at the boundary behaviour of H^p functions.

Theorem 1.4.16 *Suppose that $f \in H^1$ and f is not identically zero. Then*

$$\frac{1}{2\pi} \int_0^{2\pi} \log |f(e^{\imath\theta})| \, d\theta > -\infty,$$

and hence $f(e^{\imath\theta}) \neq 0$ almost everywhere.

The above result underlines the fact that membership in H^1 is far from being equivalent to the existence of a radial limit in $L^1(\mathbb{T})$. Indeed, Lusin (see Nikolski [154], p. 59) constructed a non-zero holomorphic function f with $\lim_{r \to 1} f(r\xi) = 0$ almost everywhere on $\mathbb{T}$.

Using the Poisson kernel and the existence of boundary values for the Hardy space H^1, we can prove the following result concerning complex measures on the unit circle.

Theorem 1.4.17 (F. and M. Riesz) *Let μ be a complex measure on $[0, 2\pi]$. If*

$$\int_0^{2\pi} e^{-int} d\mu(t) = 0$$

for all $n < 0$, then μ is absolutely continuous with respect to the Lebesgue measure.

The next result may be found in Nikolski [154, p. 45], for example.

Theorem 1.4.18 (Frostman) *Let u be an inner function and $\xi \in \mathbb{T}$. Then the composition $b_{t\xi} \circ u$ is a Blaschke product with simple zeroes for almost all $t \in (0, 1)$, where $b_\lambda(z) = \frac{\lambda - z}{1 - \bar{\lambda} z}$ for $\lambda \in \mathbb{D}$. In particular u is a uniform limit of Blaschke products with simple zeroes.*

For $1 \leq p \leq \infty$ it is useful to consider the closed subspace H_0^p of H^p, defined by

$$H_0^p = \{f \in H^p(\mathbb{D}) : f(0) = 0\}.$$

A basic fact about H^p spaces is that they are dual spaces. For $1 < p \leq \infty$ we have

$$H^p \quad \text{is isomorphically isometric to} \quad \left(L^q(\mathbb{T})/H_0^q\right)^*, \tag{1.11}$$

where $\frac{1}{p} + \frac{1}{q} = 1$. This is because H^p is closed in the the weak* topology of $L^p(\mathbb{T}) = (L^q(\mathbb{T}))^*$ and its *pre-annihilator* is

$$^{\perp}H^p(\mathbb{D}) := \{f \in L^q(\mathbb{T}) : \int_{\mathbb{D}} fg\,dm = 0 \quad \text{for all } g \in H^p(\mathbb{D})\} = H_0^q(\mathbb{D}).$$

In fact, the H^p space is reflexive for $1 < p < \infty$, being a closed subspace of the reflexive space L^p, and thus its dual space $L^q(\mathbb{T})/H_0^q$ is also reflexive. We shall discuss the dual space of H^1 in Section 1.4.6.

The next lemma characterizes the weak*-convergence of a sequence in H^∞.

Lemma 1.4.19 *A sequence $(f_n)_n$ in H^∞ weak*-converges to f in H^∞ if and only if*

$$\sup_n \|f_n\|_\infty < \infty \quad \textit{and} \quad \lim_{n\to\infty} f_n(z) = f(z) \quad \textit{for all } z \in \mathbb{D}.$$

1.4.3 The theorems of Beurling and Wiener

At this point it is appropriate to consider some very important examples of operators for which the complete lattices of invariant subspaces are known. We begin with the *bilateral shift* on $L^2(\mathbb{T})$, defined by

$$(Sf)(e^{it}) = e^{it} f(e^{it}), \qquad (f \in L^2(\mathbb{T}), \quad e^{it} \in \mathbb{T}),$$

which is a unitary operator. Note that it restricts to an operator on the invariant subspace $H^2(\mathbb{D})$, where it can be written as

$$(Sf)(z) = zf(z), \qquad (f \in H^2(\mathbb{D}), \quad z \in \mathbb{D}).$$

This is the *unilateral shift.*

We begin with the *doubly-invariant subspaces*, invariant under S and S^{-1}.

Theorem 1.4.20 (Wiener) *A closed subspace $\mathcal{M} \subset L^2(\mathbb{T})$ satisfies $S\mathcal{M} = \mathcal{M}$ if and only if $\mathcal{M} = \chi_E L^2(\mathbb{T})$ for some measurable subset $E \subset \mathbb{T}$, unique to within a set of measure zero.*

For the *singly-invariant* subspaces, we have the following classification.

Theorem 1.4.21 (Beurling–Helson) *A non-zero closed subspace $\mathcal{M} \subset L^2(\mathbb{T})$ satisfies $S\mathcal{M} \subset \mathcal{M}$ with $S\mathcal{M} \neq \mathcal{M}$ if and only if $\mathcal{M} = \phi H^2(\mathbb{D})$ for some measurable function $\phi \in L^\infty(\mathbb{T})$ such that $|\phi(e^{it})| = 1$ a.e. The function ϕ is unique to within a unimodular constant.*

The original Beurling theorem may easily be deduced from this.

Theorem 1.4.22 (Beurling) *A non-zero closed subspace $\mathcal{M} \subset H^2(\mathbb{D})$ satisfies $S\mathcal{M} \subset \mathcal{M}$ if and only if $\mathcal{M} = \phi H^2(\mathbb{D})$ for some inner function ϕ, unique to within a unimodular constant.*

These results may be found in many places, for example Garnett [112], Nikolski [154] and Partington [161].

We have a similar result for $H^\infty(\mathbb{D})$, but now we consider only the weak*-closed invariant subspaces (there is no really satisfactory classification of norm-closed invariant subspaces).

Theorem 1.4.23 *A non-zero weak*-closed subspace* $\mathcal{M} \subset H^\infty(\mathbb{D})$ *satisfies* $S\mathcal{M} \subset \mathcal{M}$ *if and only if* $\mathcal{M}$ *is an ideal in the ring* $H^\infty(\mathbb{D})$. *This holds if and only if* $\mathcal{M} = \phi H^\infty(\mathbb{D})$ *for some inner function* ϕ, *unique to within a unimodular constant.*

1.4.4 The disc algebra

The *disc algebra* $A(\mathbb{D})$ is the set of holomorphic functions $f : \mathbb{D} \to \mathbb{C}$, such that f extends to a continuous function on the closure of $\mathbb{D}$, denoted by $\overline{\mathbb{D}}$. That is,

$$A(\mathbb{D}) = \mathrm{Hol}(\mathbb{D}) \cap C(\overline{\mathbb{D}}).$$

When endowed with the pointwise addition, $(f+g)(z) = f(z) + g(z)$, and pointwise multiplication, $(fg)(z) = f(z)g(z)$, this set becomes an algebra over $\mathbb{C}$, since if f and g belong to the disc algebra then so do $f+g$ and fg.

Given the uniform norm,

$$\|f\| = \max\{|f(z)| : z \in \overline{\mathbb{D}}\},$$

it becomes a Banach algebra, and it is a sub-algebra of the Hardy space $H^\infty(\mathbb{D})$.

Note that $A(\mathbb{D})$ can also be regarded as a closed sub-algebra of $C(\mathbb{T})$, and that from the fact that $C(\mathbb{T})^* = \mathcal{M}(\mathbb{T})$ we may conclude that

$$A(\mathbb{D})^* = \mathcal{M}(\mathbb{T})/A(\mathbb{D})^\perp, \tag{1.12}$$

where $A(\mathbb{D})^\perp$ is the annihilator of $A(\mathbb{D})$ in $\mathcal{M}(\mathbb{T})$.

1.4.5 Reproducing kernels, Riesz bases and Carleson sequences

A reproducing kernel Hilbert space is a Hilbert space of functions in which pointwise evaluation is a continuous linear functional. Equivalently, it is a space that can be defined by a reproducing kernel. The subject was originally and simultaneously developed by Nachman Aronszajn and Stefan Bergman in around 1950.

Definition 1.4.24 Let $\mathcal{X}$ be an arbitrary set and $\mathcal{H}$ a Hilbert space of complex-valued functions on $\mathcal{X}$. We say that $\mathcal{H}$ is a *reproducing kernel Hilbert space* if the linear map

$$L_x : f \mapsto f(x),$$

from $\mathcal{H}$ to the complex numbers, is continuous for every x in $\mathcal{X}$. By the Riesz representation theorem, this implies that for every x in $\mathcal{X}$ there exists a unique element k_x of $\mathcal{H}$ with the property that

$$f(x) = \langle f, k_x \rangle \quad \text{for all} \quad f \in \mathcal{H}. \tag{1.13}$$

The function k_x is called the *point-evaluation functional* at the point x. Since $\mathcal{H}$ is a space of functions, the element k_x is itself a function and can therefore be evaluated at every point. We define the function

$$K : \mathcal{X} \times \mathcal{X} \to \mathbb{C}$$

by

$$K(x, y) := \overline{K_x(y)}.$$

This function is called the *reproducing kernel* for the Hilbert space $\mathcal{H}$, and it is determined entirely by $\mathcal{H}$ because the Riesz representation theorem guarantees, for every x in $\mathcal{X}$, that the element k_x satisfying (1.13) is unique.

Example 1.4.25 In the Hardy space $H^2(\mathbb{D})$, the reproducing kernels are given by

$$k_w(z) = \frac{1}{1 - \overline{w}z} \qquad (z \in \mathbb{D})$$

for $w \in \mathbb{D}$. Note that

$$\|k_w\|^2 = \langle k_w, k_w \rangle = k_w(w) = \frac{1}{1 - |w|^2}.$$

For $H^2(\mathbb{C}_+)$ the reproducing kernels are given by

$$k_w(z) = \frac{1}{2\pi} \frac{1}{\overline{w} + z} \qquad (z \in \mathbb{C}_+)$$

for $w \in \mathbb{C}_+$.

Definition 1.4.26 A sequence of vectors $(x_n)_n$ in a Hilbert space $\mathcal{H}$ is called a *Riesz sequence* if there exist constants $0 < c \le C < +\infty$ such that

$$c\left(\sum_n |a_n|^2\right) \le \left\|\sum_n a_n x_n\right\|^2 \le C\left(\sum_n |a_n|^2\right)$$

for all sequences of scalars $(a_n)_n$ in ℓ^2. A Riesz sequence $(x_n)_n$ is called a *Riesz basis* if the closed linear hull of $(x_n)_n$ is $\mathcal{H}$.

Definition 1.4.27 Let $(z_n)_n \subset \mathbb{D}$ and suppose that there exists $\delta > 0$ such that

$$\inf_m \prod_{n \neq m} \left| \frac{z_n - z_m}{1 - \overline{z_m} z_n} \right| \geq \delta. \tag{1.14}$$

Then $(z_n)_n$ is called a *Carleson sequence* in $\mathbb{D}$.

A link between the three notions, namely, reproducing kernel, Riesz basis and Carleson sequence, follows from the next result (see, for example, Nikolski [155, section C.3.3.3] or Partington [160, theorem 2.5.11]).

Theorem 1.4.28 *A sequence $\left((1 - |z_n|^2)^{1/2} k_{z_n}\right)_n$ of normalized reproducing kernels in $H^2(\mathbb{D})$ is a Riesz basis for its closed span if and only if the sequence $(z_n)_n$ is a Carleson sequence.*

We shall also require the following well-known interpolation result [56, 183].

Theorem 1.4.29 (Carleson) *Let $(z_n)_n \subset \mathbb{D}$ be a Carleson sequence satisfying (1.14). Then there exists $b > 0$ such that for all $(c_n)_n \in \ell^\infty$, there exists $f \in H^\infty(\mathbb{D})$ with $\|f\|_\infty \leq b \sup_n |c_n|$ such that $f(z_n) = c_n$ for all n.*

Remark 1.4.30 More precisely, in the above theorem, it is possible to take

$$b = \frac{2}{\delta^5}(1 - 2 \log \delta). \tag{1.15}$$

Theorem 1.4.31 *Let $z_1, \ldots, z_n$ be a finite set of distinct points in $\mathbb{D}$ satisfying (1.14). For all $c_1, \ldots, c_n \in \mathbb{C}$, there exists a rational function $f \in A(\mathbb{D})$ with $\|f\|_\infty \leq b \max_{k=1,\ldots,n} |c_k|$ such that $f(z_k) = c_k$ for all $k = 1, \ldots, n$, and where b is given by (1.15).*

It is also true (and easier to show) that (1.14) is a necessary condition for H^∞ interpolation.

Closely linked with the notion of a Carleson sequence is the following concept.

Definition 1.4.32 A positive Borel measure μ on the unit disc $\mathbb{D}$ is called a *Carleson measure* if there is a well-defined and bounded natural injection $J : H^2(\mathbb{D}) \to L^2(\mu)$.

Indeed, if $(z_n)_n \subset \mathbb{D}$ is a Carleson sequence, then it follows that the measure $\mu := \sum_{n=1}^{\infty}(1-|z_n|^2)\delta_{z_n}$ is a Carleson measure. Here again, δ_{z_n} denotes a Dirac measure (point mass) at z_n. Conversely, if this μ is a Carleson measure, then $(z_n)_n$ is a finite union of Carleson sequences.

Carleson measures can be characterized by the following result, also due to Carleson.

Theorem 1.4.33 *A positive measure μ on $\mathbb{D}$ is a Carleson measure if and only if there is a positive constant c_μ such that*

$$\mu(S_{\theta,h}) \leq c_\mu h$$

for every sector of the form $S_{\theta,h} = \{re^{it} : 1-h \leq r < 1, |t-\theta| \leq h\}$.

1.4.6 Functions of bounded mean oscillation

The space of functions of bounded mean oscillation (BMO), introduced in 1961 by John and Nirenberg [126], plays the same role in the theory of Hardy spaces that the space L^∞ of bounded functions plays in the theory of L^p-spaces.

A function u is said to be a BMO *function* if it is in $L^1_{\text{loc}}(\mathbb{R}^n)$ (the space of locally integrable functions) and the *mean oscillation* satisfies

$$\sup_Q \frac{1}{|Q|}\int_Q |u(y) - u_Q|\,dy < \infty, \tag{1.16}$$

where Q ranges over the set of all balls in $\mathbb{R}^n$. The supremum is called the BMO *norm* of u and is denoted by $\|u\|_{\text{BMO}}$. Here $|Q|$ is the volume of Q and

$$u_Q = \frac{1}{|Q|}\int_Q u(y)\,dy$$

is the average value of u on the ball Q.

The function $\|u\|_{\text{BMO}}$ becomes a norm on BMO functions on passing to the quotient by the space of constant functions (which would have norm 0).

Examples of BMO functions are all the bounded (measurable) functions (indeed, the natural inclusion of L^∞ into BMO is bounded), and $\log(|u|)$ for any polynomial u that is not identically zero. The functions in BMO are locally L^p if $0 < p < \infty$, but need not be locally bounded.

In fact, we shall more commonly consider the space BMO in the context of functions defined on $\mathbb{T}$. An analogous definition to (1.16) holds, with $|Q|$ now being interpreted as the length (or Lebesgue measure) of an arc Q.

The set of functions which are both BMO and analytic is denoted by BMOA. Fefferman [104, 105] showed that the space BMOA is the dual space of H^1, the Hardy space with $p = 1$. A pairing between $f \in H^1$ and $g \in$ BMOA is given by

$$\langle f, g\rangle = \frac{1}{2\pi}\int_0^{2\pi} f(e^{i\theta})\overline{g(e^{i\theta})}\,\mathrm{d}\theta.$$

We may identify BMOA as the image of $L^\infty(\mathbb{T})$ under the orthogonal projection (*Riesz projection*) from $L^2(\mathbb{T})$ onto $H^2(\mathbb{D})$, and this leads to the relation

$$H^\infty \subset \mathrm{BMOA} \subset \bigcap_{1\leq p<\infty} H^p,$$

with both inclusions strict.

As we have seen, functions in BMO do not need to be bounded, but the following result gives some control on their growth.

Theorem 1.4.34 (John–Nirenberg inequality) *Let $g \in BMO$, and for $\delta > 0$ write*

$$E_\delta = \{t \in (-\pi, \pi] : |f(e^{it})| > \delta\}.$$

Then there exists a constant $k > 0$ depending only on g such that

$$m(E_\delta) \leq k\exp(-k\delta) \qquad \textit{for all} \quad \delta > 0,$$

where m denotes Lebesgue measure.

1.4.7 The Hilbert transform on the unit circle

To every real trigonometric polynomial u, regarded as a function on the unit circle, one associates the real conjugate polynomial v such that $u + iv$ extends to a holomorphic function in the unit disc, where

$$u(\mathrm{e}^{\mathrm{i}\theta}) = a_0 + \sum_{k\geq 1}\big(a_k\cos(k\theta) + b_k\sin(k\theta)\big)$$

and

$$v(\mathrm{e}^{\mathrm{i}\theta}) = \sum_{k\geq 1}\big(a_k\sin(k\theta) - b_k\cos(k\theta)\big).$$

This mapping $u \mapsto v$ extends to a bounded linear operator H, the *Hilbert transform*, on $L^p(\mathbb{T}, \mathbb{R})$, when $1 < p < \infty$. However, H maps $L^1(\mathbb{T})$ into a strictly larger space called *weak-$L^1(\mathbb{T})$*. When $1 \leq p \leq \infty$, the following are equivalent for a real-valued integrable function f on the unit circle:

1. The function f is the real part of some function $g \in H^p(\mathbb{T})$.
2. The function f and its conjugate $H(f)$ belong to $L^p(\mathbb{T})$.

Theorem 1.4.35 (M. Riesz) *If $1 < p < \infty$, then the Hilbert transform H on $L^p(\mathbb{T})$ is a bounded linear operator, so that there exists a constant C_p such that*

$$\|Hu\|_p \leq C_p \|u\|_p,$$

for all $u \in L^p(\mathbb{T})$. The best constant C_p is given by

$$C_p = \begin{cases} \tan \frac{\pi}{2p} & \text{for } 1 < p \leq 2, \\ \cot \frac{\pi}{2p} & \text{for } 2 \leq p < \infty. \end{cases}$$

In particular the Hilbert transform on $L^2(\mathbb{T})$ is a contraction.

The Hilbert transform is used in Chapter 3; in particular it is applied to the argument function, arg, which by convention takes values in $(-\pi, \pi]$.

1.5 Number Theory

We shall require some results from number theory, in particular the Diophantine approximation of irrational numbers by rationals. Our main source here is Hardy and Wright [119].

Definition 1.5.1 A real number ξ is *approximable by rationals to order n* if there is a constant $K(\xi) > 0$ such that the equation

$$\left| \xi - \frac{p}{q} \right| < \frac{K(\xi)}{q^n}$$

has infinitely many solutions $p, q \in \mathbb{Z}$.

Theorem 1.5.2 (Dirichlet) *Every irrational number ξ is approximable to order 2 with $K(\xi) = 1$; however, a rational number is approximable to order 1 and no higher order.*

At the other extreme, we have a class of very well-approximable numbers.

Definition 1.5.3 A (necessarily transcendental) number is called a *Liouville number* if it is approximable to every order n.

2

The operator-valued Poisson kernel and its applications

In later chapters we shall use spectral measures and H^∞ functional calculus in order to derive results on the existence of invariant subspaces. Thus, we begin by assembling the basic tools necessary for this programme.

We begin by presenting in an elementary and self-contained way some of the properties of the so-called operator-valued Poisson kernel corresponding to a Hilbert space operator whose spectrum is contained in the closed unit disc. We study ρ-contractions with the aid of this kernel: we first show without using dilations that they are polynomially bounded, and then present the machinery of elementary spectral measures, which gives the construction of an H^∞ functional calculus in the case that the spectral measures are absolutely continuous.

In the more general context of a polynomially bounded operator T on a complex Banach space $\mathcal{X}$, when the operator-valued Poisson kernel cannot be defined in general, we construct an H^∞ functional calculus under the hypothesis that $\|T^n x\| \to 0$ for all $x \in \mathcal{X}$. We also give an alternative method, using the Valskii decomposition of a Henkin measure to obtain the absolute continuity of the spectral measures in this case.

2.1 The operator-valued Poisson kernel

Recall from Section 1.3 that the scalar Poisson kernel $P_{r,t}$ is defined by

$$P_{r,t}(e^{i\theta}) = \frac{1}{1 - re^{it}e^{-i\theta}} + \frac{1}{1 - re^{-it}e^{i\theta}} - 1.$$

Now, for $T \in \mathcal{L}(\mathcal{H})$ such that $\sigma(T) \subset \overline{\mathbb{D}}$ and for $re^{it} \in \mathbb{D}$, define the *operator-valued Poisson kernel* $K_{r,t}(T) \in \mathcal{L}(\mathcal{H})$ in the following way:

$$K_{r,t}(T) = (\mathrm{I} - re^{it}T^*)^{-1} + (\mathrm{I} - re^{-it}T)^{-1} - \mathrm{I}.$$

Remark 2.1.1 Note that $(K_{r,t}(T))^* = K_{r,t}(T)$ and $K_{r,-t}(T) = K_{r,t}(T^*)$.

As for the (scalar) Poisson kernel, we have the following identities.

Lemma 2.1.2 *For $T \in \mathcal{L}(\mathcal{H})$ such that $\sigma(T) \subset \overline{\mathbb{D}}$, we have*

$$\begin{aligned} K_{r,t}(T) &= (\mathrm{I} - re^{it}T^*)^{-1}(\mathrm{I} - r^2T^*T)(\mathrm{I} - re^{-it}T)^{-1} && (2.1)\\ &= (\mathrm{I} - re^{-it}T)^{-1}(\mathrm{I} - r^2TT^*)(\mathrm{I} - re^{it}T^*)^{-1} && (2.2)\\ &= \sum_{n\geq 0} r^n e^{int}T^{*n} + \sum_{n\geq 0} r^n e^{-int}T^n - \mathrm{I}. && (2.3) \end{aligned}$$

Proof The two first identities follow from the fact that

$$(\mathrm{I} - re^{it}T^*)K_{r,t}(T)(\mathrm{I} - re^{-it}T) = \mathrm{I} - r^2T^*T \quad \text{and}$$
$$(\mathrm{I} - re^{-it}T)K_{r,t}(T)(\mathrm{I} - re^{it}T^*) = \mathrm{I} - r^2TT^*.$$

Moreover, since $\sigma(T) \subset \overline{\mathbb{D}}$, we have $\lim_{n\to\infty} \|T^n\|^{1/n} \leq 1$. Given $r \in (0, 1)$, let $\varepsilon > 0$ be such that $r(1 + \varepsilon) < 1$. Then there exists an $N \geq 1$ such that $\|T^n\| \leq (1 + \varepsilon)^n$ for all $n \geq N$. Now, we get the convergence in norm of

$$\sum_{n\geq 0} r^n e^{int}T^{*n} + \sum_{n\geq 0} r^n e^{-int}T^n - \mathrm{I},$$

which proves the last assertion of the lemma. □

For $T \in \mathcal{L}(\mathcal{H})$ and $p \in \mathbb{C}[z]$ a polynomial, it is natural to define $p(T) \in \mathcal{L}(\mathcal{H})$ in the following way:

$$p(T) = \sum_{k=0}^{n} a_k T^k \qquad \text{if} \qquad p(z) = \sum_{k=0}^{n} a_k z^k.$$

Actually, using the operator-valued Poisson kernel, there is another way to define $p(rT)$ for $0 \leq r < 1$.

Lemma 2.1.3 *Let $T \in \mathcal{L}(\mathcal{H})$ such that $\sigma(T) \subset \overline{\mathbb{D}}$. For all $r \in [0, 1)$, we have*

$$p(rT) = \frac{1}{2\pi}\int_0^{2\pi} p(e^{it})K_{r,t}(T)\,dt, \quad p \in \mathbb{C}[z].$$

Proof Let $p(z) = \sum_{k=0}^{N} a_k z^k$. By (2.3), we have

$$K_{r,t}(T) = \sum_{n\geq 0} r^n e^{int}T^{*n} + \sum_{n\geq 0} r^n e^{-int}T^n - \mathrm{I}.$$

Since $\int_0^{2\pi} e^{ilt}\,dt = 0$ for $l \in \mathbb{Z} \setminus \{0\}$, we get

$$\begin{aligned}\int_0^{2\pi} p(e^{it})K_{r,t}(T)\,dt &= \sum_{k=0}^{N} \int_0^{2\pi} a_k r^k T^k\,dt \\ &= 2\pi \sum_{k=0}^{N} a_k r^k T^k \\ &= 2\pi p(rT).\end{aligned}$$

□

Moreover, using the operator-valued Poisson kernel, there is also another way to characterize the fact that an operator $T \in \mathcal{L}(\mathcal{H})$ is contractive.

Lemma 2.1.4 $\|T\| \le 1 \Longleftrightarrow \sigma(T) \subset \overline{\mathbb{D}}$ *and* $K_{r,t}(T) \ge 0$ *for all* $re^{it} \in \mathbb{D}$.

Proof Suppose that $T \in \mathcal{L}(\mathcal{H})$ is such that $\sigma(T) \subset \overline{\mathbb{D}}$ (automatically satisfied if $\|T\| \le 1$). Recall that $K_{r,t}(T) \ge 0$ if and only if $\langle K_{r,t}(T)h, h\rangle \ge 0$ for all $h \in \mathcal{H}$. Then, using (2.1), we have

$$\langle K_{r,t}(T)h, h\rangle = \langle (\mathrm{I} - r^2T^*T)\tilde{h}, \tilde{h}\rangle = \|\tilde{h}\|^2 - r^2\|T\tilde{h}\|^2,$$

with $\tilde{h} = (\mathrm{I} - re^{-it}T)^{-1}h$. The surjectivity of $(\mathrm{I} - re^{-it}T)^{-1}$ implies that $K_{r,t}(T) \ge 0$ if and only if $\|\tilde{h}\|^2 - r^2\|T\tilde{h}\|^2 \ge 0$ for all $\tilde{h} \in \mathcal{H}$. Now the assertion of the lemma is clear. □

More generally, we consider ρ-contractions, defined below.

Definition 2.1.5 Let $\rho \ge 1$. An operator $T \in \mathcal{L}(\mathcal{H})$ is said to be a ρ-*contraction* if its spectrum is contained in the closed unit disc and if, for all $re^{it} \in \mathbb{D}$,

$$K^{\rho}_{r,t}(T) := K_{r,t}(T) + (\rho - 1)\mathrm{I} \ge 0.$$

Obviously, by Lemma 2.1.4, the 1-contractions are the contractions in the usual sense, and the classes of ρ-contractions increase with ρ. We can also identify the 2-contractions explicitly, as follows.

Proposition 2.1.6 *An operator* $T \in \mathcal{L}(\mathcal{H})$ *is a* 2-*contraction if and only if its* numerical range $W(T) := \{\langle Tx, x\rangle : \|x\| = 1\}$ *is contained in the closed unit disc.*

Proof We first observe the classical fact that the spectrum of T is always contained in the closure of the numerical range of T. Indeed, considering $T - \lambda\mathrm{I}$, for $\lambda \in \mathbb{C}$, if necessary, it is enough to prove that if $0 \in \sigma(T)$ then $0 \in \overline{W(T)}$. So suppose $0 \in \sigma(T)$, i.e., that T is not invertible. There are two cases to consider:

- If T is not bounded below, then there exist unit vectors $x_n \in \mathcal{H}$ such that $\langle Tx_n, x_n\rangle \to 0$, and thus 0 is a limit of points in $W(T)$, and so lies in $\overline{W(T)}$.
- If T is bounded below but not onto, then $\{0\} \neq (\operatorname{Im} T)^\perp = \operatorname{Ker} T^*$, hence $0 \in W(T^*)$, and therefore $0 \in W(T)$.

Suppose that $T \in \mathcal{L}(\mathcal{H})$ satisfies $\sigma(T) \subset \overline{\mathbb{D}}$ (this is automatically the case if $W(T) \subset \overline{\mathbb{D}}$, from the above observation). Then $W(T) \subset \overline{\mathbb{D}}$ is clearly equivalent to the condition that

$$\operatorname{Re}\langle (I - re^{-it}T)x, x\rangle \geq 0 \qquad \text{for all} \quad x \in \mathcal{H} \quad \text{and} \quad re^{-it} \in \mathbb{D}.$$

This is equivalent, since $I - re^{-it}T$ is surjective, to the condition that

$$\operatorname{Re}\langle y, (I - re^{-it}T)^{-1}y\rangle \geq 0 \qquad \text{for all} \quad y \in \mathcal{H} \quad \text{and} \quad re^{-it} \in \mathbb{D},$$

which in turn is equivalent to the condition that the operator $(I - re^{-it}T)^{-1} + [(I - re^{-it}T)^{-1}]^*$ is positive. □

Using this result it is easy to give explicit examples of 2-contractions which are not contractions (see the Exercises).

Now, let us present some applications of Lemmas 2.1.3 and 2.1.4. The first one has as a corollary a dilation-free proof of the well-known *von Neumann inequality*; the proof was discovered by Heinz [121].

Theorem 2.1.7 *If T is a ρ-contraction, then, for all $p \in \mathbb{C}[z]$, we have*

$$\|p(T)\| \leq (2\rho - 1)\|p\|_{\infty,\mathbb{D}}. \tag{2.4}$$

Proof Let $x, y \in \mathcal{H}$. By Lemma 2.1.3, we have

$$\langle p(rT)x, y\rangle = \frac{1}{2\pi}\int_0^{2\pi} p(e^{it})\big\langle K^\rho_{r,t}(T)x, y\big\rangle\, dt + (1-\rho)p(0)\langle x, y\rangle$$

for $r \in [0, 1)$.

Since T is a ρ-contraction, we have also

$$\big\langle K^\rho_{r,t}(T)x, y\big\rangle = \langle \tilde{x}_{r,t}, \tilde{y}_{r,t}\rangle,$$

where $\tilde{x}_{r,t} = \sqrt{K^\rho_{r,t}(T)}x$ and $\tilde{y}_{r,t} = \sqrt{K^\rho_{r,t}(T)}y$. Therefore we get

$$|\langle p(rT)x, y\rangle| \leq \frac{\|p\|_\infty}{2\pi}\int_0^{2\pi} |\langle \tilde{x}_{r,t}, \tilde{y}_{r,t}\rangle|\, dt + (\rho - 1)|p(0)||\langle x, y\rangle|.$$

Now, applying the Cauchy–Schwarz inequality twice, we get

$$\int_0^{2\pi} |\langle \tilde{x}_{r,t}, \tilde{y}_{r,t}\rangle|\, dt \le \int_0^{2\pi} \|\tilde{x}_{r,t}\|\,\|\tilde{y}_{r,t}\|\, dt$$
$$\le \left(\int_0^{2\pi} \|\tilde{x}_{r,t}\|^2\, dt\right)^{1/2} \left(\int_0^{2\pi} \|\tilde{y}_{r,t}\|^2\, dt\right)^{1/2}.$$

Using Lemma 2.1.3 with p identically equal to 1, we get

$$\int_0^{2\pi} \|\tilde{x}_{r,t}\|^2\, dt = \int_0^{2\pi} \left\langle K^{\rho}_{r,t}(T)x, x\right\rangle dt = 2\pi\rho\|x\|^2.$$

Finally, we get

$$|\langle p(rT)x, y\rangle| \le (2\rho - 1)\|p\|_\infty \|x\|\,\|y\|,$$

which implies that $\|p(rT)\| \le (2\rho - 1)\|p\|_\infty$. Since

$$\|p(rT) - p(T)\| \le \sum_{k=0}^{N} (1 - r^k)|a_k| \qquad \text{for} \quad p(z) = \sum_{k=0}^{N} a_k z^k,$$

in particular, we obtain $\lim_{r\to 1} \|p(rT)\| = \|p(T)\|$ and thus (2.4) holds. □

Corollary 2.1.8 (von Neumann inequality) *Let $T \in \mathcal{L}(\mathcal{H})$ be a contraction. Then for all $p \in \mathbb{C}[z]$ we have*

$$\|p(T)\| \le \|p\|_{\infty,\mathbb{D}}.$$

The second application will be used in the construction of a functional calculus.

Theorem 2.1.9 *Let T be a ρ-contraction. For all $x, y \in \mathcal{H}$, there exists a unique complex measure $\mu^T_{x,y} \in \mathcal{M}(\mathbb{T})$ such that*

$$\langle K_{r,t}(T)x, y\rangle = \int_0^{2\pi} P_{r,t}(e^{i\theta})\, d\mu^T_{x,y}(e^{i\theta}),$$

and there exists a unique function in $L^1(\mathbb{T})$, denoted by $x \overset{T}{\cdot} y$, such that

$$(x \overset{T}{\cdot} y)(e^{it}) = \lim_{r\to 1^-} \langle K_{r,t}(T)x, y\rangle \tag{2.5}$$

and $\|x \overset{T}{\cdot} y\|_1 \le (2\rho - 1)\|x\|\,\|y\|$.

Proof For $z = re^{it} \in \mathbb{D}$, set $\varphi(z) = \langle K_{r,t}(T)x, y\rangle$. By (2.3), we have

$$\varphi(z) = \sum_{n\ge 0} \langle T^{*n}x, y\rangle z^n + \sum_{n\ge 0} \langle T^n x, y\rangle \bar{z}^n - \langle x, y\rangle,$$

and thus φ is a harmonic function on $\mathbb{D}$. In the proof of Theorem 2.1.7, we have seen that

$$\int_0^{2\pi} |\langle K_{r,t}(T)x, y\rangle|\, dt \leq 2\pi(2\rho - 1)\|x\|\,\|y\|.$$

Therefore we have

$$\sup_{0\leq r<1} \int_0^{2\pi} |\varphi(re^{it})|\, dt \leq 2\pi(2\rho - 1)\|x\|\,\|y\| < \infty. \tag{2.6}$$

Applying Theorem 1.3.1, there exists a unique complex measure $\mu_{x,y}^T$ on $\mathbb{T}$ such that

$$\varphi(re^{it}) = \int_0^{2\pi} P_{r,t}(e^{i\theta})\, d\mu_{x,y}^T(e^{i\theta})$$

and, moreover, $\lim_{r\to 1}\varphi(re^{it})$ is well-defined almost everywhere on $\mathbb{T}$ and is equal to an $L^1(\mathbb{T})$-function, to be denoted by $x \overset{T}{\cdot} y$, which is the Radon–Nikodým derivative of the elementary measure $\mu_{x,y}^T$. By Theorem 1.3.1, we have also $\|\mu_{x,y}^T\| = \frac{1}{2\pi}\sup_{0\leq r<1}\int_0^{2\pi}|\varphi(re^{it})|\, dt$. Since $\|x \overset{T}{\cdot} y\|_1 \leq \left\|\mu_{x,y}^T\right\|$, we obtain by (2.6) that $\|x \overset{T}{\cdot} y\|_1 \leq (2\rho - 1)\|x\|\,\|y\|$. □

Remark 2.1.10 Since $K_{r,t}(T)^* = K_{r,t}(T)$ and $K_{r,t}(T^*) = K_{r,-t}(T)$, we have

$$x \overset{T}{\cdot} y = \overline{y \overset{T}{\cdot} x}, \quad (x \overset{T^*}{\cdot} y)(e^{it}) = (x \overset{T}{\cdot} y)(e^{-it})$$

and so

$$(y \overset{T^*}{\cdot} x)(e^{it}) = \overline{(x \overset{T}{\cdot} y)(e^{-it})}.$$

The last result of this section concerns the possibility of extending analytically the density $x \overset{T}{\cdot} y$.

Proposition 2.1.11 *Let $T \in \mathcal{L}(\mathcal{H})$ be a ρ-contraction such that $\Gamma(T) := \sigma(T) \cap \mathbb{T} \neq \mathbb{T}$. For any $x, y \in \mathcal{H}$ and any closed arc $I \in \mathbb{T} \setminus \Gamma(T)$, the function $x \overset{T}{\cdot} y$ extends analytically in a neighbourhood of I.*

Proof By (2.5), for $e^{it} \notin \Gamma(T)$, we get exactly

$$(x \overset{T}{\cdot} y)(e^{it}) = \langle (\mathrm{I} - e^{it}T^*)^{-1}(\mathrm{I} - T^*T)(\mathrm{I} - e^{-it}T)^{-1}x, y\rangle.$$

Hence, the function defined by

$$z \longmapsto z\langle (\mathrm{I} - zT^*)^{-1}(\mathrm{I} - T^*T)(z\mathrm{I} - T)^{-1}x, y\rangle$$

is an analytic extension of $x \overset{T}{\cdot} y$ in a neighbourhood of I. □

2.2 The H^∞ functional calculus for absolutely continuous ρ-contractions

A reformulation of Theorem 2.1.7 is the following: The mapping $\Phi_T : \mathbb{C}[z] \to \mathcal{L}(\mathcal{H})$ defined by $\Phi_T(p) = p(T)$ is norm-continuous with $\|\Phi_T\| \leq (2\rho - 1)$ and thus it extends continuously to $\widetilde{\Phi_T}$ (similarly satisfying $\|\widetilde{\Phi_T}\| \leq (2\rho - 1)$) on $A(\mathbb{D})$. In other words,

$$\widetilde{\Phi_T}(f) = \lim_{n\to\infty} p_n(T) \quad \text{(norm limit)},$$

where $(p_n)_{n\geq 1}$ is a sequence of polynomials that converges uniformly to f on $\mathbb{D}$.

Proposition 2.2.1 *Let $T \in \mathcal{L}(\mathcal{H})$ be a ρ-contraction. Then for all $x, y \in \mathcal{H}$, there exists a unique complex measure $\mu^T_{x,y}$ on $\mathbb{T}$ such that*

$$\langle f(T)x, y\rangle = \int_0^{2\pi} f(e^{it})\, d\mu^T_{x,y}(e^{it}), \tag{2.7}$$

for $f \in A(\mathbb{D})$, and

$$\langle K_{r,t}(T)x, y\rangle = \int_0^{2\pi} P_{r,t}(e^{i\theta})\, d\mu^T_{x,y}(e^{i\theta}).$$

Proof Let $\mu^T_{x,y}$ be the measure defined in Theorem 2.1.9. It remains to check that (2.7) holds. Let f be a function in $A(\mathbb{D})$ and let $(p_n)_{n\geq 1}$ be a sequence of polynomials that converges uniformly to f on $\mathbb{D}$. By definition, we have

$$\lim_{n\to\infty} \|f(rT) - p_n(rT)\| = 0.$$

By Lemma 2.1.3, we have also

$$p_n(rT) = \frac{1}{2\pi}\int_0^{2\pi} p_n(e^{it}) K_{r,t}(T)\, dt.$$

Since $\lim_{n\to\infty} \|p_n - f\|_\infty = 0$, it follows that

$$f(rT) = \frac{1}{2\pi}\int_0^{2\pi} f(e^{it}) K_{r,t}(T)\, dt,$$

and therefore

$$\langle f(rT)x, y\rangle = \frac{1}{2\pi}\int_0^{2\pi} f(e^{it}) \langle K_{r,t}(T)x, y\rangle\, dt.$$

Now, since we have

$$\langle K_{r,t}(T)x, y\rangle = \int_0^{2\pi} P_{r,t}(e^{i\theta})\, d\mu^T_{x,y}(e^{i\theta}),$$

by Fubini's theorem (see (1.1)) and the fact that

$$\frac{1}{2\pi}\int_0^{2\pi} f(e^{it})P_{r,\theta}(e^{i\theta})\,dt = f(re^{i\theta}),$$

it follows that

$$\langle f(rT)x, y\rangle = \int_0^{2\pi} f(re^{i\theta})\,d\mu^T_{x,y}(e^{i\theta}).$$

Define $f_r \in A(\mathbb{D})$ by $f_r(z) = f(rz)$. Since $\lim_{r\to 1}\|f_r - f\|_\infty = 0$, the continuity of the $A(\mathbb{D})$ functional calculus $\widetilde{\Phi_T}$ implies that (2.7) holds. □

Now, we would like to extend the $A(\mathbb{D})$ functional calculus into an H^∞ functional calculus; however, the above construction does not produce a norm-continuous functional calculus, and, indeed, it is more useful to work with a weak*-continuous functional calculus. Recall from (1.11) that H^∞ is isometrically isomorphic to the dual of the Banach space $L^1(\mathbb{T})/H^1_0$. Moreover, $\mathcal{L}(\mathcal{H})$ is isometrically the dual of the Banach space $\mathcal{S}_1(\mathcal{H})$ as in (1.4). Thus there is a natural definition of a weak*-weak* continuous functional calculus, obtained by giving each of these spaces its weak* topology. (We shall see a Banach-space version of this in Section 2.3.)

Definition 2.2.2 A ρ-contraction $T \in \mathcal{L}(\mathcal{H})$ is said to be *absolutely continuous* if, for all $x, y \in \mathcal{H}$, the elementary spectral measure $\mu^T_{x,y}$ satisfying

$$\langle K_{r,t}(T)x, y\rangle = \int_0^{2\pi} P_{r,t}(e^{i\theta})\,d\mu^T_{x,y}(e^{i\theta})$$

is absolutely continuous with respect to Lebesgue measure. In other words, using the notation of Theorem 2.1.9, T is an absolutely continuous ρ-contraction if $d\mu^T_{x,y}(e^{it}) = (x \overset{T}{\cdot} y)(e^{it})\,\frac{dt}{2\pi}$ for all $x, y \in \mathcal{H}$.

Proposition 2.2.3 *Let $T \in \mathcal{L}(\mathcal{H})$ be an absolutely continuous ρ-contraction. Then T has a weak*-weak* continuous, norm-contractive and multiplicative H^∞ functional calculus Ψ_T extending its $A(\mathbb{D})$ functional calculus and defined by*

$$\Psi_T(f) := f(T) \qquad (f \in H^\infty),$$

where, for all $x, y \in \mathcal{H}$,

$$\langle f(T)x, y\rangle = \frac{1}{2\pi}\int_0^{2\pi} f(e^{it})(x \overset{T}{\cdot} y)(e^{it})\,dt, \tag{2.8}$$

with $(x \overset{T}{\cdot} y)(e^{it}) = \lim_{r\to 1^-}\langle K_{r,t}(T)x, y\rangle$.

We have an explicit expression of the Fourier coefficients of the density of the elementary spectral measures.

Lemma 2.2.4 *Let $T \in \mathcal{L}(\mathcal{H})$ be an absolutely continuous ρ-contraction. For all $x, y \in \mathcal{H}$, we have*

$$\widehat{x \overset{T}{\cdot} y}(n) = \begin{cases} \langle T^{*n}x, y\rangle & \text{if} \quad n \geq 0, \\ \langle T^{|n|}x, y\rangle & \text{if} \quad n \leq 0. \end{cases}$$

Proof Using (2.8) with $f(z) = z^n$, $n \geq 0$, we get

$$\langle T^n x, y\rangle = \frac{1}{2\pi}\int_0^{2\pi} e^{int}(x \overset{T}{\cdot} y)(e^{it})\,dt = \widehat{x \overset{T}{\cdot} y}(-n).$$

Moreover, replacing T by T^*, we have also

$$\langle T^{*n}x, y\rangle = \frac{1}{2\pi}\int_0^{2\pi} e^{int}(x \overset{T^*}{\cdot} y)(e^{it})\,dt.$$

Now, by Remark 2.1.10, we get

$$\begin{aligned} \langle T^{*n}x, y\rangle &= \frac{1}{2\pi}\int_0^{2\pi} e^{int}(x \overset{T}{\cdot} y)(e^{-it})\,dt \\ &= \frac{1}{2\pi}\int_0^{2\pi} e^{-int}(x \overset{T}{\cdot} y)(e^{it})\,dt = \widehat{x \overset{T}{\cdot} y}(n). \end{aligned}$$

□

Remark 2.2.5 Obviously, when $T \in \mathcal{L}(\mathcal{H})$ is an absolutely continuous contraction, T has also a weak*-weak* continuous, norm-contractive $L^\infty(\mathbb{T})$ functional calculus (extending its H^∞ functional calculus), defined as follows for $f \in L^\infty$ and for all $x, y \in \mathcal{H}$:

$$\langle f(T)x, y\rangle = \frac{1}{2\pi}\int_0^{2\pi} f(e^{it})(x \overset{T}{\cdot} y)(e^{it})\,dt, \tag{2.9}$$

where $(x \overset{T}{\cdot} y)(e^{it}) = \lim_{r\to 1^-}\langle K_{r,t}(T)x, y\rangle$. Nevertheless, this L^∞ functional calculus is multiplicative if and only if T is unitary, which is a very restrictive condition.

We now mention one application to invariant subspaces, which makes use of the H^∞ functional calculus established in Proposition 2.2.3.

Definition 2.2.6 A completely non-unitary contraction $T \in \mathcal{L}(\mathcal{H})$ is said to be of *class* C_0 if there exists a non-trivial function $h \in H^\infty$ such that $h(T) = 0$.

Theorem 2.2.7 *Let $T \in \mathcal{L}(\mathcal{H})$ be a completely non-unitary contraction of class C_0. Then either T is a scalar multiple of the identity or it has a non-trivial hyperinvariant subspace.*

Proof Consider the set

$$J := \{h \in H^\infty : h(T) = 0\}.$$

This is a non-zero ideal in H^∞, and it is weak*-closed, since the functional calculus is weak*-weak* continuous. Thus, by Theorem 1.4.23, there is an inner function $m_T \in H^\infty$, the *minimal function* of T, such that $J = m_T H^\infty$.

If m_T is an elementary Blaschke factor, then T is a scalar multiple of the identity. Otherwise, it follows from the discussion of inner functions in Section 1.4.1 that m_T has a non-trivial factorization into two inner functions $m_T = \theta\phi$, for if m_T has both a Blaschke factor and a singular factor, we may take one of them for θ; if $m(T)$ is singular, then we may take θ to be its square root; if m_T is a Blaschke product with more than one factor, we may take θ to be a Blaschke factor.

Now $\mathcal{M} = \operatorname{Ker}\theta(T)$ is a hyperinvariant subspace for T. Since m_T does not divide θ, we have $\mathcal{M} \neq \mathcal{H}$, and similarly $y = \phi(T)x \neq 0$ for some $x \in \mathcal{H}$ and $y \in \mathcal{M}$; hence $\mathcal{M}$ is non-trivial. □

2.3 H^∞ functional calculus in a complex Banach space

Let $\mathcal{X}$ be a complex Banach space, and denote by $\mathcal{X}^*$ its topological dual. The aim of this section is to study under which simple conditions on an operator $T \in \mathcal{L}(\mathcal{X})$ one can construct a weak*-continuous H^∞ functional calculus for T.

Definition 2.3.1 Let $T \in \mathcal{L}(\mathcal{X})$. We say that T has an H^∞ *functional calculus* if there exists a homomorphism of algebras $\Phi_T : H^\infty \to \mathcal{L}(\mathcal{X})$ such that $\Phi_T(p) = p(T)$ for any polynomial p. Whenever T has an H^∞ functional calculus, $\Phi_T(f)$ is denoted by $f(T)$.

Note that a necessary condition for the existence of a norm-continuous H^∞ functional calculus for T is obviously that T satisfies the following condition.

Definition 2.3.2 An operator $T \in \mathcal{L}(\mathcal{X})$ is said to be *polynomially bounded* if there exists $C > 0$ such that

$$\|p(T)\| \leq C \sup_{z \in \mathbb{D}} |p(z)|$$

for all polynomials p.

The set of polynomially bounded operators is denoted $PB(\mathcal{X})$.

Given that this necessary condition is satisfied, the density of the polynomials in the disc algebra $A(\mathbb{D})$ implies that every $T \in PB(\mathcal{X})$ has a continuous $A(\mathbb{D})$ functional calculus. Indeed, for any sequence $(p_n)_n$ uniformly converging to f, we may define $f(T)$ as the norm limit of the Cauchy sequence $(p_n(T))_n$.

The above construction does not provide a norm-continuous H^∞ functional calculus for $T \in PB(\mathcal{X})$, and, as in Section 2.2, it is more useful to work with a weak*-continuous functional calculus. Recalling once more that H^∞ is isometrically isomorphic to the dual space of a Banach space, namely $L^1(\mathbb{T})/H_0^1$ as in (1.11), we shall also require the result of Lemma 1.4.19, that a sequence $(f_n)_n$ in H^∞ is weak*-convergent to f in H^∞ if and only if $\sup_n \|f_n\|_\infty < \infty$ and $f_n(z) \to f(z)$ for all $z \in \mathbb{D}$. Such a sequence is called a *Montel sequence.*

We now recall the definition of some useful and standard topologies on $\mathcal{L}(\mathcal{X})$.

Definition 2.3.3 A sequence $(T_n)_n$ in $\mathcal{L}(\mathcal{X})$ converges to $T \in \mathcal{L}(\mathcal{X})$ in the *weak operator topology (wot)* if, for all $(x, x^*) \in \mathcal{X} \times \mathcal{X}^*$,

$$\lim_{n\to\infty} \langle T_n x, x^* \rangle = \langle Tx, x^* \rangle.$$

A sequence $(T_n)_n$ in $\mathcal{L}(\mathcal{X})$ converges to $T \in \mathcal{L}(\mathcal{X})$ in the *strong operator topology (sot)* if, for all $x \in \mathcal{X}$,

$$\lim_{n\to\infty} \|T_n x - Tx\| = 0.$$

We are now ready to define a weak*-continuous H^∞ functional calculus in a Banach-space setting.

Definition 2.3.4 We say that T has a *weak*-continuous* H^∞ *functional calculus* Φ_T if T has an H^∞ functional calculus Φ_T and if, whenever a sequence $(f_n)_n$ tends to zero in the weak* topology, $(f_n(T))_n$ tends to zero in the weak operator topology.

One of the motivations for constructing a weak*-continuous H^∞ functional calculus is that, given $(x, x^*) \in \mathcal{X} \times \mathcal{X}^*$, the mapping

$$f \mapsto \langle f(T)x, x^* \rangle$$

is a weak*-continuous linear form and therefore is an element of the predual of H^∞, namely $L^1(\mathbb{T})/H_0^1$, which we denote by $[x \overset{T}{\cdot} x^*]$.

Note also that if T has a weak*-continuous H^∞ functional calculus, it follows that

$$\begin{array}{rcl} B_T : \mathcal{X} \times \mathcal{X}^* & \to & L^1(\mathbb{T})/H_0^1 \\ (x, x^*) & \mapsto & [x \overset{T}{\cdot} x^*] \end{array}$$

is a continuous bilinear map, since $\|[x \overset{T}{\cdot} x^*]\| \leq \|\Phi_T\|\|x\|\|x^*\|$. In Section 3.5, we will study the properties of such bilinear mappings.

Here is a result providing a simple condition on $T \in \mathcal{L}(\mathcal{X})$ for the existence of a weak*-continuous H^∞ functional calculus associated with T.

Theorem 2.3.5 *Let $T \in PB(\mathcal{X})$. If*

$$\lim_{n\to\infty} \|T^n x\| = 0 \qquad \textit{for all} \quad x \in \mathcal{X}, \tag{2.10}$$

then T has a norm-continuous and a weak-continuous H^∞ functional calculus such that for every sequence $(f_n)_n$ in H^∞ converging to zero in the weak* topology, the sequence $(f_n(T))_n$ tends to zero in the strong operator topology.*

In order to prove this theorem, we need the following observation.

Lemma 2.3.6 *Let $x \in \mathcal{X}$, $f \in H^\infty$ and $T \in PB(\mathcal{X})$. Let $(f_n)_n \subset A(\mathbb{D})$ be such that $(f_n)_n$ weak*-converges to f. Then $(f_n(T)x)_n$ is a Cauchy sequence in $\mathcal{X}$ whenever (2.10) is satisfied.*

In the particular case where $(f_n)_n$ weak-converges to 0, (2.10) implies that $(f_n(T))_n$ sot converges to 0.*

Proof Without loss of generality we may assume that $\|x\| \leq 1$, $\|f\|_\infty \leq 1$ and $\|f_n\|_\infty \leq M$ with $M \geq 1$. Let $C > 0$ such that $\|p(T)\| \leq C\|p\|_\infty$ for all polynomials p. Let $0 < \varepsilon < 2C$ and $F_{n,p} := f_n - f_{n+p}$ for any non-negative integers n and p. If (2.10) is satisfied, let k be such that

$$\|T^k x\| \leq \frac{\varepsilon}{6CM}.$$

Using the Cauchy formula, for all $n, j \geq 0$ and $0 < r < 1$ we have

$$\frac{f_n^{(j)}(0)}{j!} = \frac{1}{2i\pi} \int_{|\xi|=r} \frac{f_n(\xi)}{\xi^{j+1}} d\xi.$$

Since $(f_n)_n$ is bounded and converges pointwise to f, using the dominated-convergence theorem, $\left(f_n^{(j)}(0)\right)_n$ converges to $f^{(j)}(0)$ and thus $\left(f_n^{(j)}(0)\right)_n$

is a Cauchy sequence. Therefore, there exists an integer N such that, for all $n \geq N$, $p \geq 0$ and $j < k$,

$$\left| \frac{F_{n,p}^{(j)}(0)}{j!} \right| < \frac{\varepsilon}{2Ck}.$$

Now we write $F_{n,p}(z) = g_{n,p}(z) + z^k h_{n,p}(z)$, where $g_{n,p}(z) = \sum_{j<k} \frac{F_{n,p}^{(j)}(0)}{j!} z^j$. So, we have

$$\|g_{n,p}\|_\infty \leq \frac{\varepsilon}{2C} \quad \text{and} \quad \|h_{n,p}\|_\infty \leq \|F_{n,p}\|_\infty + \|g_{n,p}\|_\infty \leq 3M.$$

Under assumption (2.10), we get, for all $n \geq N$ and $p \geq 0$,

$$\begin{aligned} \|f_n(T)x - f_{n+p}(T)x\| &\leq \|g_{n,p}(T)x\| + C\|h_{n,p}\|_\infty \|T^k x\| \\ &\leq C\|g_{n,p}\|_\infty \|x\| + 3CM \frac{\varepsilon}{6CM} \\ &\leq \frac{\varepsilon}{2} + \frac{\varepsilon}{2} = \varepsilon. \end{aligned}$$

In the particular case where $(f_n)_n$ weak*-converges to 0, the assertion of the lemma follows along the same lines, decomposing f_n instead of $F_{n,p}$. □

Proof of Theorem 2.3.5 By Lemma 2.3.6, under assumption (2.10), for each $x \in \mathcal{X}$ the sequence $(f_n(T)x)_n$ converges and its limit is independent of the choice of the sequence (f_n) weak*-convergent to f. Denote by $f(T)$ the sot limit of $(f_n(T))_n$, where (f_n) is an arbitrary sequence of $A(\mathbb{D})$ weak*-convergent to f. Note that such sequences $(f_n)_n$ always exist and we can even guarantee that $\|f_n\|_\infty \leq \|f\|_\infty$, taking for example $f_n(z) := f\left(\left(1 - \frac{1}{1+n}\right) z\right)$. It is clear that $f \mapsto f(T)$ is linear and, moreover, since

$$\begin{aligned} \|f(T)\| &= \sup_{\|x\|\leq 1} \|f(T)x\| \\ &= \sup_{\|x\|\leq 1} \lim_n \|f_n(T)x\|| \\ &\leq \sup_{\|x\|\leq 1} \limsup_n C\|f_n\|_\infty \|x\| \\ &\leq C\|f\|_\infty, \end{aligned}$$

it follows that $f \mapsto f(T)$ is also bounded on H^∞.

Now let us check that condition (2.10) implies that T has in fact a weak*-continuous H^∞ functional calculus. Let $(f_n)_n$ in H^∞ be such that $(f_n)_n$ weak*-converges to 0. We may assume that $\|f_n\|_\infty \leq 1$. Set

$$g_n(z) = f_n((1 - 2^{-n-1})z).$$

Then $g_n \in A(\mathbb{D})$, $\|g_n\|_\infty \le \|f_n\|_\infty$ and $(f_n - g_n)$ weak*-converges to 0. By definition of $f_n(T)$, condition (2.10) implies that, given $x \in \mathcal{X}$,

$$\lim_{n\to\infty} \|(f_n(T) - g_n(T))x\| = 0. \tag{2.11}$$

Since $(f_n)_n$ weak*-converges to 0, the same is true for $(g_n)_n$. By Lemma 2.3.6, condition (2.10) implies that $(g_n(T))_n$ sot converges to 0, and thus $(f_n(T))_n$ sot (and thus wot) converges to 0.

Finally, it remains to check that $fg(T) = f(T)g(T)$ for all $f, g \in H^\infty$. Let $f \in A(\mathbb{D})$ and $g \in H^\infty$. Take $(g_n)_n \in A(\mathbb{D})$ with $(g_n - g)$ weak*-convergent to 0. In the following discussion, all the limits in $\mathcal{X}$ are norm limits. We have

$$\begin{aligned}(fg)(T)x &= \lim_{n\to\infty} (fg_n)(T)x \\ &= \lim_{n\to\infty} f(T)g_n(T)x \\ &= f(T)g(T)x.\end{aligned}$$

Now, for $f, g \in H^\infty$, taking $(g_n) \in A(\mathbb{D})$ with $(g_n - g)$ weak*-convergent to 0, and using the identity $fg_n(T)x = f(T)g_n(T)x$, we have

$$\begin{aligned}\lim_{n\to\infty} fg_n(T)x &= \lim_{n\to\infty} f(T)g_n(T)x \\ &= f(T)g(T)x.\end{aligned}$$

Since $(fg_n - fg)$ is weak*-convergent to 0 and since $f \mapsto f(T)$ is weak*-continuous on H^∞, it follows that

$$\lim_{n\to\infty} fg_n(T)x = fg(T)x.$$

Therefore, for all $x \in \mathcal{X}$, $fg(T)x = f(T)g(T)x$, which proves that $f \mapsto f(T)$ is an algebra homomorphism. □

2.4 Absolutely continuous elementary spectral measures

We now present an alternative approach to the construction of an H^∞ functional calculus for polynomially bounded operators on a Banach space. Provided $\lim_{n\to\infty} \|T^n u\| = 0$ for all $u \in \mathcal{X}$, we can represent $g(T)$, with $g \in H^\infty$, by means of absolutely continuous elementary spectral measures.

Definition 2.4.1 A measure $\mu \in \mathcal{M}(\mathbb{T})$ is a *Henkin measure* if

$$\lim_{n\to\infty} \int_{\mathbb{T}} f_n \, d\mu = 0$$

for every Montel sequence $(f_n)_n$ in $A(\mathbb{D})$ weak*-converging to 0.

Theorem 2.4.2 (Valskii's decomposition) *If μ is a Henkin measure on $\mathbb{T}$, then there exists $f \in L^1(\mathbb{T})$ such that $\mu = fm$ (i.e., $d\mu = f\,dm$), where m is the Lebesgue measure on the unit circle $\mathbb{T}$.*

Proof Let $A(\mathbb{D})^*$ be the dual space of $A(\mathbb{D})$, which is isometrically isomorphic to the quotient space $\mathcal{M}(\mathbb{T})/A(\mathbb{D})^{\perp}$ (see (1.12)). For $\mu \in \mathcal{M}(\mathbb{T})$, let $\|\mu\|_{A(\mathbb{D})^*}$ denote its norm as a linear functional on $A(\mathbb{D})$, and let $\|\mu\|$ be its total variation, as usual. The proof breaks into two steps.

Step 1. If σ is a Henkin measure and $\varepsilon > 0$, then there exists $h \in L^1(m)$ such that $\|h\|_1 \leq \|\sigma\|$ and $\|\sigma - hm\|_{A(\mathbb{D})^*} < \varepsilon$.

To prove this, put $u = P(\sigma)$, the Poisson integral of σ, and let $u_r(\xi) := u(r\xi)$ where $\xi \in \mathbb{T}$ and $0 < r < 1$. We claim that

$$\lim_{r\to 1} \|\sigma - u_r m\|_{A(\mathbb{D})^*} = 0. \tag{2.12}$$

Since $\|u_r\|_1 \leq \|\sigma\|$, (2.12) implies that $h = u_r$ has the desired properties if r is close enough to 1. Assume, to reach a contradiction, that (2.12) fails. Then there exist $\delta > 0$, a positive sequence $(r_i)_i$ increasing to 1, and $f_i \in A(\mathbb{D})$ with $\|f_i\|_\infty \leq 1$, such that

$$\left| \int_{\mathbb{T}} f_i d\sigma - \int_{\mathbb{T}} f_i u_{r_i} dm \right| \geq \delta, \tag{2.13}$$

for $i = 1, 2, \ldots$. Since $u_r = P(\sigma)_r$, Fubini's theorem gives

$$\int_{\mathbb{T}} f u_r\, dm = \int_{\mathbb{T}} f_r\, d\sigma,$$

for $f \in A(\mathbb{D})$ and $0 < r < 1$. Thus (2.13) becomes

$$\left| \int_{\mathbb{T}} (f_i(\xi) - f_i(r_i\xi)) d\sigma(\xi)\, dm \right| \geq \delta. \tag{2.14}$$

But if $g_i(z) = f_i(z) - f_i(r_i z)$ for $z \in \overline{\mathbb{D}}$, then $(g_i)_i$ is a Montel sequence (since $(f_i)_i$ is equicontinuous on every compact subset of $\mathbb{D}$) and the assumption that σ is a Henkin measure shows therefore that $\int_{\mathbb{T}} g_i d\sigma$ tends to 0 as $i \to \infty$, contradicting (2.14).

Step 2. We now complete the proof of the theorem. Choose $\varepsilon_i > 0$ so that $\varepsilon_1 > \|\mu\|_{A(\mathbb{D})^*}$ and $\sum_{i=1}^{\infty} \varepsilon_i < \infty$. Put $\mu_1 = \mu$ and make the induction hypothesis that $k \geq 1$ and that μ_k is a Henkin measure with $\|\mu_k\|_{A(\mathbb{D})^*} < \varepsilon_k$. By the Hahn–Banach theorem, this means that $\|\mu_k - \nu_k\| < \varepsilon_k$ for some

$\nu_k \in A(\mathbb{D})^\perp$. Apply Step 1 to $\mu_k - \nu_k$: there is an $h_k \in L^1(m)$ such that $\|h_k\|_1 < \varepsilon_k$ and

$$\|\mu_k - \nu_k - h_k m\|_{A(\mathbb{D})^*} < \varepsilon_{k+1}.$$

Define $\mu_{k+1} = \mu_k - \nu_k - h_k m$, and proceed. It follows, for $k = 1, 2, \ldots$, that

$$\mu = \mu_{k+1} + \sum_{i=1}^{k} \nu_i + \sum_{i=1}^{k} h_i m.$$

Put $g = \sum_{i=1}^{k} h_i$. Then $g \in L^1(m)$, $\|g\|_1 \leq \sum_{i=1}^{\infty} \varepsilon_i$ and

$$\mu - gm = \mu_{k+1} + \sum_{i=1}^{k} \nu_i - \sum_{i=k+1}^{\infty} h_i m.$$

Since $\nu_i \in A(\mathbb{D})^\perp$, it follows that

$$\|\mu - gm\|_{A(\mathbb{D})^*} \leq \|\mu_{k+1}\|_{A(\mathbb{D})^*} + \sum_{i=k+1}^{\infty} \|h_i\|_1. \tag{2.15}$$

Since the right side of (2.15) tends to 0 as $k \to \infty$, $\mu - gm$ is in $A(\mathbb{D})^\perp$. Now recall that the F. and M. Riesz theorem (Theorem 1.4.17) implies that $A(\mathbb{D})^\perp \subset L^1(m)$. Therefore there exists an $h \in L^1(m)$ such that $\mu - gm = hm$, and then the proof is complete taking $f = g + h$. □

We now have all the tools to get the following result on elementary spectral measures.

Theorem 2.4.3 *Let T be a polynomially bounded operator on a Banach space $\mathcal{X}$. Suppose that $\lim_{n\to\infty} \|T^n u\| = 0$ for all $u \in \mathcal{X}$. Let $x \in \mathcal{X}$ and $x^* \in \mathcal{X}^*$. Then there exists $f \in L^1(\mathbb{T})$ such that*

$$\langle g(T)x, x^* \rangle = \frac{1}{2\pi} \int_0^{2\pi} g(e^{it}) f(e^{it})\, dt,$$

for all $g \in H^\infty$.

Proof Let μ be a measure representing the functional $x \otimes x^*$ such that $\|\mu\| = \|x \otimes x^*\|$. By Lemma 2.3.6, μ is a Henkin measure. Applying Theorem 2.4.2, it follows that there exists a function $f \in L^1(\mathbb{T})$ such that $\mu = fm$. It follows that

$$\langle p(T)x, x^* \rangle = \frac{1}{2\pi} \int_0^{2\pi} p(e^{it}) f(e^{it})\, dt,$$

for all polynomials p. The conclusion follows from Theorem 2.3.5. □

Exercises

1. Let $T = \begin{pmatrix} 0 & 2 \\ 0 & 0 \end{pmatrix}$, regarded as an operator on a two-dimensional Hilbert space. Show that T is a ρ-contraction for $\rho = 2$, but is not a contraction.
2. Construct a 2×2 matrix M whose spectrum is in the closed unit disc and such that $\|M^n\| \to \infty$ (i.e., M is not *power-bounded* and thus, obviously, not polynomially bounded).
3. Suppose that $(e_n)_n$ is the usual basis in $\ell^2(\mathbb{N})$, S the shift ($Se_n = e_{n+1}$), $\mathcal{H} = \ell^2(\mathbb{N}) \oplus \ell^2(\mathbb{N})$ and n_k a sequence such that $n_k > n_{k+1}$. We define Q to be the orthogonal projection onto the closed linear hull of $(e_{n_k})_k$, and finally
$$T = \begin{pmatrix} S^* & Q \\ 0 & S \end{pmatrix}.$$
Prove that T is power-bounded.
(Hint: check that
$$T^n = \begin{pmatrix} S^{*n} & Q_n \\ 0 & S^n \end{pmatrix},$$
where $Q_{n+1} = \sum_{i=0}^{n} S^{*n-i} Q S^i$, and then check that
$$Q_{n+1} e_m = \begin{cases} e_{m+2i-n} & \text{if } \ \exists i \le n, \quad \text{such that} \quad m + i = n_k, \\ & \text{and} \quad m + 2i - n \ge 0, \\ 0 & \text{otherwise}, \end{cases}$$
which implies that Q_{n+1} is a contraction.)
4. Suppose that $T \in \mathcal{L}(\mathcal{H})$ satisfies $\sigma(T) \subset \overline{\mathbb{D}}$ and there exists a self-adjoint $A \in \mathcal{L}(\mathcal{H})$ such that $K_{r,t}(T) + A \ge 0$ for all $re^{it} \in \mathbb{D}$. Show that T is a ρ-contraction for some $\rho \ge 1$.
5. Let $T \in \mathcal{L}(\mathcal{H})$ be a ρ-contraction for $\rho \ge 1$. Check that, for all $f \in A(\mathbb{D})$, we have
$$d\mu^T_{f(T)x,y} = d\mu^T_{x,f(T)^*y}.$$
6. Show that the unilateral shift S is an absolutely continuous contraction. (Hint: calculate $\langle K_{r,t}(S)x, y\rangle$.)
7. Suppose that $\Phi : A(\mathbb{D}) \to \mathcal{L}(\mathcal{H})$ is a contractive linear map; then for any $x, y \in \mathcal{H}$, the map $f \mapsto \langle \Phi(f)x, y\rangle$ is a continuous linear functional on $A(\mathbb{D})$, of norm less than $\|x\|\|y\|$. It can be extended to $C(\mathbb{T})$ and

represented by a measure $\mu_{x,y}$. We say that $\mu_{x,y}$ is a *representing measure* (for x, y) if

$$\int_{\mathbb{T}} f(z)\, d\mu_{x,y} = \langle \Phi(f)x, y\rangle,$$

for all $f \in A(\mathbb{D})$.

(a) Check that a representing measure is determined modulo $A(\mathbb{D})^{\perp}$.
(b) Check that $(x, y) \mapsto \mu_{x,y}$ is sesquilinear modulo $A(\mathbb{D})^{\perp}$.
(c) Check that Φ is multiplicative if and only if

$$g\, d\mu_{x,y} - d\mu_{\Phi(g)x,y} \in A(\mathbb{D})^{\perp},$$

for all $g \in A(\mathbb{D})$ and $x, y \in \mathcal{H}$.

8. Let $\theta \in H^2$ be an inner function, let K_θ denote the orthogonal complement $H^2 \ominus \theta H^2$ and let $S_\theta \in \mathcal{L}(K_\theta)$ denote the *restricted shift operator* $S_\theta = P_{K_\theta} S_{|K_\theta}$, where $S \in \mathcal{L}(H^2)$ is the usual shift.

(a) Verify that $S_\theta^* = S^*_{|K_\theta}$.
(b) Show that S_θ is of class C_0 and that its minimal function m_{S_θ} equals θ.

Comments

The dilation-free proof of the well-known von Neumann inequality (Corollary 2.1.8) was discovered by Heinz [121]. This proof is reproduced by Riesz and Sz.-Nagy [175, section 153]. For a more recent account, see also Paulsen [163]. Recall that von Neumann's inequality characterizes the Hilbert spaces among the complex Banach spaces. Indeed, it is proved by Foiaş [109] that, if $\mathcal{X}$ is a complex Banach space such that von Neumann's inequality holds for all contractions $T \in \mathcal{L}(\mathcal{X})$, then $\mathcal{X}$ is necessarily a complex Hilbert space.

A part of Section 2.1 is also presented in Chalendar [61].

Originally, the ρ-contractions were defined to be the class of Hilbert space operators with unitary ρ-*dilations*; that is, operators $T \in \mathcal{L}(\mathcal{H})$ for which there is a Hilbert space $\mathcal{K}$ containing $\mathcal{H}$ isometrically and a unitary operator $U \in \mathcal{L}(\mathcal{K})$ such that, for all positive integers n,

$$T^n = \rho P_{\mathcal{H}} U^n_{|\mathcal{H}},$$

where $P_{\mathcal{H}}$ is the orthogonal projection from $\mathcal{K}$ onto $\mathcal{H}$. The ρ-contractions were introduced by Holbrook ([124]) and Sz.-Nagy and Foiaş ([190, 191]) as a natural generalization of the usual contractions. The equivalence between the definition using the positivity of the kernel $K^{\rho}_{r,t}(T)$ and that based on

ρ-dilations follows from a result by Sz.-Nagy and Foiaş ([190, p. 17] or [191, section 1.11]) and is detailed [60, p. 315] by Cassier and Fack.

The fact that the set of 2-contractions coincides with the set of operators for which the numerical range is included in the closed unit disc was first proved by Berger (see Berger and Stampfli [38]).

For further information see also Cassier and Fack [60], who study the behaviour of contractions, and operators similar to contractions, in the context of von Neumann algebras. We also mention the works of Curto and Vasilescu [198, 86, 199], where the authors use the operator-valued Poisson kernel to define an $A(\mathbb{D}^n)$ functional calculus for an n-tuple of commuting operators $T = (T_1, \dots, T_n)$ with $\sigma(T) \subset \mathbb{D}^n$. Vasilescu [199] obtained a von Neumann inequality and a series of dilation-theory objects related to T when

$$T_1^* T_1 + \cdots + T_n^* T_n \leq \mathrm{I}.$$

Proposition 2.1.11 is Lemma 5.1 in Cassier, Chalendar and Chevreau [58].

Our presentation of the construction of a H^∞ functional calculus follows Corollary 2.5 in Réjasse [173]. For further information on the construction of an H^∞-functional calculus in various contexts, see also Apostol [13], Bercovici [32], Eschmeier [100, 101], Octavio and Kosiek [158] and Szafraniec [192].

Theorem 2.2.7, concerning the properties of C_0 operators, is given in Sz. Nagy and Foiaş [191] and Bercovici [32, chapter II].

Theorem 2.4.2 can be found in Rudin [178].

The operator defined in Exercise 3 is called *Foguel's operator*, and Lebow [141] proved that it is a power-bounded operator (as defined in Exercise 2) but not polynomially bounded. Pisier [165] proved that there exists also a polynomially bounded operator on Hilbert space which is not similar to a contraction.

While contractions on Hilbert spaces are, by von Neumann's inequality, polynomially bounded, this result does not extend to the case of general Banach spaces. For example, Zarrabi [205] characterized all the polynomially bounded isometries with spectrum contained in a given closed subset of the unit circle.

3

Properties ($\mathbb{A}_{n,m}$) and factorization of integrable functions

We begin by outlining the ideas behind the S. Brown method for constructing invariant subspaces; this technique makes it natural to consider certain classes of absolutely continuous contractions, which will be denoted $\mathbb{A}_{n,m}$.

In the particular case where T is the unilateral shift on H^2, its membership of $\mathbb{A}_{1,1}$ is linked to the Douglas–Rudin problem of characterizing functions $f \in L^1$ such that $f = g\overline{h}$, for $g, h \in H^2$. We give full details of the solution of this problem and derive related factorization results.

The original application of the S. Brown method was in proving the existence of invariant subspaces for subnormal operators. We shall give this result too, although deriving it by a simpler method due to Thomson.

The full power of the method relies on an approximation scheme presented initially in the context of contractions on a Hilbert space. The key idea is to show the surjectivity of certain bilinear (or sesquilinear) mappings satisfying various vanishing conditions, and we describe this in a very general context (this will be taken further in Chapter 4).

3.1 The basis of the S. Brown method

3.1.1 The starting point

Our aim is now to give an idea of S. Brown's method, and its connection with Chapter 2. The starting point of the method is the following lemma.

Lemma 3.1.1 *Let $T \in \mathcal{L}(\mathcal{H})$, $\|T\| \leq 1$ and $\lambda \in \mathbb{D}$. Suppose that there exist $x, y \in \mathcal{H}$ such that*

$$\langle x, y\rangle = 1 \qquad \text{and} \quad \langle T^n x, y\rangle = \lambda^n \quad \text{for} \quad n \geq 1. \tag{3.1}$$

Then T has a non-trivial invariant subspace.

Proof Let $\mathcal{M}$ the closed linear hull of $\{(T - \lambda I)T^n x : n \geq 0\}$. Then $T\mathcal{M} \subset \mathcal{M}$. The equality (3.1) implies that $y \perp \mathcal{M}$ with $y \neq 0$. Then $\mathcal{M}$ or $\mathbb{C}x$ is a non-trivial invariant subspace (the second case arising if $\mathcal{M} = \{0\}$). □

In view of the previous section, one can reformulate Lemma 3.1.1. We have seen in Lemma 2.2.4 that if $T \in \mathcal{L}(\mathcal{H})$ is an absolutely continuous contraction, then for all $x, y \in \mathcal{H}$ there exists a unique function $x \overset{T}{\cdot} y \in L^1(\mathbb{T})$ such that

$$\begin{cases} (T^n x, y) &= \widehat{(x \overset{T}{\cdot} y)}(-n), \quad n \geq 0, \\ (T^{*n} x, y) &= \widehat{(x \overset{T}{\cdot} y)}(n), \quad n > 0. \end{cases}$$

We shall require the following terminology.

Definition 3.1.2 Let $T \in \mathcal{L}(\mathcal{H})$ be an absolutely continuous contraction and let $f \in L^1(\mathbb{T})$. If there exist $x, y \in \mathcal{H}$ such that $f = x \overset{T}{\cdot} y$, we say that T *factorizes* f.

Note that for the purposes of constructing invariant subspaces it is generally sufficient to consider factorizations in the quotient space $L^1(\mathbb{T})/H_0^1$. Now $[f]_{L^1(\mathbb{T})/H_0^1} = [g]_{L^1(\mathbb{T})/H_0^1}$ if and only if $\hat{f}(n) = \hat{g}(n)$ for all $n \leq 0$, and this is equivalent to the condition that $f - g \in H_0^1$. With this in mind, we now consider the factorization of Poisson kernels P_λ.

A reformulation of S. Brown's starting point is this:

If there exist $x, y \in \mathcal{H}$ and $\lambda \in \mathbb{D}$ such that $[x \overset{T}{\cdot} y]_{L^1(\mathbb{T})/H_0^1} = [P_\lambda]_{L^1(\mathbb{T})/H_0^1}$, then T has non-trivial invariant subspaces.

We now give a direct extension of this observation. Let $L^1(\mathbb{T}) \cap \dfrac{H^1(\mathbb{D})}{H^\infty(\mathbb{D})}$ denote the set of functions $f \in L^1(\mathbb{T})$ that can be written as $f = h_1/h_2$ for some $h_1 \in H^1$ and $h_2 \in H^\infty$.

Theorem 3.1.3 *Let $T \in \mathcal{L}(\mathcal{H})$ be an absolutely continuous contraction and suppose that there exist non-zero vectors $x, y \in H$, a function $f \in L^1(\mathbb{T}) \cap \dfrac{H^1(\mathbb{D})}{H^\infty(\mathbb{D})}$ and an $n_0 \in \mathbb{N}$ such that*

$$\langle T^n x, y \rangle = \hat{f}(-n), \qquad \text{for all } n \geq n_0.$$

Then T has a non-trivial closed invariant subspace.

Proof We write $f = h_1/h_2$, as in the definition of $L^1(\mathbb{T}) \cap \dfrac{H^1(\mathbb{D})}{H^\infty(\mathbb{D})}$, and consider the subspace $\mathcal{M} = \langle h_2(T)T^{n_0+1}x \rangle$. This is clearly T-invariant. Moreover, for $k \geq 1$,

$$\begin{aligned}\langle h_2(T)T^{n_0+k}x, y\rangle &= \langle h_2(z)z^{n_0+k}, x \overset{T}{\cdot} y\rangle \\ &= \frac{1}{2\pi}\int_0^{2\pi} h_2(e^{i\theta})e^{i(n_0+k)\theta}(x \overset{T}{\cdot} y)(e^{i\theta})\,d\theta \\ &= \frac{1}{2\pi}\int_0^{2\pi} h_2(e^{i\theta})e^{i(n_0+k)\theta}\frac{h_1(e^{i\theta})}{h_2(e^{i\theta})}\,d\theta \\ &\qquad\qquad \text{since } z^{n_0}\left(x \overset{T}{\cdot} y - \frac{h_1}{h_2}\right) \in H^1 \\ &= \frac{1}{2\pi}\int_0^{2\pi} e^{i(n_0+k)\theta}h_1(e^{i\theta})\,d\theta \\ &= \hat{h}_1(-n_0-k) = 0.\end{aligned}$$

Hence $\mathcal{M} \neq \mathcal{H}$, and if $\mathcal{M} \neq \{0\}$ then T has a non-trivial closed invariant subspace.

Suppose now that $\mathcal{M} = \{0\}$. Then $h_2(T)T^{n_0+1}x = 0$. Let

$$\mathcal{N} = \operatorname{Ker}(h_2(T)T^{n_0+1}) \neq \{0\}.$$

If $\mathcal{N} \neq \mathcal{H}$, then it is a non-trivial closed invariant subspace for T.

There remains the possibility that $\mathcal{N} = \mathcal{H}$, i.e., that $h_2(T)T^{n_0+1} = 0$. Then T is a C_0 operator, i.e., there exists a non-trivial function $h \in H^\infty$ such that $h(T) = 0$. In this case T has a non-trivial invariant subspace, by Theorem 2.2.7. □

Corollary 3.1.4 *Let T be an absolutely continuous contraction on $\mathcal{H}$ and suppose that there exist $x, y \in \mathcal{H}$ such that $[x \overset{T}{\cdot} y]_{L^1(\mathbb{T})/H_0^1} = [g]_{L^1(\mathbb{T})/H_0^1}$ holds for a non-zero function $g = \sum_{n\geq 1} c_n P_{\lambda_n}$, where $(\lambda_n)_{n\geq 1}$ is a Blaschke sequence (possibly finite) and $\sum_{n\geq 1} |c_n| < \infty$. Then T has a non-trivial invariant subspace.*

Proof If f is a single Poisson kernel P_λ, then taking $b \in H^\infty$ as the Blaschke product $b(z) = \frac{z-\lambda}{1-\overline{\lambda}z}$, and $|z| = 1$, we have

$$f(z)b(z) = \frac{1-|\lambda|^2}{(1-\overline{\lambda}z)(1-\lambda\overline{z})}\frac{z-\lambda}{1-\overline{\lambda}z} = \frac{(1-|\lambda|^2)z}{(1-\overline{\lambda}z)^2} \in H^\infty.$$

The general case follows on noting that $gB \in H^\infty$, where B is a Blaschke product with zeroes $(\lambda_n)_{n\geq 1}$, as is now easily verified. □

It is not possible to extend the above methods to arbitrary sequences $(\lambda_n)_{n\geq 1} \subset \mathbb{D}$, since any function in $C(\mathbb{T})$ can be written as an ℓ^1 sum of Poisson kernels (see, for example, Bonsall [45] and Chalendar and Esterle [62]), including some functions which are not in H^1/H^∞, since they are zero on sub-intervals of positive measure but not identically zero.

It is clear that the class $L^1(\mathbb{T}) \cap \frac{H^1(\mathbb{D})}{H^\infty(\mathbb{D})}$ is dense in $L^1(\mathbb{T})$, since it contains all rational functions without poles on the unit circle. However, we do not know an explicit characterization of it. If $f \in L^1(\mathbb{T}) \cap \frac{H^1(\mathbb{D})}{H^\infty(\mathbb{D})}$, then certainly $\log|f| \in L^1(\mathbb{T})$, but this is not a sufficient condition, as the example $1 + \chi_\Omega$ shows, taking Ω to be a closed subset of normalized measure strictly between 0 and 1. For if $1 + \chi_\Omega = h_1/h_2$, then $h_1 = h_2$ on a set of uniqueness, and hence almost everywhere, which is absurd.

By using the operator-valued Poisson kernel presented in Chapter 2, we may consider the possibility of factorizing other functions in $L^1(\mathbb{T})$.

The next proposition collects together some standard facts, which will show that the hypotheses in the theorem that follows it are very natural.

Proposition 3.1.5 *(i) Suppose that, for some $\lambda \in \mathbb{D}$ and vectors x, $y \in \mathcal{H}$, we have $[P_\lambda] = [x \overset{T}{\cdot} y]$. Then there is a vector $\tilde{x} \in \mathcal{H}$ such that $[P_\lambda] = [\tilde{x} \overset{T}{\cdot} \tilde{x}]$.*

(ii) Let $f \in L^1(\mathbb{T})$ and $x \in \mathcal{H}$; then the following assertions are equivalent:

1. $f \geq 0$ a.e. and $[f] = [x \overset{T}{\cdot} x]$.

2. $f = x \overset{T}{\cdot} x$.

Proof (i) Without loss of generality we may suppose that $y \in \langle x \rangle$. By the functional calculus we have

$$\langle f(T)x, y \rangle = \frac{1}{2\pi} \int_0^{2\pi} f(e^{it}) P_\lambda(e^{it})\, dt = f(\lambda)$$

for all $f \in H^\infty$, and so $y \perp \mathcal{V} := \langle (T - \lambda \mathrm{I})x \rangle$. Let $\tilde{x}$ be a unit vector in the one-dimensional space $\langle x \rangle \ominus \mathcal{V}$; then one can easily check that $[P_\lambda] = [\tilde{x} \overset{T}{\cdot} \tilde{x}]$ (see Exercise 1).

(ii) Since the operator-valued Poisson kernel is positive, we have

$$(x \overset{T}{\cdot} x)(e^{it}) = \lim_{r \to 1-} \|K_{r,t}^{1/2}(T)x\|^2 \geq 0,$$

and thus condition 2 implies condition 1.

Conversely, if condition 1 is satisfied, then $\langle T^n x, x \rangle = \hat{f}(-n)$ for all $n \geq 0$, and since f is real-valued we have

$$\hat{f}(n) = \overline{\hat{f}(-n)} = \langle T^{*n} x, x \rangle = (x \overset{T}{\cdot} x)\hat{}(n), \qquad \text{for } n \geq 0,$$

i.e., $f = x \overset{T}{\cdot} x$. □

As we shall see later (Theorem 4.6.8), if T is a contraction and $\sigma(\mathbb{T}) \supseteq \mathbb{T}$, then T has a non-trivial invariant subspace. We now analyse some of the possibilities when $\sigma(T) \cap \mathbb{T}$ is not the whole circle.

Theorem 3.1.6 *Let $T \in \mathcal{L}(\mathcal{H})$ be an absolutely continuous contraction. Suppose that there is an open arc $I \subset \mathbb{T} \setminus \sigma(T)$ and $x \in \mathcal{H}$ such that one has $[x \stackrel{T}{\cdot} x] = [f]$, where f is a non-zero real function vanishing on I. Then T has a non-trivial invariant subspace.*

Proof Using Proposition 3.1.5, we have $f = x \stackrel{T}{\cdot} x$ and $f \geq 0$. By Proposition 2.1.11, since $I \subset \mathbb{T} \setminus \sigma(T)$, the function $x \stackrel{T}{\cdot} x$ extends analytically in a neighbourhood of I. For $e^{it} \in I$ we have

$$\begin{aligned} 0 = (x \stackrel{T}{\cdot} x)(e^{it}) &= \left\langle (\mathrm{I} - T^*T)(\mathrm{I} - e^{-it}T)^{-1}x, (\mathrm{I} - e^{-it}T)^{-1}x \right\rangle \\ &= \|(\mathrm{I} - T^*T)^{1/2}(\mathrm{I} - e^{-it}T)^{-1}x\|^2 \\ &= \|(\mathrm{I} - T^*T)^{1/2}(e^{it}\mathrm{I} - T)^{-1}x\|^2. \end{aligned}$$

Hence

$$(\mathrm{I} - T^*T)^{1/2}(e^{it}\mathrm{I} - T)^{-1}x = 0 \qquad \text{for all } t \text{ with } e^{it} \in I.$$

On taking successive derivatives with respect to t, we obtain

$$(\mathrm{I} - T^*T)^{1/2}(e^{it}\mathrm{I} - T)^{-n}x = 0 \qquad \text{for all } n \geq 1, \text{ and } t \text{ with } e^{it} \in I.$$

By multiplying T by a unimodular constant, we may reduce to the case when

$$(\mathrm{I} - T^*T)^{1/2}(\mathrm{I} - T)^{-n}x = 0 \qquad \text{for all } n \geq 1.$$

Let $\mathcal{M} = \operatorname{span}\{(\mathrm{I} - T)^{-n}x : n \geq 1\}$. Clearly $\mathcal{M} \neq \{0\}$, since it contains the non-zero vector $(\mathrm{I} - T)^{-1}x$. If $\mathcal{M} = \mathcal{H}$, then it follows that $\mathrm{I} - T^*T = 0$ and thus T is an isometry, and thus has a rich lattice of invariant subspaces. Otherwise, let us set $A = (\mathrm{I} - T)^{-1}$. Then $\mathcal{M}$ is a non-trivial invariant subspace for A. Now $\sigma(A) \subset \{z \in \mathbb{C} : \operatorname{Re} z > 0\}$, and so, by the holomorphic functional calculus, there is a sequence $(p_n)_{n\geq 1}$ of complex polynomials converging uniformly to the function $z \mapsto 1 - 1/z$ on a neighbourhood of $\sigma(A)$. Then $\|p_n(A) - T\| \to 0$ and hence $\mathcal{M}$ is a non-trivial invariant subspace for T as well. □

A simple example of the above situation is obtained by taking $\mathcal{H} = L^2(\Omega)$, where Ω is a proper closed subset of $\mathbb{T}$. Let T be the operator of multiplication by the independent variable. Then for $x, y \in \mathcal{H}$ one has $(x \stackrel{T}{\cdot} y)(e^{it}) = (x\overline{y})(e^{it})$, where x and y are regarded as functions in $L^2(\mathbb{T})$ that vanish on the complement of Ω.

3.1.2 The class $\mathbb{A}$

In Chapter 2, we mentioned that an absolutely continuous contraction $T \in \mathcal{L}(\mathcal{H})$ has a contractive H^∞ functional calculus Ψ_T. The case where Ψ_T is an isometry is of particular interest, and so we make the following definition.

Definition 3.1.7 Let $T \in \mathcal{L}(\mathcal{H})$. We say that T is in the class $\mathbb{A}$ if T is an absolutely continuous contraction for which the H^∞ functional calculus is isometric.

The following result by Apostol [12] emphasizes the importance of the class $\mathbb{A}$.

Theorem 3.1.8 *Let $T \in \mathcal{L}(\mathcal{H})$ be a contraction such that $\sigma(T) \supset \mathbb{T}$. Then either T has a non-trivial invariant subspace or T is in the class $\mathbb{A}$.*

We shall not prove this now, but note that a more powerful result will be given in Theorem 4.6.8. However, we now present some sufficient conditions for a contraction to be in the class $\mathbb{A}$. To that end, we introduce the following definition.

Definition 3.1.9 A subset Γ of $\mathbb{D}$ is a *dominating set* of $\mathbb{D}$ (or a dominating set for H^∞) if, for every $f \in H^\infty$, we have

$$\|f\|_\infty = \sup\{|f(\lambda)| : \lambda \in \Gamma\}.$$

The dominating sets of $\mathbb{D}$ were characterized by Brown, Shields and Zeller [49, Theorem 3], as follows: a subset Γ of $\mathbb{D}$ is a dominating set of $\mathbb{D}$ if and only if almost every point $\xi \in \mathbb{T}$ is a non-tangential limit point of Γ (that is, ξ is the limit of a sequence in Γ that lies inside some proper angular opening with vertex at ξ).

Lemma 3.1.10 *If T is an absolutely continuous contraction such that $\sigma(T)$ is a dominating set of $\mathbb{D}$, then $T \in \mathbb{A}$.*

Proof Recall that if T is an absolutely continuous contraction, its H^∞ functional calculus is contractive, and thus $\|f(T)\| \leq \|f\|_\infty$, $f \in H^\infty$. Using the spectral mapping inclusion $f(\sigma(T)) \subset \sigma(f(T))$ and the fact that $\|f(T)\| \geq \sup\{|\beta| : \beta \in \sigma(f(T))\}$, we have

$$\sup\{|f(\lambda)| : \lambda \in \sigma(T) \cap \mathbb{D}\} \leq \|f(T)\|.$$

Therefore, if $\sigma(T)$ is a dominating set of $\mathbb{D}$, then $T \in \mathbb{A}$. □

3.1.3 Classes $\mathbb{A}_{n,m}$

In 1983, Bercovici, Foias and Pearcy [34] introduced subclasses of the class $\mathbb{A}$, defined as follows.

Definition 3.1.11 Let $1 \le n, m \le \aleph_0$.

$$\mathbb{A}_{n,m} = \Big\{T \in \mathbb{A} : \textit{ for all } (f_{p,q}) \subset L^1(\mathbb{T}), \exists (x_p)_{0 \le p < n}, (y_q)_{0 \le q < m} \subset \mathcal{H} \\ \text{such that } [f_{p,q}]_{L^1(\mathbb{T})/H_0^1} = \big[x_p \overset{T}{\cdot} y_q\big]_{L^1(\mathbb{T})/H_0^1}\Big\}.$$

By definition, we have the following scheme of inclusion for the classes $\mathbb{A}_{n,m}$:

$$\begin{array}{ccccccc}
\mathbb{A}_{1,1} & \supset & \mathbb{A}_{1,2} & \supset & \cdots & \supset & \mathbb{A}_{1,\aleph_0} \\
\cup & & \cup & & & & \cup \\
\mathbb{A}_{2,1} & \supset & \mathbb{A}_{2,2} & \supset & \cdots & \supset & \mathbb{A}_{2,\aleph_0} \\
\cup & & \cup & & & & \cup \\
\vdots & & \vdots & & & & \vdots \\
\cup & & \cup & & & & \cup \\
\mathbb{A}_{\aleph_0,1} & \supset & \mathbb{A}_{\aleph_0,2} & \supset & \cdots & \supset & \mathbb{A}_{\aleph_0,\aleph_0}
\end{array}$$

Membership of a class $\mathbb{A}_{n,m}$ implies the existence of a rich lattice of invariant subspaces. Some of the properties of the classes $\mathbb{A}_{n,m}$ are presented in the next proposition.

Proposition 3.1.12 *Let $0 \le n, m \le \aleph_0$ and $T \in \mathcal{L}(\mathcal{H})$ be an absolutely continuous contraction.*

1. *$T \in \mathbb{A}_{n,m}$ if and only if $T^* \in \mathbb{A}_{m,n}$.*
2. *If $T \in \mathbb{A}_{n,n}$ then, for all $\lambda \in \mathbb{D}$, there exists $\mathcal{M}_\lambda \in \mathrm{Lat}(T)$ such that $\dim(\mathcal{M}_\lambda \ominus (T - \lambda \mathrm{I})\mathcal{M}_\lambda) = n$.*
3. *If $T \in \mathbb{A}_{\aleph_0,\aleph_0}$, then $\mathrm{Lat}(T)$ contains a sublattice isomorphic to the set of all closed subspaces of $\mathcal{H}$.*

Proof 1. This is an obvious consequence of the equality

$$(y \overset{T^*}{\cdot} x)(e^{it}) = \overline{(x \overset{T}{\cdot} y)(e^{-it})}$$

(see Remark 2.1.10) and the fact that for all $f \in L^1(\mathbb{T})$ there exists $g \in L^1(\mathbb{T})$ such that $f(e^{it}) = \overline{g(e^{-it})}$.

2. If $T \in \mathbb{A}_{n,n}$, taking $f_{p,p} = P_{r,t}$ (with $\lambda = re^{it}$) and $f_{p,q} = 0$ for $0 \le p, q < n$ with $p \ne q$, there exist $(x_p)_{0 \le p < n}$ and $(y_q)_{0 \le q < n}$ in $\mathcal{H}$ such that

$$\langle T^k x_p, y_q \rangle = \delta_{p,q} \lambda^k. \tag{3.2}$$

Now consider $\mathcal{M}_\lambda$, the invariant subspace for T generated by $(x_p)_{0\le p<n}$. Then $\dim(\mathcal{M}_\lambda \ominus (T-\lambda I)\mathcal{M}_\lambda)$ is necessarily at most n and is at least n because, by (3.2), the vectors y'_q given by the orthogonal projection of y_q onto $\mathcal{M}_\lambda$ are linearly independent and belong to $\mathcal{M}_\lambda \ominus (T-\lambda I)\mathcal{M}_\lambda$.

3. If $T \in \mathbb{A}_{\aleph_0,\aleph_0}$, taking $f_{p,p} = P_{0,0} = 1$ and $f_{p,q} = 0$ for $0 \le p, q < \aleph_0$ with $p \ne q$, along the same lines of the proof of Part 2, one can prove that there exists $\mathcal{M} \in \mathrm{Lat}(T)$ such that $\dim \mathcal{M} \ominus T\mathcal{M} = \aleph_0$. Set $\mathcal{K} = \mathcal{M} \ominus T\mathcal{M}$. Obviously we have $P_\mathcal{K} T_{|\mathcal{K}} = 0$, where $P_\mathcal{K}$ is the orthogonal projection of $\mathcal{H}$ onto $\mathcal{K}$. By construction, we have $\{\mathcal{K}' \oplus T\mathcal{M} : \mathcal{K}' \text{ a closed subspace of } \mathcal{K}\}$ contained in $\mathrm{Lat}(T)$. □

In what follows we will give an idea of the proof of the fact that classes $\mathbb{A}_{n,m}$ are distinct.

Let $L = \{z \in \mathbb{C} : (|z| = 1,\ \mathrm{Re}(z) \le 0) \quad \text{or} \quad (\mathrm{Re}(z) = 0,\ |\mathrm{Im}(z)| \le 1)\}$. Thus L is the boundary of the open left half unit disc; let $\mathbb{D}_L$ denote this left half disc (so $\mathbb{D}_L$ is the simply connected component of $\mathbb{C}\backslash L$). Put arc-length measure ℓ on L, and define $L^2(L, d\ell)$ to be the space of (equivalence classes of) square integrable complex functions on L. Now, define $E^2(L, d\ell)$ as the closure of the polynomials in $L^2(L, d\ell)$. Similarly, let us define

$$R = \{z \in \mathbb{C} :\ |z| = 1,\ \mathrm{Re}(z) \ge 0\} \cup \{z \in \mathbb{C} :\ \mathrm{Re}(z) = 0,\ |\mathrm{Im}(z)| \le 1)\}.$$

So R is the boundary of the open right half unit disc $\mathbb{D}_R$, and we define $L^2(R, d\ell)$ and $E^2(R, d\ell)$ in the analogous way.

Let N_L be the (normal) operator of multiplication by z on $L^2(L, d\ell)$, and T_L its (subnormal) restriction to $E^2(L, d\ell)$. Let N_R and T_R be defined similarly relative to R.

Let $w = \phi(z)$ be a map that takes the disc to the left semi-disc $\mathbb{D}_L$; for example,

$$z = \phi^{-1}(w) = (w^2 - 2w - 1)/(w^2 + 2w - 1).$$

Note that

$$\phi'(z) = (w^2 + 2w - 1)^2/(4(w^2 + 1)).$$

Then we have an unitary equivalence between T_L and the operator M_ϕ on H^2, where $M_\phi(f) = \phi f$.

The following formula will be of use in interpreting factorizations with M_ϕ. In what follows, $\phi : \mathbb{D} \to \mathbb{D}_L$ is a conformal bijection, and we shall choose it so that the left-hand arc $[e^{i\pi/2}, e^{3i\pi/2}]$ is mapped to itself (to achieve this, consider instead $\phi \circ \mu$ for a suitable Möbius map μ).

Lemma 3.1.13 *For $\pi/2 < t < 3\pi/2$ and $f, g \in H^2(\mathbb{T})$, we have*

$$f \overset{M_\phi}{\cdot} g(e^{it}) = \frac{f\overline{g}(\phi^{-1}(e^{it}))e^{it}}{\phi'(\phi^{-1}(e^{it}))\phi^{-1}(e^{it})} + \int_{-\pi/2}^{\pi/2} f(e^{i\theta})\overline{g(e^{i\theta})}\frac{1-|\phi(e^{i\theta})|^2}{|1-e^{-it}\phi(e^{i\theta})|^2}\frac{d\theta}{2\pi},$$

and the second term has an analytic extension to the left-hand half-plane $\{\operatorname{Re} z < 0\}$.

Proof Using Theorem 2.1.9, we have, for $\pi/2 < t < 3\pi/2$,

$$\begin{aligned} f \overset{M_\phi}{\cdot} g(e^{it}) &= \lim_{r\to 1^-}\int_0^{2\pi} f(e^{i\theta})\overline{g(e^{i\theta})}\frac{1-r^2|\phi(e^{i\theta})|^2}{|1-re^{-it}\phi(e^{i\theta})|^2}\frac{d\theta}{2\pi} \\ &= \lim_{r\to 1^-}\int_{\pi/2}^{3\pi/2} f(e^{i\theta})\overline{g(e^{i\theta})}\frac{1-r^2}{|1-re^{-it}\phi(e^{i\theta})|^2}\frac{d\theta}{2\pi} \\ &\quad + \int_{-\pi/2}^{\pi/2} f(e^{i\theta})\overline{g(e^{i\theta})}\frac{1-|\phi(e^{i\theta})|^2}{|1-e^{-it}\phi(e^{i\theta})|^2}\frac{d\theta}{2\pi}. \end{aligned}$$

Define a new function $\psi : \mathbb{T} \to \mathbb{T}$ by

$$\psi(e^{i\theta}) = \begin{cases} \phi(e^{i\theta}) & \text{if } \pi/2 < \theta < 3\pi/2, \\ e^{i\theta} & \text{otherwise.} \end{cases}$$

Then, since $\lim_{r\to 1^-} \frac{1-r^2}{|1-re^{-it}\psi(e^{i\theta})|^2} = 0$ for $\theta \in (0, 2\pi) \setminus \left[\frac{\pi}{2}, \frac{3\pi}{2}\right]$, we get

$$\begin{aligned} f \overset{M_\phi}{\cdot} g(e^{it}) &= \lim_{r\to 1^-}\int_0^{2\pi} f(e^{i\theta})\overline{g(e^{i\theta})}\frac{1-r^2}{|1-re^{-it}\psi(e^{i\theta})|^2}\frac{d\theta}{2\pi} \\ &\quad + \int_{-\pi/2}^{\pi/2} f(e^{i\theta})\overline{g(e^{i\theta})}\frac{1-|\phi(e^{i\theta})|^2}{|1-e^{-it}\phi(e^{i\theta})|^2}\frac{d\theta}{2\pi} \\ &= \frac{f\overline{g}(\phi^{-1}(e^{it}))e^{it}}{\phi'(\phi^{-1}(e^{it}))\phi^{-1}(e^{it})} + \int_{-\pi/2}^{\pi/2} f(e^{i\theta})\overline{g(e^{i\theta})}\frac{1-|\phi(e^{i\theta})|^2}{|1-e^{-it}\phi(e^{i\theta})|^2}\frac{d\theta}{2\pi}, \end{aligned}$$

using a change of variables and the standard properties of the Poisson kernel. Note that the first term involves the value of $f\overline{g}$ at just one point of the left-hand semi-circle, and the second term involves values only on the right-hand semi-circle.

Since, for e^{it} on the unit circle,

$$\begin{aligned} |1-e^{-it}w|^{-2} &= |e^{it}-w|^{-2} \\ &= (e^{it}-w)^{-1}(e^{-it}-\overline{w})^{-1} = \frac{e^{it}}{(e^{it}-w)(1-\overline{w}e^{it})}, \end{aligned}$$

it is clear that the second term has an analytic extension to the left-hand half-plane $\{\operatorname{Re} z < 0\}$. □

Theorem 3.1.14 $T_L \oplus T_R^* \in \mathbb{A} \setminus (\mathbb{A}_{1,2} \cup \mathbb{A}_{2,1})$.

Proof Clearly $\sigma\left(T_L \oplus T_R^*\right) = \overline{\mathbb{D}}$, which, using Lemma 3.1.10, is a sufficient condition for membership in $\mathbb{A}$. Using the first assertion of Proposition 3.1.12, it is sufficient to prove that $M_\phi \oplus T_R^* \notin \mathbb{A}_{2,1}$. Let $\delta = 1/2$ and $C_\delta = \{z \in \mathbb{C} : \operatorname{Re} z < -\delta\}$. Now consider the function γ defined on $\mathbb{T}$ by

$$\gamma(e^{it}) = \frac{e^{it}}{\phi'(\phi^{-1}(e^{it}))\phi^{-1}(e^{it})}.$$

Take Ω_1 and Ω_2 to be closed sub-arcs of $C_\delta \cap \mathbb{T}$ such that $\ell\left(\Omega_1^c \cap \Omega_2^c \cap C_\delta\right) > 0$, $\ell(\Omega_1 \cap \Omega_2) = 0$, and $\ell(\Omega_j) > 0$ for $j = 1, 2$. Suppose that $M_\phi \oplus T_R^* \in \mathbb{A}_{2,1}$. It follows that there exist functions f_1, f_2, g in H^2 and x_1, x_2, y in $E^2(R, d\ell)$ such that

$$\begin{cases} [f_1 \overset{M_\phi}{\cdot} g] + [x_1 \overset{T_R{}^*}{\cdot} y] &= [\chi_{\Omega_1}], \\ [f_2 \overset{M_\phi}{\cdot} g] + [x_2 \overset{T_R{}^*}{\cdot} y] &= [\chi_{\Omega_2}]. \end{cases}$$

Since $\sigma\left(T_R^*\right) \cap \mathbb{T} = R \cap \mathbb{T}$, it follows from the proof of Proposition 2.1.11 that $x_1 \overset{T_R{}^*}{\cdot} y$ and $x_2 \overset{T_R{}^*}{\cdot} y$ have analytic extensions to C_δ. Using Lemma 3.1.13, there exist h_1 and h_2 in H^2 and k_1, k_2 analytic on C_δ, such that, on $C_\delta \cap \mathbb{T}$,

$$\begin{cases} \chi_{\Omega_1}(e^{it}) + e^{it}h_1(e^{it}) &= f_1(\phi^{-1}(e^{it}))\overline{g(\phi^{-1}(e^{it}))}\gamma(e^{it}) + k_1(e^{it}), \\ \chi_{\Omega_2}(e^{it}) + e^{it}h_2(e^{it}) &= f_2(\phi^{-1}(e^{it}))\overline{g(\phi^{-1}(e^{it}))}\gamma(e^{it}) + k_2(e^{it}). \end{cases} \tag{3.3}$$

On $\Omega_1^c \cap \Omega_2^c \cap C_\delta$ we have

$$\left[e^{it}h_1(e^{it}) - k_1(e^{it})\right]f_2(\phi^{-1}(e^{it})) = \left[e^{it}h_2(e^{it}) - k_2(e^{it})\right]f_1(\phi^{-1}(e^{it})),$$

and hence the same inequality holds on $C_\delta \cap \mathbb{T}$ since both sides of the equation are E^1 (Hardy class) functions on $C_\delta \cap \mathbb{D}$. Multiplying the first equation of (3.3) by $f_2(\phi^{-1}(e^{it}))$ and the second by $f_1(\phi^{-1}(e^{it}))$ and subtracting, we see that

$$\chi_{\Omega_1}(e^{it})f_2(\phi^{-1}(e^{it})) = \chi_{\Omega_2}(e^{it})f_1(\phi^{-1}(e^{it})) \qquad \text{on} \quad C_\delta \cap \mathbb{T},$$

which implies that f_1 and f_2 are identically zero, since they vanish on subsets of positive measure. This is absurd as χ_{Ω_1} and χ_{Ω_2} are not the restrictions of analytic functions. □

For every pair of positive integers n and m, denote by $T_L^{(n)} \oplus T_R^{*(m)}$ the operator given by the direct sum of n copies of T_L and m copies of T_R^*.

One can extend the methods of this section to obtain the following result.

Theorem 3.1.15 $T_L^{(n)} \oplus T_R^{*(m)} \in \mathbb{A}_{n,m} \setminus (\mathbb{A}_{n+1,1} \cup \mathbb{A}_{1,m+1})$.

3.2 Factorization of log-integrable functions

We now consider some questions in harmonic analysis; these are important in their own right, and they will later have applications in operator theory.

Since there is no bounded projection of $L^\infty(\mathbb{T})$ onto H^∞, there is no exact decomposition of a general function in $L^\infty(\mathbb{T})$ into the sum of a function in H^∞ and a function in $\overline{H_0^\infty}$. Nevertheless, it is possible to obtain an approximate decomposition with estimates on the norms of the functions.

We denote by $\mathcal{N}^+ = \mathcal{N}^+(\mathbb{D})$ the *Smirnoff class*, which can be defined as the algebra of all holomorphic functions f on $\mathbb{D}$ such that $f = u/g$, where $u, g \in H^\infty$ and g is an outer function (if g is not restricted to being outer, we obtain the *Nevanlinna class*, $\mathcal{N}(\mathbb{D})$).

Lemma 3.2.1 *Let f be a function in $\mathcal{N}^+$ and let $\delta \geq 1$. Then there exists a function $g \in H^\infty$ with $\|g\|_\infty \leq \delta$, such that*

$$\|f - g\|_2 \leq \sqrt{\frac{2+\delta^2}{\pi}} \left(\int_{E_\delta} |f(e^{it})|^2 \, dt \right)^{1/2},$$

where

$$E_\delta = \{t \in (-\pi, \pi] : |f(e^{it})| > \delta\}.$$

Proof Consider the analytic function g in $\mathbb{D}$ defined by $g = fG$, where G is the outer function

$$G(z) = \exp\left(\frac{1}{2\pi} \int_{E_\delta} (\log \delta - \log |f(e^{it})|) \frac{e^{it} + z}{e^{it} - z} \, dt \right).$$

If we set $F_\delta = \{t \in (-\pi, \pi] : |f(e^{it})| \leq \delta\}$, then we obtain

$$|g(e^{it})| = \begin{cases} |f(e^{it})| & \text{if } t \in F_\delta, \\ \delta & \text{otherwise.} \end{cases}$$

Thus, the function g belongs to H^∞ and $\|g\|_\infty \leq \delta$. It is clear that

$$\begin{aligned} 2\pi \|f - g\|_2^2 &= \int_{E_\delta} |(f - g)(e^{it})|^2 \, dt + \int_{F_\delta} |(f - g)(e^{it})|^2 \, dt \\ &= \int_{E_\delta} |(f - g)(e^{it})|^2 \, dt + \int_{F_\delta} |f(e^{it})|^2 |1 - G(e^{it})|^2 \, dt. \end{aligned}$$

If we define the function ϕ on $\mathbb{T}$ by

$$\phi(e^{it}) = \begin{cases} 0 & \text{if } t \in F_\delta, \\ \log \delta - \log |f(e^{it})| & \text{otherwise,} \end{cases}$$

then $G(e^{it}) = \exp(\phi(e^{it}) + i(H\phi)(e^{it}))$, for $t \in (-\pi, \pi]$, where H denotes the Hilbert transform. Since $G(e^{it}) = \exp(i(H\phi)(e^{it}))$ for $t \in F_\delta$, we get

$$\int_{F_\delta} |(f-g)(e^{it})|^2\, dt \le \delta^2 \int_{F_\delta} |1 - \exp(i(H\phi)(e^{it}))|^2\, dt.$$

Using the inequality $|1 - e^{ix}|^2 \le 2x^2$ for $x \in \mathbb{R}$, we obtain

$$\int_{F_\delta} |(f-g)(e^{it})|^2\, dt \le 2\delta^2 \int_{\mathbb{T}} |(H\phi)(e^{it})|^2\, dt.$$

Since the Hilbert transform is a contraction with respect to the $L^2(\mathbb{T})$-norm, and since $\phi(e^{it}) = 0$ for $t \in F_\delta$, it follows that

$$\int_{F_\delta} |(f-g)(e^{it})|^2\, dt \le 2\delta^2 \int_{E_\delta} |\phi(e^{it})|^2\, dt.$$

Since $\delta \ge 1$, it follows that, for $t \in E_\delta$,

$$|\phi(e^{it})| = \log |f(e^{it})/\delta| \le |f(e^{it})|.$$

Thus we have

$$\int_{F_\delta} |(f-g)(e^{it})|^2\, dt \le 2\delta^2 \int_{E_\delta} |f(e^{it})|^2\, dt.$$

Moreover, we easily get that

$$\int_{E_\delta} |(f-g)(e^{it})|^2\, dt \le 4 \int_{E_\delta} |f(e^{it})|^2\, dt,$$

and the result follows. □

The next preliminary result gives us information on how well we can approximate unbounded H^p and BMOA functions by H^∞ functions.

Proposition 3.2.2 *(i) Let $p > 4$ and let $f \in H^p$. For every $\varepsilon \in (0, 1]$ there exists a function $g \in H^\infty$ such that*

$$\begin{cases} \|f - g\|_2 \le \varepsilon \|f\|_p, & \\ \|g\|_\infty \le c_p \varepsilon^{2/(4-p)} \|f\|_p, & \text{where} \quad 0 < c_p \le 6^{1/(p-4)}. \end{cases}$$

(ii) Let $\varepsilon \in (0, \frac{1}{2}]$ and let $f \in$ BMOA. Then there exists a function $g \in H^\infty$ and a constant $d > 0$ such that

$$\begin{cases} \|f - g\|_2 \le \varepsilon \|f\|_{\mathrm{BMO}}, \\ \|g\|_\infty \le d \log\left(\frac{1}{\varepsilon}\right) \|f\|_{\mathrm{BMO}}. \end{cases}$$

Proof (i) We may suppose that $\|f\|_p \leq 1$. Since $H^p \subset \mathcal{N}^+$ for $p > 0$, we may apply Lemma 3.2.1 and deduce that for each $\delta \geq 1$ there exists a function $g \in H^\infty$, with $\|g\|_\infty \leq \delta$, such that

$$\|f-g\|_2 \leq \sqrt{\frac{2+\delta^2}{\pi}} \left(\int_{E_\delta} |f(e^{it})|^2 \, dt \right)^{1/2}, \tag{3.4}$$

where $E_\delta = \{t \in (-\pi, \pi] : |f(e^{it})| > \delta\}$. Applying Hölder's inequality (1.2), we obtain

$$\int_{E_\delta} |f(e^{it})|^2 \, dt \leq \left(\int_{E_\delta} |f(e^{it})|^p \, dt \right)^{2/p} (2\pi \mu(E_\delta))^{1-2/p}. \tag{3.5}$$

Moreover, since

$$\|f\|_p^p \geq \frac{1}{2\pi} \int_{E_\delta} |f(e^{it})|^p \, dt \geq \delta^p \mu(E_\delta),$$

we obtain $\mu(E_\delta) \leq 1/\delta^p$. Hence we get, for $p > 4$,

$$\|f-g\|_2 \leq \frac{\sqrt{4+2\delta^2}}{\delta^{p/2-1}} \leq \frac{\sqrt{6}}{\delta^{(p-4)/2}}.$$

(ii) We may now suppose that $\|f\|_{\mathrm{BMO}} \leq 1$. By using (3.4) and (3.5) for $p = 3$ (for example), noting that $\mathrm{BMO} \subset \bigcap_{p>0} H^p$, we get

$$\|f-g\|_2 \leq K\delta\mu(E_\delta)^{1/6}$$

for some constant $K > 0$. Moreover, by Theorem 1.4.34 (John–Nirenberg), there exists a constant $k > 0$ such that $\mu(E_\delta) \leq k \exp(-k\delta)$. Hence we have, for some constant $c_0 > 0$,

$$\begin{cases} \|f-g\|_2 \leq c_0 \delta \exp(-k\delta/6), \\ \|g\|_\infty \leq \delta. \end{cases}$$

For ε small enough, set $\delta = \frac{12}{k} \log \frac{1}{\varepsilon}$. We easily get that

$$\begin{cases} \|f-g\|_2 \leq \varepsilon, \\ \|g\|_\infty \leq d \log\left(\frac{1}{\varepsilon}\right) \end{cases}$$

for some positive constant d, which completes the proof of the lemma. □

We now come to the main decomposition theorem, which will be useful in several contexts.

Theorem 3.2.3 *Let $\varepsilon \in (0, 1/2]$ and $f \in L^\infty(\mathbb{T})$. Then there exist $g^+ \in H^\infty$ and $g^- \in \overline{H_0^\infty}$ such that*

$$\begin{cases} \|f - (g^+ + g^-)\|_2 \le \varepsilon \|f\|_\infty, \\ \|g^+\|_\infty + \|g^-\|_\infty \le c \log\left(\frac{1}{\varepsilon}\right) \|f\|_\infty, \end{cases}$$

where c is an absolute constant.

Proof Since the Riesz projection P_+ maps $L^\infty(\mathbb{T})$ into BMOA and since $L^\infty(\mathbb{T})$ embeds continuously in BMO (see Section 1.4.6), the corollary follows immediately from the second assertion of Proposition 3.2.2, applied to the functions $P_+ f$ and $\overline{f - P_+ f}$. □

A first corollary is a result on approximation of real bounded functions by the real parts of H^∞ functions.

Corollary 3.2.4 *Let $\varepsilon \in (0, 1/2]$ and f be a real function in $L^\infty(\mathbb{T})$. Then there exists $h \in H^\infty$ such that*

$$\begin{cases} \|f - \operatorname{Re} h\|_2 \le \varepsilon \|f\|_\infty, \\ \|h\|_\infty \le c \log\left(\frac{1}{\varepsilon}\right) \|f\|_\infty, \end{cases}$$

where c is the absolute constant in Theorem 3.2.3.

Proof Let g^+ and g^- be associated with f as in Theorem 3.2.3. Set

$$h = g^+ + \overline{g^-}.$$

Then h is in H^∞ and it is easy to see that $\operatorname{Re}(h) = \operatorname{Re}(g^+ + g^-)$. The result follows immediately. □

The key to the results that follow lies in a study of the properties of the argument of a ratio B_1/B_2 of two Blaschke products. We begin by finding ratios with arguments approximating step functions, and then consider more complicated functions. Recall that m denotes the normalized Lebesgue measure on $\mathbb{T}$.

Lemma 3.2.5 *Let $\varepsilon, \theta, \delta \in (0, \pi)$, $\eta \in (0, \theta/\pi)$, $\alpha \in (\varepsilon - \pi, \pi - \varepsilon)$ and $\Omega = \{e^{iw} : -\theta < w < \theta\}$. Then there exist finite Blaschke products B_1 and B_2 having simple zeroes such that*

$$m(\{z \in \Omega : |\alpha - \arg(B_1(z)/B_2(z))| > \varepsilon\}) < \eta, \tag{3.6}$$

while

$$-\delta < \arg(B_1(z)/B_2(z)) < \delta \qquad \textit{for all} \quad z \in \mathbb{T} \setminus \Omega, \tag{3.7}$$

and

$$\sum_{\{z:\, B_1(z)B_2(z)=0\}} (1-|z|) \le 2m(\Omega)\log\left(\frac{4\pi}{\varepsilon}\right). \tag{3.8}$$

Proof We fix some $r \in (0,1)$, N a positive integer and $0 < \beta \le \theta - \pi/N$, which will be determined later. Then the zero sets of B_1 and B_2 are given by

$$Z_1 = \{r^{1/N}e^{2\pi ik/N} : 0 \le k \le N-1\} \cap \{r^{1/N}e^{iw} : -\beta < w < \beta\},$$
$$Z_2 = \{e^{i\alpha/N}z : z \in Z_1\}.$$

Note that

$$\sum_{z\in Z_j}(1-|z|) \le (1-r^{1/N})N\frac{\theta}{\pi} \le \log(1/r)m(\Omega) \qquad \text{for} \quad j=1,2. \tag{3.9}$$

Let $q : \mathbb{T} \to \mathbb{C}$ be the ratio of two finite Blaschke products whose zeroes are respectively $Z_1' = \{r^{1/N}e^{2\pi ik/N} : 0 \le k \le N-1\}$ and $Z_2' = \{e^{i\alpha/N}z : z \in Z_1'\}$. Then

$$q(z) = \frac{(r-z^N)(e^{i\alpha}-rz^N)}{(1-rz^N)(re^{i\alpha}-z^N)} = e^{i\alpha}\frac{(1-rz^{-N})(1-rz^Ne^{-i\alpha})}{(1-rz^N)(1-rz^{-N}e^{i\alpha})}.$$

Since for all $\xi \in \mathbb{T}$, $|\arg(1-r\xi)| \le \frac{\pi}{2}r$, it follows that

$$|\arg(e^{-i\alpha}q(z))| \le 4\left(\frac{\pi r}{2}\right) = 2\pi r,$$

for all $z \in \mathbb{T}$. Now we set $r = \frac{\varepsilon}{4\pi}$. Using (3.9), this ensures that (3.8) holds.

Since $|\alpha| + 2\pi r < \pi$, we see that

$$|\alpha - \arg(q(z))| \le \varepsilon/2, \tag{3.10}$$

for all $z \in \mathbb{T}$. Then we have, for $z \in \mathbb{T}$,

$$B_1(z)/B_2(z) = \prod_{w\in Z_1} e^{i\alpha/N}\frac{(w-z)(1-\overline{we^{i\alpha/N}}z)}{(1-\overline{w}z)(we^{i\alpha/N}-z)} \tag{3.11}$$

$$= q(z)\Big/\prod_{w\in Z_1'\setminus Z_1} e^{i\alpha/N}\frac{(w-z)(1-\overline{we^{i\alpha/N}}z)}{(1-\overline{w}z)(we^{i\alpha/N}-z)}. \tag{3.12}$$

We shall use (3.11) to estimate $\arg(B_1(z)/B_2(z))$ on the complement of Ω, whereas we apply (3.12) on Ω.

For $w \in Z_1$ and $z \in \mathbb{T}$, we have

$$\left| e^{i\alpha/N} \frac{(w-z)(1-\overline{we^{i\alpha/N}}z)}{(1-\overline{w}z)(we^{i\alpha/N}-z)} - 1 \right| = \left| \frac{(1-|w|^2)(1-e^{i\alpha/N})}{(1-\overline{w}z)(we^{i\alpha/N}-z)} \right|$$
$$\leq \frac{(2N^{-1}\log(1/r))(\alpha/N)}{|z-w||z-we^{i\alpha/N}|},$$

since $|1-e^{i\alpha/N}| \leq \alpha/N$ and $1-r^{2/N} \leq 2N^{-1}\log(1/r)$. Therefore, from (3.11), we have

$$|\arg(B_1(z)/B_2(z))| \leq \sum_{w \in Z_1} \left(\frac{\pi}{2} \frac{2\alpha N^{-2}\log(1/r)}{|z-w||z-we^{i\alpha/N}|} \right)$$
$$\leq \frac{\pi\alpha N^{-1}\log(1/r)}{\operatorname{dist}(z, Z_1)\operatorname{dist}(z, Z_2)}, \tag{3.13}$$

for all $z \in \mathbb{T}$. Similarly, it follows from (3.12) that

$$|\arg(q(z)^{-1}B_1(z)/B_2(z))| \leq \frac{\pi\alpha N^{-1}\log(1/r)}{\operatorname{dist}\left(z, Z_1' \setminus Z_1\right)\operatorname{dist}\left(z, Z_2' \setminus Z_2\right)}, \tag{3.14}$$

for all $z \in \mathbb{T}$.

From now on we may assume that $N \geq 2/\eta$ and we take $\beta = \theta - \pi\eta/2$. Then, for all $z \in \mathbb{T}$ with $|\arg(z)| \geq \theta$, the quantities $\operatorname{dist}(z, Z_1)$ and $\operatorname{dist}(z, Z_2)$ are bounded below, independently of N. Likewise, if $|\arg(z)| \geq \theta - \pi\eta$, the quantities $\operatorname{dist}\left(z, Z_1' \setminus Z_1\right)$ and $\operatorname{dist}\left(z, Z_2' \setminus Z_2\right)$ are also bounded below, independently of N. In the first case the estimate (3.13) is less than δ for all N sufficiently large, and in the second case the estimate (3.14) is less than $\varepsilon/2$ for all N sufficiently large. Therefore we get (3.7). Moreover, since

$$|\alpha - \arg(B_1(z)/B_2(z))| \leq |\alpha - \arg(q(z))| + |\arg(q^{-1}(z)B_1(z)/B_2(z))|,$$

using (3.10) and the last estimate on (3.14), we have

$$|\alpha - \arg(B_1(z)/B_2(z))| < \varepsilon,$$

for all z in some neighbourhood of the arc $[e^{-i(\theta-\pi\eta)}, e^{i(\theta-\pi\eta)}]$ in $\mathbb{T}$. This implies (3.6), and concludes the proof. □

Using Lemma 3.2.5 and an induction argument we can derive a similar approximation result for an arbitrary relatively open subset of $\mathbb{T}$.

Lemma 3.2.6 *Let $\varepsilon, \delta \in (0, \pi)$, $\eta \in (0, 1)$, Ω be a relatively open subset of $\mathbb{T}$ and $\psi : \Omega \to (-\pi, \pi]$ a measurable function. Then there exist finite Blaschke products B_1 and B_2 having simple zeroes such that*

$$m(\{z \in \Omega : |\psi(z) - \arg(B_1(z)/B_2(z))| > \varepsilon\}) < \eta,$$

while

$$-\delta < \arg(B_1(z)/B_2(z)) < \delta \qquad \textit{for all} \quad z \in \mathbb{T} \setminus \Omega$$

and

$$\sum_{\{z:\, B_1(z)B_2(z)=0\}} (1-|z|) \le 2m(\Omega)\log\left(\frac{12\pi}{\varepsilon}\right).$$

Proof Since Ω is open and $\psi : \Omega \to (-\pi, \pi]$ is measurable, there exist a positive integer k, real numbers $\alpha_1, \ldots, \alpha_k \in [\varepsilon/3 - \pi, \pi - \varepsilon/3]$ and pairwise disjoint proper relatively open arcs $I_1, \ldots, I_k \subset \Omega$, such that

$$m(\{z \in I_j : |\psi(z) - \alpha_j| > \varepsilon/3\}) < \eta/(3k) \tag{3.15}$$

and

$$\sum_{j=1}^{k} m(I_j) > m(\Omega) - \eta/3, \tag{3.16}$$

for all $j \in \{1, \ldots, k\}$. Then, applying Lemma 3.2.5, by replacing ε with $\varepsilon/3$, η with $\eta/3$ and δ with $\min(\eta, \varepsilon/3)/k$ we obtain Blaschke products $\left(B_1^{(j)}\right)_{j=1}^{k}$ and $\left(B_2^{(j)}\right)_{j=1}^{k}$ such that

$$m\{z \in I_j : \left|\alpha_j - \arg\left(B_1^{(j)}(z)/B_2^{(j)}(z)\right)\right| > \varepsilon/3\} < \nu/3, \tag{3.17}$$

with

$$\left|\arg\left(B_1^{(j)}(z)/B_2^{(j)}(z)\right)\right| < \min(\eta, \varepsilon/3)/k \tag{3.18}$$

and

$$\sum_{\left\{z:B_1^{(j)}(z)B_2^{(j)}(z)=0\right\}} (1-|z|) \le 2m(I_j)\log(12\pi/\varepsilon), \tag{3.19}$$

for all $j \in \{1, \ldots, k\}$ and $z \in I_j$. Now, let B_1 and B_2 be the Blaschke products given by

$$B_l = \prod_{j=1}^{k} B_l^{(j)}, \qquad \text{for all} \quad l = 1, 2.$$

By construction and the fact that the arcs I_j are pairwise disjoint, we find that B_1 and B_2 have simple zeroes. By summing the estimates (3.18) and (3.19) over all indices j, we obtain

$$|\arg(B_1(z)/B_2(z))| < \delta$$

and

$$\sum_{\{z:B_1(z)B_2(z)=0\}} (1-|z|) \leq 2m(\Omega)\log(12\pi/\varepsilon),$$

as required. Now, using (3.17) and (3.18), we have

$$m(\{z \in I_j : |\alpha_j - \arg(B_1(z)/B_2(z))| > 2\varepsilon/3\}) < \eta/(3k),$$

for all $j \in \{1,\ldots,k\}$. Then, by applying (3.15), summing the resultant measures over all indices j and finally applying (3.16), we obtain

$$m(\{z \in \Omega : |\psi(z) - \arg(B_1(z)/B_2(z))| > \varepsilon\}) < \eta,$$

completing the proof. □

We are now ready to prove a strong scalar inner approximation result.

Lemma 3.2.7 *Let $\varepsilon \in (0,\pi]$, $\Delta \subset \mathbb{T}$ measurable and let $\phi : \mathbb{T} \to (-\pi,\pi]$ be a measurable function vanishing outside Δ. Then there exist Blaschke products B_1 and B_2 such that*

$$\|\phi - \arg(B_1/B_2)\|_\infty < \varepsilon,$$

with

$$\sum_{\{z:\,B_1(z)B_2(z)=0\}} (1-|z|) \leq 2m(\Delta)\log\left(\frac{50\pi}{\varepsilon}\right).$$

Proof We may assume, without loss of generality, that $m(\Delta) > 0$. Let $(\eta_k)_{k\geq 0}$ and $(\delta_k)_{k\geq 0}$ be any sequences of positive reals such that

$$2\sum_{k=0}^{\infty} \eta_k \leq m(\Delta)\log(25/24) \qquad \text{and} \qquad \sum_{k=0}^{\infty} \delta_k < \varepsilon/4.$$

Let $\Omega_1 \subset \mathbb{T}$ be any relatively open set containing Δ and such that $m(\Omega_1) < m(\Delta) + \eta_0$; let $\psi_1 : \mathbb{T} \to (-\pi,\pi]$ be given by

$$\psi_1(z) = \begin{cases} \phi(z) - \varepsilon/2 & \text{if } \phi(z) > 0, \\ \phi(z) + \varepsilon/2 & \text{if } \phi(z) \leq 0, \end{cases}$$

for all $z \in \mathbb{T}$. We shall inductively find sequences of sets $(\Omega_k)_{k=2}^{\infty}$, relatively open in $\mathbb{T}$, measurable functions $(\phi_k)_{k=2}^{\infty}$ from $\mathbb{T}$ to $(-\pi,\pi]$, and Blaschke products $\left(B_1^{(k)}\right)_{k=1}^{\infty}$ and $\left(B_2^{(k)}\right)_{k=1}^{\infty}$ such that

1. $\sum_{\{z:\,B_1^{(k)}(z)B_2^{(k)}(z)=0\}} (1-|z|) \leq 2m(\Omega_k)\log\left(\frac{48\pi}{\varepsilon}\right)$
2. $\left|\arg(B_1^{(k)}/B_2^{(k)})\right| < \delta_k$ for all $z \in \mathbb{T}\setminus\Omega_k$

3. Ω_{k+1} contains the set

$$\Delta_{k+1} = \{z \in \Omega_k : |\psi_k(z) - \arg(B_1^{(k)}(z)/B_2^{(k)}(z))| > \varepsilon/4\},$$

and

$$m(\Omega_{k+1}) < m(\Delta_{k+1}) + \eta_k < 2\eta_k$$

4. For all $z \in \mathbb{T}$ we have

$$\psi_{k+1}(z) = \begin{cases} \arg\left(e^{i(\phi(z)-\varepsilon/2)} / \prod_{j=1}^{k}(B_1^{(j)}(z)/B_2^{(j)}(z))\right) & \text{if } f(z) > 0, \\ \arg\left(e^{i(\phi(z)+\varepsilon/2)} / \prod_{j=1}^{k}(B_1^{(j)}(z)/B_2^{(j)}(z))\right) & \text{if } f(z) \leq 0, \end{cases}$$

for all $k \geq 1$.

Let l be a positive integer, and suppose by induction that we have found Ω_{k+1}, ψ_{k+1}, $B_1^{(k)}$ and $B_2^{(k)}$ for all integers $k < l$, satisfying the above hypotheses for all such k. Then, by setting $\eta = \eta_k$, $\delta = \delta_k$, $\Omega = \Omega_k$ and $\psi = \psi_{k|\Omega}$, and applying Lemma 3.2.6 with $\varepsilon/4$ in place of ε, we obtain Blaschke products B_1 and B_2 such that

$$\begin{aligned} m(\{z \in \Omega : |\psi(z) - \arg(B_1(z)/B_2(z))| > \varepsilon/4\}) &< \eta, \\ |\arg(B_1(z)/B_2(z))| &< \delta, \quad \text{for all} \quad z \in \mathbb{T} \setminus \Omega, \\ \sum_{\{z:\, B_1(z)B_2(z)=0\}} (1 - |z|) &\leq 2m(\Omega) \log\left(\frac{48\pi}{\varepsilon}\right). \end{aligned}$$

Now, by setting $B_1^{(l)} = B_1$ and $B_2^{(l)} = B_2$, and choosing a suitable relatively open set $\Omega_{l+1} \subset \mathbb{T}$, Hypotheses 1, 2 and 3 above are satisfied for $k = l$. Then Hypothesis 4 uniquely defines the function $\psi_{l+1} : \mathbb{T} \to (-\pi, \pi]$. By induction on l, we obtain thereby sequences $(\Omega_k)_k$, $(\psi_k)_k$, $\left(B_1^{(k)}\right)_k$ and $\left(B_2^{(k)}\right)_k$ which satisfy the above hypotheses for all positive integers k.

Now, using Hypothesis 1, we have

$$\begin{aligned} \sum_{k=1}^{\infty} \sum_{\{z:\, B_1^{(k)}(z)B_2^{(k)}(z)=0\}} (1 - |z|) &\leq 2\sum_{k=1}^{\infty} m(\Omega_k) \log\left(\frac{48\pi}{\varepsilon}\right) \\ &\leq 2\left(m(\Delta) + 2\sum_{k=1}^{\infty} \eta_k\right) \log\left(\frac{48\pi}{\varepsilon}\right) \\ &\leq 2m(\Delta)\left(\log\left(\frac{48\pi}{\varepsilon}\right) + \log\left(\frac{25}{24}\right)\right) \\ &= 2m(\Delta) \log\left(\frac{50\pi}{\varepsilon}\right) < \infty. \end{aligned}$$

In other words, the infinite products

$$B_1 = \prod_{k=1}^{\infty} B_1^{(k)} \qquad \text{and} \qquad B_2 = \prod_{k=1}^{\infty} B_2^{(k)}$$

are Blaschke products satisfying

$$\sum_{\{z:\, B_1(z)B_2(z)=0\}} (1-|z|) \le 2m(\Delta)\log\left(\frac{50\pi}{\varepsilon}\right),$$

as required.

Finally, for any $z \in \Omega_1 \setminus \bigcap_{k=2}^{\infty} \Omega_k$ there exists some $\ell \ge 1$ such that $z \in \Omega_\ell$ and $z \notin \Omega_k$ for all $k > \ell$.

Therefore, by Hypothesis 4, we have

$$\psi_\ell(z) + \arg\left(\prod_{k=1}^{\ell-1}\left(B_1^{(k)}(z)/B_2^{(k)}(z)\right)\right) \equiv \alpha \qquad (\operatorname{mod} 2\pi)$$

for some $\alpha \in (\varepsilon/2 - \pi, \pi - \varepsilon/2]$ satisfying $|\phi(z) - \alpha| \le \varepsilon/2$. But $\psi_k(z)$ lies within $\varepsilon/4$ of $\arg\left(B_1^{(k)}(z)/B_2^{(k)}(z)\right)$, by Hypothesis 3. Hence,

$$\arg\left(\prod_{k=1}^{\ell}\left(B_1^{(k)}(z)/B_2^{(k)}(z)\right)\right) = \beta$$

for some $\beta \in (\varepsilon/4 - \pi, \pi - \varepsilon/4]$ satisfying $|\phi(z) - \beta| \le 3\varepsilon/4$. Now by Hypothesis 2 we have $\left|\arg\left(B_1^{(k)}/B_2^{(k)}\right)\right| < \delta_k$ for all $k > \ell$, and the δ_k sum to less than $\varepsilon/4$. Hence

$$\limsup_{m\to\infty}\left|\phi(z) - \arg\left(\prod_{k=1}^{m}\left(B_1^{(k)}(z)/B_2^{(k)}(z)\right)\right)\right| < \varepsilon.$$

This product is bounded away from $\pm\pi$, and thus we may pass to the limit and deduce that

$$|\phi(z) - \arg(B_1(z)/B_2(z))| < \varepsilon, \tag{3.20}$$

for almost all $z \in \Omega_1 \setminus \bigcap_{k=2}^{\infty} \Omega_k$. However, $m(\Omega_k) \to 0$ as $k \to \infty$, and so the intersection has measure zero. Applying Hypothesis 2 once more, we see that

$$|\arg(B_1(z)/B_2(z))| \le \sum_{k=1}^{\infty} \delta_k < \varepsilon$$

for all $z \in \mathbb{T} \setminus \Omega_1$. Thus (3.20) holds for almost all $z \in \mathbb{T}$, as claimed. □

Theorem 3.2.8 *Let $\phi \in L^{\infty}(\mathbb{T}, (-\pi, \pi])$ and $\varepsilon > 0$. Then there exist Blaschke products B_1 and B_2 such that*

$$\|H(\phi - \arg(B_1/B_2))\|_{\infty} < \varepsilon,$$

where H is the Hilbert transform.

Proof The proof is based on an induction.

Let δ be a positive constant which depends on ε in a way to be specified later. Set $\phi_0 = \phi$ and $h_1 = 0$. By Lemma 3.2.7, there exists $\phi_1 \in L^{\infty}(\mathbb{T}, (-\pi, \pi])$ with $\|\phi_1\|_{\infty} \leq \frac{1}{2}\delta$ and

$$\phi_1 = \phi_0 - \operatorname{Re}(h_1) - \arg\left(\frac{B_1^1}{B_1^2}\right),$$

where B_1^1 and B_1^2 are finite Blaschke products with

$$\sum_{\{z:\, B_1^1(z) B_2^1(z)=0\}} (1 - |z|) \leq \log\left(\frac{100\pi}{\delta}\right).$$

Suppose that, for $k \geq 1$, we have constructed ϕ_k with $\|\phi_k\|_{\infty} \leq 2^{-k}\delta$. Using Corollary 3.2.4, there is a function $h_{k+1} \in H^{\infty}$ such that

$$\begin{cases} \|\phi_k - \operatorname{Re} h_{k+1}\|_2 \leq 2^{-k}\|\phi_k\|_{\infty} \leq 2^{-2k}\delta, \\ \|h_{k+1}\|_{\infty} \leq c \log\left(2^k\right) \|\phi_k\|_{\infty} \leq \frac{ck \log 2}{2^k}\delta. \end{cases}$$

Now define

$$\Delta_k := \{\xi \in \mathbb{T} : |\phi_k(\xi) - \operatorname{Re}(h_{k+1}(\xi))| > 2^{-(k+2)}\delta\}.$$

Then, by Chebyshev's inequality,

$$m(\Delta_k) \leq \frac{\|\phi_k - \operatorname{Re} h_{k+1}\|_2^2}{2^{-2(k+2)}\delta^2} \leq \frac{16}{2^{2k}}.$$

Then take

$$f_k = (\phi_k - \operatorname{Re} h_{k+1})\chi_{\Delta_k}.$$

Using Lemma 3.2.7, we may find Blaschke products B_1^{k+1} and B_2^{k+1} such that

$$\left\| f_k - \arg\left(B_1^{k+1}/B_2^{k+1}\right)\right\|_{\infty} < 2^{-(k+2)}\delta,$$

with

$$\begin{aligned} \sum_{\{z:\, B_1(z) B_2(z)=0\}} (1 - |z|) &\leq 2m(\Delta_k) \log\left(\frac{50\pi}{2^{-(k+2)}\delta}\right) \\ &\leq \frac{32}{2^{2k}} \log\left(\frac{200\pi 2^k}{\delta}\right). \end{aligned}$$

We now set

$$\phi_{k+1} = \phi_k - \operatorname{Re} h_{k+1} - \arg\left(B_1^{k+1}/B_2^{k+1}\right).$$

Note that, on Δ_k, $\phi_{k+1} = f_k - \arg\left(B_1^{k+1}/B_2^{k+1}\right)$, and thus

$$\|\phi_{k+1}\|_{\infty,\Delta_k} \leq 2^{-(k+2)}\delta.$$

On the other hand, on $\mathbb{T} \setminus \Delta_k$, since $|\phi_k - \operatorname{Re} h_{k+1}| \leq 2^{-(k+2)}\delta$ and since $\left|\arg\left(B_1^{k+1}/B_2^{k+1}\right)\right| \leq 2^{-(k+2)}\delta$, it follows that

$$\|\phi_{k+1}\|_{\infty,\ \mathbb{T}\setminus\Delta_k} \leq 2^{-(k+1)}\delta.$$

Therefore we have

$$\|\phi_{k+1}\|_{\infty} \leq 2^{-(k+1)}\delta.$$

Note that

$$\begin{aligned}\sum_{k=0}^{\infty} \|\phi_k - \arg\left(B_1^{k+1}/B_2^{k+1}\right) - \phi_{k+1}\|_{\infty} &\leq \sum_{k=0}^{\infty} \|\operatorname{Re} h_{k+1}\|_{\infty} \\ &\leq \sum_{k=0}^{\infty} \|h_{k+1}\|_{\infty} \leq \sum_{k=0}^{\infty} \frac{ck\log 2}{2^k}\delta < \varepsilon,\end{aligned}$$

for δ sufficiently small. So

$$\sum_{k=0}^{\infty}\left(\phi_k - \arg\left(B_1^{k+1}/B_2^{k+1}\right) - \phi_{k+1}\right)$$

converges in $L^{\infty}(\mathbb{T})$ and is equal to

$$\phi_0 - \arg\left(\prod_{k=0}^{\infty} B_1^{k+1}/B_2^{k+1}\right).$$

Moreover,

$$\left\| H\left(\phi_0 - \arg\left(\prod_{k=0}^{\infty} B_1^{k+1}/B_2^{k+1}\right)\right)\right\|_{\infty} \leq \sum_{k=0}^{\infty} \|h_{k+1}\|_{\infty} < \varepsilon.$$

Finally, since

$$\sum_{k=0}^{\infty} \sum_{\left\{z:\, B_1^{k+1}(z)B_2^{k+1}(z)=0\right\}} (1-|z|) < \infty,$$

the infinite products $\prod_{k=0}^{\infty} B_1^{k+1}$ and $\prod_{k=0}^{\infty} B_2^{k+1}$ converge to give Blaschke products. □

We now arrive at the necessary and sufficient condition for a function f to have a Douglas–Rudin factorization $f = g\overline{h}$, with $f, g \in H^\infty$.

Corollary 3.2.9 (Barclay) *Let $f \in L^\infty(\mathbb{T})$ such that $\log|f| \in L^1(\mathbb{T})$. Then, for each $\varepsilon > 0$, there exist $g, h \in H^\infty$ such that*

$$f = g\overline{h} \qquad \text{and} \qquad \|f\|_\infty \leq \|g\|_\infty \|h\|_\infty < (1+\varepsilon)\|f\|_\infty.$$

Proof Clearly, for any factorization $f = g\overline{h}$, we have $\|f\|_\infty \leq \|g\|_\infty \|h\|_\infty$.

Let v be the outer function defined by

$$v(z) = \exp\left(\frac{1}{2\pi}\int_{-\pi}^{\pi} \log|f(e^{it})| \frac{e^{it}+z}{e^{it}-z}\, dt\right).$$

Then $v \in H^\infty$ and f/v is unimodular on $\mathbb{T}$. We write it as $f/v = e^{i\phi}$ where $\phi \in L^\infty(\mathbb{T}, (-\pi, \pi])$. Using Theorem 3.2.8, there exist Blaschke products B_1 and B_2 such that $\gamma = \phi - \arg(B_1/B_2)$ satisfies

$$\|H(\gamma)\|_\infty < \log(1+\varepsilon).$$

It follows that $F := \frac{1}{2}(\gamma + iH(\gamma)) \in H^\infty$. By definition of F, we have

$$\exp(i\phi) = B_1 \exp(iF)\overline{B_2 \exp(-iF)}.$$

Thus we have the factorization $f = g\overline{h}$ with

$$g = vB_1 \exp(iF) \qquad \text{and} \qquad h = B_2 \exp(-iF),$$

and $\|g\|_\infty < \|f\|_\infty \sqrt{1+\varepsilon}$, while $\|h\|_\infty < \sqrt{1+\varepsilon}$, as required. □

Corollary 3.2.10 (Bourgain, Barclay) *Let $p, q, r \in [1, \infty]$ such that $\frac{1}{p} = \frac{1}{q} + \frac{1}{r}$, and let $f \in L^p(\mathbb{T})$ such that $\log|f| \in L^1(\mathbb{T})$. Then, for each $\varepsilon > 0$, there exist $g \in H^q$ and $h \in H^r$ such that*

$$f = g\overline{h} \qquad \text{and} \qquad \|f\|_p \leq \|g\|_q \|h\|_r < (1+\varepsilon)\|f\|_p.$$

Proof Suppose first that $q, r < \infty$. Let $\ell = \log|f| \in L^1(\mathbb{T})$. Now define

$$g_0 = \exp\left(\frac{p}{q}(\ell + iH(\ell))\right) \quad \text{and} \quad h_0 = \exp\left(\frac{p}{r}(\ell + iH(\ell))\right).$$

Note that $g_0 \in H^q$ and $h_0 \in H^r$, since $f \in L^p(\mathbb{T})$. Moreover,

$$|g_0\overline{h_0}| = \exp\left(\left(\frac{p}{q} + \frac{p}{r}\right)\ell\right) = |f|.$$

So, $f = g_0\overline{h_0}u$, where $u \in L^\infty(\mathbb{T})$, $|u| = 1$ almost everywhere and obviously $\log|u| \in L^1(\mathbb{T})$. Using Corollary 3.2.9, we may write $u = g_1\overline{h_1}$ where $g_1, h_1 \in H^\infty$ and $\|g_1\|_\infty \|h_1\|_\infty \leq (1+\varepsilon)$. Since

$$\|g_0\|_q = \|f\|_p^{p/q} \quad \text{and} \quad \|h_0\|_r = \|f\|_p^{p/r},$$

we get $f = (g_0g_1)\overline{(h_0h_1)}$, where $g_0g_1 \in H^q$, $h_0h_1 \in H^r$, with

$$\|g_0g_1\|_q \|h_0h_1\|_r < (1+\varepsilon)\|f\|_p.$$

On the other hand, Hölder's inequality gives $\|f\|_p \leq \|g_0g_1\|_q \|h_0h_1\|_r$.

For $q = r = \infty$, the result is given in Corollary 3.2.9.

The remaining case, when only one of q and r is infinite, is similar and is left as an exercise. □

Remark 3.2.11 Corollaries 3.2.9 and 3.2.10 are sharp, in the sense that it is not in general possible to find g and h such that $f = g\overline{h}$ and $\|f\|_p = \|g\|_q \|h\|_r$. See Exercises 2 and 3 for examples in $L^\infty(\mathbb{T})$.

We shall require the following consequence of Corollary 3.2.10 in the case $q = r = 2$.

Corollary 3.2.12 *Let $f_0 \in L^\infty(\mathbb{T})$ with $\frac{1}{f_0} \in L^\infty(\mathbb{T})$. Then*

$$L^1(\mathbb{T}) = H^2\overline{H^2} + \mathbb{C}f_0.$$

Proof We begin with the case $f_0 = 1$. Let $f \in L^1(\mathbb{T})$ and for $\theta \in (-\pi, \pi]$, consider $g_\theta = f + e^{i\theta}$. Applying Jensen's inequality in the θ variable, we get

$$\begin{aligned}\int_{-\pi}^{\pi} \frac{1}{2\pi} \int_{-\pi}^{\pi} \log|g_\theta(e^{i\omega})|\, d\omega\, d\theta &= \int_{-\pi}^{\pi} \frac{1}{2\pi} \int_{-\pi}^{\pi} \log|f(e^{i\omega})e^{-i\theta} + 1|\, d\theta\, d\omega \\ &\geq \int_{-\pi}^{\pi} \log 1\, d\omega \geq 0.\end{aligned}$$

Hence there is a θ for which

$$\frac{1}{2\pi} \int_{-\pi}^{\pi} \log|g_\theta(e^{i\omega})|\, d\omega > 0,$$

i.e., $f + e^{i\theta} \in H^2\overline{H^2}$ for some θ.

We now consider the general case. It is clear that $\log|f_0| \in L^1$. Hence, by Corollary 3.2.9, since f_0 is also bounded, it follows that $f_0 \in H^\infty\overline{H^\infty}$; that is, we can write $f_0 = g_0\overline{h_0}$, where $g_0, h_0 \in H^\infty$. As noted above, we know that, for $f \in L^1(\mathbb{T})$, there exist $g, h \in H^2$ and $c \in \mathbb{C}$, such that $f/f_0 = g\overline{h} + c$, and so finally

$$f = gg_0\overline{hh_0} + cf_0 \in H^2\overline{H^2} + \mathbb{C}f_0.$$ □

3.3 Applications in harmonic analysis

The *degree* of a finite Blaschke product b is defined to be the cardinality of its zero set Zero(b) (taking into account multiplicity).

We shall prove a result providing a link between the density of the elementary spectral measure associated with T and that associated with $b(T)$. To do this we need information on the location of the zeroes of b'.

Theorem 3.3.1 *If b is a Blaschke product (finite or not), then*

$$\text{Zero}(b') \subset \textit{closed convex hull of } \text{Zero}(b) \cup \{0\}.$$

Proof First let us prove the theorem when b is a finite Blaschke product of degree $d < \infty$. Let $b(z) = \prod_{k=1}^{d} \frac{z-\alpha_k}{1-\overline{\alpha_k}z} = \frac{p(z)}{q(z)}$, where

$$p(z) = \prod_{k=1}^{d} (z - \alpha_k) \quad \text{and} \quad q(z) = \prod_{k=1}^{d} (1 - \overline{\alpha_k} z).$$

Then, for any $z \in \overline{\mathbb{D}}$ we have $b'(z) = \frac{p'(z)q(z)-q'(z)p(z)}{q(z)^2}$ and for any $z \in \overline{\mathbb{D}} \backslash \{\alpha_1, \ldots, \alpha_d\}$ we have $\frac{b'(z)}{b(z)} = \frac{p'(z)}{p(z)} - \frac{q'(z)}{q(z)}$. Given that

$$p'(z) = \sum_{k=1}^{d} \prod_{l \neq k, l=1}^{d} (z - \alpha_l) \text{ and } q'(z) = \sum_{k=1}^{d} -\overline{\alpha_k} \prod_{l \neq k, l=1}^{d} (1 - \overline{\alpha_l} z),$$

we obtain

$$\frac{b'(z)}{b(z)} = \sum_{k=1}^{d} \left(\frac{1}{z - \alpha_k} + \frac{\overline{\alpha_k}}{1 - \overline{\alpha_k} z} \right) \tag{3.21}$$

for any $z \in \overline{\mathbb{D}} \backslash \{\alpha_1, \ldots, \alpha_d\}$.

Let $\xi \in (\text{Zero}(b') \cap \overline{\mathbb{D}}) \backslash \{\alpha_1, \ldots, \alpha_d\}$. Then ξ satisfies

$$\sum_{k=1}^{d} \left(\frac{1}{\overline{\xi} - \overline{\alpha_k}} + \frac{\alpha_k}{1 - \alpha_k \overline{\xi}} \right) = 0, \tag{3.22}$$

that is,

$$\sum_{k=1}^{d} \left(\frac{\xi - \alpha_k}{|\xi - \alpha_k|^2} + \frac{\alpha_k (1 - \overline{\alpha_k} \xi)}{|1 - \alpha_k \overline{\xi}|^2} \right) = 0.$$

Hence ξ satisfies the identity

$$\xi \sum_{k=1}^{d} \left(\frac{1}{|\xi - \alpha_k|^2} - \frac{|\alpha_k|^2}{|1 - \alpha_k \overline{\xi}|^2} \right) = \sum_{k=1}^{d} \alpha_k \left(\frac{1}{|\xi - \alpha_k|^2} - \frac{1}{|1 - \alpha_k \overline{\xi}|^2} \right).$$

Moreover, we have

$$\frac{1}{|\xi - \alpha_k|^2} - \frac{|\alpha_k|^2}{|1 - \alpha_k \bar{\xi}|^2} = \frac{(1 - |\alpha_k|^2)(|1 - \overline{\alpha_k}\xi|^2 + |\alpha_k|^2(1 - |\xi|^2))}{|\xi - \alpha_k|^2 |1 - \alpha_k \bar{\xi}|^2}$$

and

$$\frac{1}{|\xi - \alpha_k|^2} - \frac{1}{|1 - \alpha_k \bar{\xi}|^2} = \frac{(1 - |\alpha_k|^2)(1 - |\xi|^2)}{|\xi - \alpha_k|^2 |1 - \alpha_k \bar{\xi}|^2}.$$

Thus, if $\xi \in (\mathrm{Zero}(b') \cap \overline{\mathbb{D}}) \backslash \{\alpha_1, \cdots \alpha_d\}$, we obtain

$$\begin{aligned} \xi \sum_{k=1}^{d} & \frac{(1 - |\alpha_k|^2)(|1 - \overline{\alpha_k}\xi|^2 + |\alpha_k|^2(1 - |\xi|^2))}{|\xi - \alpha_k|^2 |1 - \alpha_k \bar{\xi}|^2} \\ &= \sum_{k=1}^{d} \alpha_k \left(\frac{(1 - |\alpha_k|^2)(1 - |\xi|^2)}{|\xi - \alpha_k|^2 |1 - \alpha_k \bar{\xi}|^2} \right). \end{aligned} \tag{3.23}$$

Note that if $|\xi| = 1$ we get

$$\sum_{k=1}^{d} \frac{(1 - |\alpha_k|^2)|1 - \overline{\alpha_k}\xi|^2}{|\xi - \alpha_k|^2 |1 - \alpha_k \bar{\xi}|^2} = 0,$$

which is absurd since each term of the above sum is positive.

Thus we have proved that if $\xi \in (\mathrm{Zero}(b') \cap \overline{\mathbb{D}}) \backslash \{\alpha_1, \ldots, \alpha_d\}$, then $|\xi| < 1$ and $\xi = \sum_{k=1}^{d} \rho_k \alpha_k$, where

$$\rho_k = \frac{\frac{(1 - |\alpha_k|^2)(1 - |\xi|^2)}{|\xi - \alpha_k|^2 |1 - \alpha_k \bar{\xi}|^2}}{\sum_{l=1}^{d} \left(\frac{(1 - |\alpha_l|^2)(|1 - \overline{\alpha_l}\xi|^2 + |\alpha_l|^2(1 - |\xi|^2))}{|\xi - \alpha_l|^2 |1 - \alpha_l \bar{\xi}|^2} \right)}.$$

Since

$$|1 - \overline{\alpha_l}\xi|^2 + |\alpha_l|^2(1 - |\xi|^2) = |\xi - \alpha_l|^2 + 1 - |\xi|^2,$$

it is clear that $1 - |\xi|^2 \leq |1 - \overline{\alpha_l}\xi|^2 + |\alpha_l|^2(1 - |\xi|^2)$, which implies that $\sum_{k=1}^{d} \rho_k \leq 1$. The proof of the theorem follows in the particular case where b is a finite Blaschke product.

Consider now the case where b is Blaschke product of infinite degree. If $\alpha \in \mathbb{D}$, for $z \in \mathbb{D}$, set

$$u_\alpha(z) = \begin{cases} z & \text{if } \alpha = 0, \\ \frac{\bar{\alpha}}{\alpha} \frac{\alpha - z}{1 - \bar{\alpha} z} & \text{otherwise.} \end{cases}$$

Then b is of the form $b(z) = \prod_{k \geq 1} u_{\alpha_k}(z)$ where $(\alpha_k)_{k \geq 1} \subset \mathbb{D}$ satisfies the Blaschke condition,

$$\sum_{k \geq 1} (1 - |\alpha_k|^2) < \infty. \tag{3.24}$$

Using the Blaschke condition (3.24), we obtain the analogous equation to (3.21):

$$\frac{b'(z)}{b(z)} = \sum_{k\geq 1}\left(\frac{1}{z-\alpha_k} + \frac{\overline{\alpha_k}}{1-\overline{\alpha_k}z}\right) \tag{3.25}$$

$$= \sum_{k\geq 1}\frac{1-|\alpha_k|^2}{(z-\alpha_k)(1-\overline{\alpha_k}z)}. \tag{3.26}$$

Therefore, for every $\xi \in \mathrm{Zero}(b') \cap \overline{\mathbb{D}}\backslash\overline{\mathrm{Zero}(b)}$, following Equations (3.22)–(3.23), we get

$$\xi \sum_{k\geq 1}\frac{(1-|\alpha_k|^2)(|1-\overline{\alpha_k}\xi|^2 + |\alpha_k|^2(1-|\xi|^2))}{|\xi-\alpha_k|^2|1-\alpha_k\overline{\xi}|^2}$$

$$= \sum_{k\geq 1}\alpha_k\left(\frac{(1-|\alpha_k|^2)(1-|\xi|^2)}{|\xi-\alpha_k|^2|1-\alpha_k\overline{\xi}|^2}\right),$$

once more thanks to the Blaschke condition (3.24). As in the case of finite Blaschke products we can prove that we obtain a contradiction if $|\xi| = 1$, which implies that if $\xi \in \mathrm{Zero}(b') \cap \overline{\mathbb{D}}\backslash\overline{\mathrm{Zero}(b)}$, then $\xi = \sum_{k\geq 1}\rho_k\alpha_k$, where

$$\rho_k = \frac{\frac{(1-|\alpha_k|^2)(1-|\xi|^2)}{|\xi-\alpha_k|^2|1-\alpha_k\overline{\xi}|^2}}{\sum_{l\geq 1}\left(\frac{(1-|\alpha_l|^2)(|1-\overline{\alpha_l}\xi|^2+|\alpha_l|^2(1-|\xi|^2))}{|\xi-\alpha_l|^2|1-\alpha_l\overline{\xi}|^2}\right)}.$$

The fact that $\rho_k \leq 1$ follows from the same argument as in the case of finite Blaschke products. □

Therefore, if b is a finite Blaschke product, $b'(\xi) \neq 0$ for all $\xi \in \mathbb{T}$.

We can now provide a formula for the density of the elementary spectral measure associated with $b(T)$.

Theorem 3.3.2 *Let $T \in \mathcal{L}(\mathcal{H})$ be an absolutely continuous contraction and let b be a finite Blaschke product of degree d. Then*

$$(x \overset{b(T)}{\cdot} y)(e^{it}) = \sum_{j=1}^{d}\frac{(x \overset{T}{\cdot} y)(\xi_j)}{|b'(\xi_j)|} \qquad a.e., \tag{3.27}$$

where $\xi_1, \ldots, \xi_d$ are the solutions of $b(z) = e^{it}$.

Proof We may partition $\mathbb{T}$ into intervals $J_1, \ldots, J_d$, each of which is mapped onto $\mathbb{T}$ by b (we leave this as Exercise 6 below, but the reader may also consult Cassier and Chalendar [57]). Then we get

$$
\begin{aligned}
(x \overset{b(T)}{\cdot} y)(e^{it}) &= \lim_{r\to 1^-} \langle K_{r,t}(b(T))x, y\rangle \\
&= \lim_{r\to 1^-} \langle (\mathrm{I} - re^{-it}b(T))^{-1}x, y\rangle \\
&\quad + \langle x, (\mathrm{I} - re^{-it}b(T))^{-1}y\rangle - \langle x, y\rangle \\
&= \lim_{r\to 1^-} \langle (1 - re^{-it}b)^{-1}, x \overset{T}{\cdot} y\rangle + \langle (1 - re^{it}\overline{b})^{-1}, x \overset{T}{\cdot} y\rangle \\
&\quad - \langle 1, x \overset{T}{\cdot} y\rangle,
\end{aligned}
$$

since $\langle f(T)x, y\rangle = \langle f, x \overset{T}{\cdot} y\rangle$ for all $f \in L^\infty$. Hence

$$
\begin{aligned}
(x \overset{b(T)}{\cdot} y)(e^{it}) &= \lim_{r\to 1^-} \int_0^{2\pi} (x \overset{T}{\cdot} y)(e^{i\theta}) \frac{1 - r^2}{|1 - re^{-it}b(e^{i\theta})|^2} \frac{d\theta}{2\pi} \\
&= \lim_{r\to 1^-} \sum_{j=1}^{d} \int_{J_j} (x \overset{T}{\cdot} y)(e^{i\theta}) \frac{1 - r^2}{|1 - re^{-it}b(e^{i\theta})|^2} \frac{d\theta}{2\pi} \\
&= \lim_{r\to 1^-} \sum_{j=1}^{d} \int_0^{2\pi} \frac{(x \overset{T}{\cdot} y)(b^{-1}(e^{i\alpha}))e^{i\alpha}}{b'(b^{-1}(e^{i\alpha}))b^{-1}(e^{i\alpha})} \frac{1 - r^2}{|1 - re^{-it}e^{i\alpha}|^2} \frac{d\alpha}{2\pi},
\end{aligned}
$$

where $e^{i\alpha} = b(e^{i\theta})$. Since $\frac{d\theta}{d\alpha} \geq 0$, we have

$$
\begin{aligned}
(x \overset{b(T)}{\cdot} y)(e^{it}) &= \lim_{r\to 1^-} \sum_{j=1}^{d} \int_0^{2\pi} \frac{(x \overset{T}{\cdot} y)(b^{-1}(e^{i\alpha}))}{|b'(b^{-1}(e^{i\alpha}))|} \frac{1 - r^2}{|1 - re^{-it}e^{i\alpha}|^2} \frac{d\alpha}{2\pi} \\
&= \sum_{j=1}^{d} \frac{(x \overset{T}{\cdot} y)(\xi_j)}{|b'(\xi_j)|}.
\end{aligned}
$$

□

Here is a reformulation of Bourgain's original result, the case $p = 1, q = r = 2$ of Corollary 3.2.10.

Lemma 3.3.3 *Let $S \in \mathcal{L}(\mathcal{H})$ with $\mathcal{H} = H^2(\mathbb{T})$ defined by*

$$Sf(e^{it}) = e^{it} f(e^{it}).$$

Then S factorizes $f \in L^1(\mathbb{T})$ if and only if $f \equiv 0$ or $\log|f| \in L^1(\mathbb{T})$.

Proof Recall that S is an absolutely continuous contraction (see Exercise 6 of Chapter 2). The only thing to check is that, for $g, h \in H^2(\mathbb{T})$, we have $g \overset{S}{\cdot} h = g\overline{h}$. By Lemma 2.2.4, we get

$$
\widehat{g \overset{S}{\cdot} h}(n) = \begin{cases} \langle S^{*n}g, h\rangle = \frac{1}{2\pi}\int_0^{2\pi} g(e^{it})\overline{h(e^{it})}e^{-int}\,dt = \widehat{g\overline{h}}(n) & \text{if } n \geq 0, \\ \langle S^{|n|}g, h\rangle = \frac{1}{2\pi}\int_0^{2\pi} e^{-int}g(e^{it})\overline{h(e^{it})}\,dt = \widehat{g\overline{h}}(n) & \text{if } n \leq 0. \end{cases}
$$

The injectivity of the Fourier transform implies that $g \overset{S}{\cdot} h = g\overline{h}$. □

A reformulation of Corollary 3.2.12 in terms of factorization of L^1 functions is the following.

Lemma 3.3.4 *Let $f_0 \in L^\infty(\mathbb{T})$ with $\frac{1}{f_0} \in L^\infty(\mathbb{T})$. Let $A \in \mathcal{L}(\mathcal{H})$ be an absolutely continuous contraction which factorizes f_0. Then $T := S \oplus A$ factorizes $L^1(\mathbb{T})$.*

Proof Let $f \in L^1$. Using Corollary 3.2.12, there exist $g, h \in H^2(\mathbb{T})$ and $\lambda \in \mathbb{C}$ such that $f = g\overline{h} + \lambda f_0$. We have already noticed that $g\overline{h} = g \overset{S}{\cdot} h$. Our hypothesis implies that there exist $x, y \in \mathcal{H}$ such that $\lambda f_0 = x \overset{A}{\cdot} y$, and hence $f = (g \oplus x) \overset{T}{\cdot} (h \oplus y)$. Indeed, if T_1 and T_2 are absolutely continuous contractions on $\mathcal{H}_1$ and $\mathcal{H}_2$, respectively, then for $x_1, y_1 \in \mathcal{H}_1$ and $x_2, y_2 \in \mathcal{H}_2$ it is not difficult to check that

$$(x_1 \oplus x_2) \overset{T_1 \oplus T_2}{\cdot} (y_1 \oplus y_2) = x_1 \overset{T_1}{\cdot} y_1 + x_2 \overset{T_2}{\cdot} y_2.$$ □

Now, let us give an immediate corollary of the previous lemma.

Corollary 3.3.5 *Let $A \in \mathcal{L}(\mathcal{H})$ be an absolutely continuous contraction such that $(\sigma_p(A) \cap \mathbb{D}) \cup (\sigma_p(A^*) \cap \mathbb{D}) \neq \emptyset$. Then $S \oplus A$ factorizes $L^1(\mathbb{T})$.*

Proof Let $\lambda = re^{it} \in \sigma_p(A) \cap \mathbb{D}$. Then there exists $x \in \mathcal{H}$ such that $\|x\| = 1$, $\langle A^n x, x\rangle = \lambda^n$ and $\langle A^{*n} x, x\rangle = \overline{\lambda}^n$. So A factorizes $P_{r,t}$ with $P_{r,t} \in L^\infty(\mathbb{T})$ and $\frac{1}{P_{r,t}} \in L^\infty(\mathbb{T})$. If $\lambda \in \sigma_p(A^*) \cap \mathbb{D}$, then A factorizes $P_{r,-t}$. The end of the proof follows, applying Lemma 3.3.4. □

A combination of Corollary 3.3.5 and Theorem 3.3.2 provides the following factorization result.

Theorem 3.3.6 *For all $f \in L^1(\mathbb{T})$, there exist $g, h \in H^2(\mathbb{T})$ such that*

$$f(e^{it}) = \frac{(g\overline{h})(e^{it/2}) + (g\overline{h})(-e^{it/2})}{2}.$$

Proof Since $\sigma_p(S^*) = \mathbb{D}$, the operator $S \oplus S$, which is unitarily equivalent to S^2, factorizes L^1. Consider the finite Blaschke product $b(z) = z^2$. We have seen that, for $g, h \in H^2(\mathbb{T})$,

$$(g \overset{S^2}{\cdot} h)(e^{it}) = \sum_{k=1}^{2} \frac{(g \overset{S}{\cdot} h)(\xi_k)}{2},$$

where ξ_1, ξ_2 are the solutions of $z^2 = e^{it}$ and $g \overset{S}{\cdot} h = g\overline{h}$. Now S^2 factorizes every function in L^1, so, given $f \in L^1$, there exist $g, h \in H^2(\mathbb{T})$ such that

$$f(e^{it}) = \frac{(g\overline{h})(e^{it/2}) + (g\overline{h})(-e^{it/2})}{2}.$$ □

3.4 Subnormal operators

Although the original application of the S. Brown method was in proving the existence of invariant subspaces for subnormal operators, there are now simpler proofs available. We shall begin by discussing normal operators, a functional calculus for them, and their invariant subspaces. Then we shall consider their restrictions to invariant subspaces, namely the subnormal operators.

3.4.1 Borelian functional calculus for normal operators

Let $N \in \mathcal{L}(\mathcal{H})$ be a normal operator. Let $\mathcal{P}$ be the sub-algebra of $C(\sigma(N))$, the algebra of continuous functions on $\sigma(N)$, consisting of functions of the form $f(z) = p(z, \overline{z})$, where $p \in \mathbb{C}[X, Y]$. Obviously, for $\lambda \neq \mu$ there exists $p \in \mathcal{P}$ such that $p(\lambda) \neq p(\mu)$. The complex version of the Stone–Weierstrass theorem, Theorem 1.1.6, implies that the closure of $\mathcal{P}$ is equal to $C(\sigma(N))$. For $f \in \mathcal{P}$ such that $f(z) = p(z, \overline{z})$, set $f(N) = p(N, N^*)$. Then $f(N)$ is also a normal operator. By Proposition 1.2.18, we have

$$\begin{aligned}\|f(N)\| &= \|p(N, N^*)\| \\ &= \rho(p(N, N^*)) \\ &= \max\{|p(z, \overline{z})| : z \in \sigma(N)\} \\ &= \|f\|_\infty.\end{aligned}$$

It follows that $\Psi : f \mapsto f(N)$ is an isometric isomorphism from $\mathcal{P}$ into the C^*-algebra generated by N and N^*. In particular Ψ extends uniquely and isometrically on the closure of $\mathcal{P}$, namely $C(\sigma(N))$, mapping into $\mathcal{L}(\mathcal{H})$. For $x, y \in \mathcal{H}$, define $\Phi_{x,y}$ on $C(\sigma(N))$ by $\Phi_{x,y}(f) = \langle f(N)x, y\rangle$. The Cauchy–Schwarz inequality and the identity $\|f(N)\| = \|f\|_\infty$ imply that $|\Phi_{x,y}(f)| \leq \|f\|_\infty \|x\| \|y\|$, and thus $\Phi_{x,y}$ is continuous. By Theorem 1.1.18, there exists a Borel measure $\mu_{x,y}$ on $\sigma(N)$ such that

$$\langle f(N)x, y\rangle = \int_{\sigma(N)} f \, d\mu_{x,y},$$

for all $f \in C(\sigma(N))$. Note that the bounded Borelian functions on $\sigma(N)$, of which we denote the class by $B_b(\sigma(N))$, are also integrable with respect to the measures $\mu_{x,y}$, for all $x, y \in \mathcal{H}$. Therefore, $\Phi_{x,y}$ extends continuously on $B_b(\sigma(N))$, and its norm is less than $\|x\| \|y\|$. It follows that, for all $f \in B_b(\sigma(N))$,

$$(x, y) \mapsto \int_{\sigma(N)} f \, d\mu_{x,y}$$

is a continuous sesquilinear map on $\mathcal{H} \times \mathcal{H}$ with norm less that $\|f\|_\infty$. The Riesz representation theorem implies that there exists $N_f \in \mathcal{L}(\mathcal{H})$ such that

$$\langle N_f x, y\rangle = \int_{\sigma(N)} f \, d\mu_{x,y},$$

for all $x, y \in \mathcal{H}$. The operator N_f is denoted by $f(N)$. In particular, if Ω is a Borel set of $\sigma(N)$ and if χ_Ω is the characteristic function of Ω, we get

$$\langle \chi_\Omega(N)x, y\rangle = \int_\Omega d\mu_{x,y}.$$

The operator $\chi_\Omega(N)$ is denoted by $E(\Omega)$. The function E, defined on the Borel sets of $\sigma(N)$, maps into $\mathcal{L}(\mathcal{H})$ and is a *spectral measure* on $\sigma(N)$, i.e., for each Borel subset Ω of $\sigma(N)$ the operator $E(\Omega)$ is an orthogonal projection; moreover, $E(\emptyset) = 0$, $E(\sigma(N)) = \mathrm{I}$, $E(\Omega_1 \cup \Omega_2) = E(\Omega_1) + E(\Omega_2)$ and $E(\Omega_1 \cap \Omega_2) = E(\Omega_1)E(\Omega_2)$ for all disjoint Borel subsets Ω_1, Ω_2 of $\Sigma(N)$.

An easy consequence of the existence of a Borel functional calculus for normal operators is the following.

Theorem 3.4.1 *A normal operator N that is not a multiple of the identity has a non-trivial hyperinvariant subspace.*

Proof If $\sigma(N) = \{a\}$, then $\rho(N - a\mathrm{I}) = 0$, and by Proposition 1.2.18, it follows that $N = a\mathrm{I}$. If $\sigma(N)$ is not reduced to a singleton, there exist two disjoint non-empty Borel subsets of $\sigma(N)$, say Ω_1 and Ω_2, such that $\Omega_1 \cup \Omega_2 = \sigma(N)$. It follows that $E(\Omega_1) \neq 0$, $E(\Omega_2) \neq 0$ and $E(\Omega_1)E(\Omega_2) = 0$. Therefore, Ker $E(\Omega_1)$ as well as Ker $E(\Omega_2)$ are non-trivial hyperinvariant subspaces for N. □

3.4.2 Invariant subspaces for subnormal operators

Definition 3.4.2 Suppose that $S \in \mathcal{L}(\mathcal{H})$ is a subnormal operator on $\mathcal{H}$ and $N \in \mathcal{L}(\mathcal{K})$ is a normal extension of S. Then N is a *minimal normal extension* of S if, whenever $\mathcal{M}$ reduces N and $\mathcal{H} \subset \mathcal{M} \subset \mathcal{K}$, then $\mathcal{M} = \mathcal{K}$.

Proposition 3.4.3 *Every subnormal operator S has a minimal normal extension.*

Proof Let N be a normal extension of S, and let $\mathcal{K}$ be the closed linear span of $\{N^{*k}h : h \in \mathcal{H}, k \geq 0\}$. Clearly $\mathcal{K}$ reduces N and contains $\mathcal{H}$. Moreover, $N_{|\mathcal{K}}$ is a minimal normal extension of S. □

Minimal normal extensions of subnormal operators are of particular interest because of their spectral properties.

Lemma 3.4.4 *Let $S \in \mathcal{L}(\mathcal{H})$ be a subnormal operator, and denote by $N \in \mathcal{L}(\mathcal{K})$ a minimal normal extension of S. Then $\sigma(N) \subset \sigma(S)$.*

Proof Suppose that $\lambda \in \mathbb{C}$ is not in $\sigma(S)$. We shall prove that $\lambda \notin \sigma(N)$. Without loss of generality, we may assume that $\lambda = 0$. By Theorem 1.2.16 there is a Borel set $\Omega \subset \mathbb{C}$, a positive measure μ defined on Borel subsets of Ω and a function $\Phi \in L^\infty(\Omega, \mu)$ such that N is unitarily equivalent to the operator M_Φ of multiplication by Φ on $L^2(\Omega, \mu)$. Without loss of generality, we will assume that $N = M_\Phi$, and so $N^* = M_{\overline{\Phi}}$. Let $\varepsilon > 0$ and let

$$\mathcal{M} = \chi_{\{z \in \Omega : |\Phi(z)| < \varepsilon\}} L^2(\Omega, \mu).$$

For $f \in \mathcal{M}$ and $k \geq 1$, it follows that $\|N^k f\| \leq \varepsilon^k \|f\|$. Since S is invertible, for $h \in \mathcal{H}$, we have

$$\begin{aligned} |\langle f, h\rangle| &= |\langle f, S^k S^{-k} h\rangle| = |\langle f, N^k S^{-k} h\rangle| = |\langle N^{*k} f, S^{-k} h\rangle| \\ &\leq \|N^{*k} f\| \|S^{-k} h\| = \|N^k f\| \|S^{-k} h\| \\ &\leq \varepsilon^k \|f\| \|S^{-k} h\| \leq \varepsilon^k \|S^{-1}\|^k \|f\| \|h\|. \end{aligned}$$

Letting $k \to \infty$ we see that if $\varepsilon < \|S^{-1}\|^{-1}$ then $\langle f, h\rangle = 0$. In this case $\mathcal{H} \subset \mathcal{M}^\perp$, which is a reducing subspace for N. The minimality of N implies that $\mathcal{M} = \{0\}$, and so N is invertible. □

Theorem 3.4.5 *Let $S \in \mathcal{L}(\mathcal{H})$ be a subnormal operator with* $\dim \mathcal{H} \geq 2$. *Then S has non-trivial invariant subspaces.*

The key idea of the proof is Lemma 3.1.1, the starting point of S. Brown's method.

The proof presented here relies on the three following results. The first lemma reduces the study of subnormal operators to the case of operators of multiplication by z on a closed subspace of $L^2(\mu)$, where μ is a positive Borel measure.

Lemma 3.4.6 *Let S be a subnormal operator with a cyclic vector. Then there exists a positive Borel measure μ on $\sigma(S)$ such that S is unitarily equivalent to M_z, the operator of multiplication by z on $P^2(\mu)$, where $P^2(\mu)$ is the closure in $L^2(\mu)$ of the polynomial functions restricted to $\sigma(S)$.*

Proof Let $x \in \mathcal{H} \setminus \{0\}$ be a cyclic vector for S. Let $N \in \mathcal{L}(\mathcal{K})$ be a minimal normal extension of S, i.e., suppose that $\mathcal{K}$ is the closed linear span of $\{S^{*k} x : k \geq 0\}$. Let E be the spectral measure associated with N on $\sigma(N)$ and let $\tilde{\mu}$ be the positive measure defined on the Borel subsets of $\sigma(N)$ by

$$\tilde{\mu}(\Omega) = \langle E(\Omega)x, x\rangle.$$

By Lemma 3.4.4, we have $\sigma(N) \subset \sigma(S)$. Obviously one can extend the measure $\tilde{\mu}$ on the Borel sets of $\sigma(S)$, setting, for all Borel sets $A \subset \sigma(S)$,

$$\mu(A) := \tilde{\mu}(A \cap \sigma(N)).$$

Note that, for all $p \in \mathbb{C}[z]$, we have

$$\begin{aligned}\|p(S)x\|^2 &= \langle p(S)x, p(S)x\rangle = \langle p(N)x, p(N)x\rangle = \langle |p|^2(N)x, x\rangle \\ &= \int_{\sigma(N)} |p|^2 \, d\mu_{x,x} \\ &= \int_{\sigma(N)} |p|^2 \, d\tilde{\mu} \\ &= \int_{\sigma(S)} |p|^2 \, d\mu = \|p\|^2_{L^2(\mu)}.\end{aligned}$$

In other words, ϕ defined on $\{p(S)x : p \in \mathbb{C}[z]\}$ by $\phi(p(S)x) = p$ is well-defined and isometric, and therefore extends isometrically from $\mathcal{H}$ onto $P^2(\mu)$. Note that $\phi(S(S^n x)) = \phi(S^{n+1}x) = z^{n+1} = M_z\phi(S^n x)$. The definition of $P^2(\mu)$ implies that $\phi S = M_z \phi$ on $P^2(\mu)$, where ϕ is an isometric surjection from $\mathcal{H}$ onto $P^2(\mu)$. □

An easy example of Lemma 3.4.6 in action is obtained by considering the unilateral shift S, which we may naturally regard as the operator M_z acting on the Hardy space H^2.

Lemma 3.4.7 *Let μ be a positive Borel measure on a non-trivial compact set $K \in \mathbb{C}$ and let $g \in L^q(\mu) \setminus \{0\}$ with $1 < q < 2$. Then there exists $w \in \mathbb{C}$ satisfying the following two conditions:*

1. $z \mapsto \frac{g(z)}{z-w} \in L^q(\mu)$.
2. $\int_K \frac{g(z)}{z-w} \, d\mu(z) \neq 0$.

Proof Let $R > 0$ such that $K \subset \{z : |z| \leq R\}$. The function h, defined on $K \times K$ by

$$h(z, w) = \begin{cases} \frac{|g(z)|^q}{|z-w|^q} & \text{if } z \neq w, \\ 0 & \text{if } z = w, \end{cases}$$

is measurable. Denote by m the planar Lebesgue measure. For all $z \in K$, we have

$$\begin{aligned}\int_{\sigma(S)} \frac{1}{|z-w|^q} \, dm(w) &\leq \int_{\{w:|z-w|\leq R\}} \frac{1}{|w|^q} \, dm(w) \\ &\leq \int_{\{w:|w|\leq 2R\}} \frac{1}{|w|^q} \, dm(w) < \infty,\end{aligned}$$

since $q < 2$. Applying Fubini's theorem, we get that

$$\int_k \left(\int_K h(z, w) d\mu(z) \right) dm(w) < \infty,$$

and thus there is a Borel set $\Omega \subset \mathbb{C}$ with $m(\Omega) = 0$ such that $\int_K h(z, w) d\mu(z)$ is finite for all $w \in K \setminus \Omega$. Moreover, if $w \notin K$, it is clear that the function $z \mapsto \frac{g(z)}{z-w}$ lies in $L^q(\mu)$. Therefore, $z \mapsto \frac{g(z)}{z-w} \in L^q(\mu)$ for all $w \in \mathbb{C} \setminus \Omega$, with $m(\Omega) = 0$.

Let $C_c^1(\mathbb{C})$ be the set of compactly supported functions on $\mathbb{C}$, whose derivative is continuous. In a similar way, Fubini's theorem shows that, for $\phi \in C_c^1(\mathbb{C})$, $z \mapsto \frac{g(z)}{z-w} \bar{\partial} \varphi(w) \in L^1(\mu)$ for all $w \in \mathbb{C} \setminus \Omega$, for some subset Ω with $m(\Omega) = 0$. In addition we have

$$\begin{aligned} \int_{\mathbb{C}} \left(\int_K \frac{g(z)}{z - w} \bar{\partial} \varphi(w) \, d\mu(z) \right) dm(w) &= \int_K g(z) \left(\int_{\mathbb{C}} \frac{\bar{\partial} \varphi(w)}{z - w} dm(w) \right) d\mu(z) \\ &= \pi \int_K g(z) \phi(z) \, d\mu(z). \end{aligned}$$

The last equality follows from the complex version of Green's formula, also known as the Cauchy–Pompeiu formula (see, for example, Berenstein and Gay [37, proposition 2.1.2]). By the Stone–Weierstrass theorem, $C_c^1(\mathbb{C})$ restricted to K is dense in $C(K)$. Since $C(K)$ is dense in $L^p(\mu)$, where $\frac{1}{p} + \frac{1}{q} = 1$, and since g is not identically equal to 0, there exists $\phi_0 \in C_c^1(\mathbb{C}) \setminus \{0\}$ such that

$$\int_K g(z) \phi_0(z) \, d\mu(z) \neq 0.$$

Therefore, there exists a Borel set $A \subset \mathbb{C}$ such that $m(A) > 0$ and

$$\int_K \frac{g(z)}{z - w} \bar{\partial} \varphi_0(w) \, d\mu(z) \neq 0,$$

for all $w \in A$. For any $w \in A \setminus \Omega$, both assertions of the lemma are satisfied. □

If $p > 0$ and μ is a positive Borel measure on a compact set K, denote by $P^p(\mu)$ the closure in $L^p(\mu)$ of $\{p_{|K} : p \in \mathbb{C}[z]\}$. Using Lemma 3.4.7, we will prove that for $p > 2$, provided $P^p(\mu) \neq L^p(\mu)$, there exists $w \in \mathbb{C}$ such that the evaluation at w defined on $\mathbb{C}[z]$ extends continuously on $P^p(\mu)$.

Theorem 3.4.8 *Let $p > 2$ and μ be a positive Borel measure on a compact set K, such that $P^p(\mu) \neq L^p(\mu)$. Then there exists $w \in \mathbb{C}$ and $c > 0$ such that*

$$|p(w)| \leq c \|p\|_{L^p(\mu)},$$

for all $p \in \mathbb{C}[z]$.

Proof Let q be such that $\frac{1}{p} + \frac{1}{q} = 1$. In particular note that $1 < q < 2$. Let $g \in L^q(\mu) \setminus \{0\}$ such that $g \perp P^p(\mu)$. By Lemma 3.4.7, there exists $w \in \mathbb{C}$ such that $z \mapsto \frac{g(z)}{z-w} \in L^q(\mu)$ and

$$\alpha := \int_K \frac{g(z)}{z-w}\, d\mu(z) \neq 0.$$

Since $g \perp P^p(\mu)$ and since the function $z \mapsto \frac{p(z)-p(w)}{z-w}$ is in $\mathbb{C}[z]$, it follows that

$$\int_K \frac{p(z)-p(w)}{z-w} g(z)\, d\mu(z) = 0,$$

for all $p \in \mathbb{C}[z]$. In other words, we have

$$p(w) = \frac{1}{\alpha} \int_K p(z) \frac{g(z)}{z-w}\, d\mu(z),$$

and thus, using Hölder's inequality,

$$|p(w)| \leq c \|p\|_{L^p(\mu)},$$

for all $p \in \mathbb{C}[z]$ and with $c = \frac{1}{|\alpha|} \|z \mapsto \frac{g(z)}{z-w}\|_{L^q(\mu)}$. □

We now have all the ingredients to give a complete proof of Theorem 3.4.5, which asserts the existence of non-trivial invariant subspaces for subnormal operators.

Proof Let $x \in \mathcal{H} \setminus \{0\}$. Without loss of generality concerning the existence of non-trivial invariant subspaces, we may assume that x is cyclic for S. In this case, by Lemma 3.4.6, S is unitarily equivalent to M_z, the operator of multiplication by z on $P^2(\mu)$, where μ is a positive Borel measure on $\sigma(S)$ and where $P^2(\mu)$ is the closure in $L^2(\mu)$ of the polynomial functions restricted to $\sigma(S)$. Moreover, we may assume that $P^2(\mu) \neq L^2(\mu)$, since otherwise S is normal and, by Theorem 3.4.1, S has non-trivial invariant subspaces.

It remains to check that in this case there exists $w \in \mathbb{C}$ and $x, y \in P^2(\mu)$ such that, for all non-negative integers n,

$$\langle S^n x, y \rangle = w^n.$$

By Theorem 3.4.8 with $p = 3$ and $q = 3/2$, there exists $w \in \mathbb{C}$ such that $L : f \mapsto f(w)$ is a continuous linear map on $P^3(\mu)$. The Hahn–Banach theorem implies the existence of $h \in L^{3/2}(\mu)$ such that $\|h\|_q = \|L\|$ and $L(f) = \int_{\sigma(S)} f h\, d\mu$. The reflexivity of $P^3(\mu)$ implies that there exists $r \in P^3(\mu)$ such that $\|r\|_3 = 1$ and $L(r) = \|L\|$. It follows that

$$\|h\|_q = \|L\| = L(r) = \int_K r h\, d\mu \leq \|r\|_3 \|h\|_{3/2}.$$

In other words we have an equality in Hölder's inequality. By (1.3), there exists $c > 0$ such that $|r|^3 = c|h|^{3/2}$, i.e., $|r|^2 = c^{2/3}|h|$, μ-almost everywhere.

Now, consider u defined as follows: $u(z) = h(z)/r(z)$ if $r(z) \neq 0$ and $h(z) = 0$ otherwise. Note that if $r(z) \neq 0$ then

$$|u^2(z)| = |h(z)|^2/|r(z)|^2 = |h(z)|^2/c^{2/3}.$$

Therefore, $u \in L^2(\mu)$, since $h \in L^{3/2}(\mu) \supset L^1(\mu)$.

We conclude the proof as follows. Take $x = r \in L^3(\mu) \subset L^2(\mu)$ and y equal to the orthogonal projection of $\overline{u}$ onto $P^2(\mu)$. Then, for $p \in \mathbb{C}[z]$,

$$\langle p(S)x, y\rangle = \langle p(S)r, \overline{u}\rangle = \int_{\sigma(N)} pru\, d\mu = \int_{\sigma(N)} ph\, d\mu = p(w).$$

By Lemma 3.1.1, S has non-trivial invariant subspaces. □

3.5 Surjectivity of continuous bilinear mapping

The other feature of the S. Brown method is an approximation scheme that ensures the surjectivity of certain mappings. We now explain this.

In this section, B will denote a continuous bilinear or sesquilinear mapping from $\mathcal{X} \times \mathcal{Y}$ to $\mathcal{Z}$, where $\mathcal{X}, \mathcal{Y}$ are complex Banach spaces and $\mathcal{Z}$ is a normed space. The purpose of this section is to provide sufficient conditions that imply the surjectivity of B, and even a stronger form of surjectivity called the property $(\mathbb{A}_{n,m})$, defined as follows.

Definition 3.5.1 Let $1 \leq n, m \leq \aleph_0$. We say that B possesses the property $(\mathbb{A}_{n,m})$ if, for every $n \times m$ matrix $(z_{i,j})_{1\leq i\leq n, 1\leq j\leq m}$ in $\mathcal{Z}$, there exist $x_1, \ldots, x_n \in \mathcal{X}$ and $y_1, \ldots, y_m \in \mathcal{Y}$ such that $B(x_i, y_j) = z_{i,j}$. For short, we write $(\mathbb{A}_n)$ for $(\mathbb{A}_{n,n})$.

Our next aim is to provide sufficient conditions for the properties $(\mathbb{A}_{\aleph_0})$ and $(\mathbb{A}_{1,\aleph_0})$; it will be seen that they follow from the existence of *approximate* solutions to equations of the form $B(x_i, y_j) = z_{i,j}$ together with certain 'vanishing conditions'.

3.5.1 A sufficient condition for property $(\mathbb{A}_{\aleph_0})$

Let c be a constant in $(0, 1)$, $C_1, C_2 > 0$.

We say that B has the *left and right vanishing approximation property (LRVAP)* if, for every arbitrary finite set $x_1, \ldots, x_n \in \mathcal{X}$ and $y_1, \ldots, y_n \in \mathcal{Y}$, $\varepsilon > 0$ and L in $\mathcal{Z}$, there exist $x \in \mathcal{X}$ and $y \in \mathcal{Y}$ such that

(i) $\|B(x, y) - L\| \leq c\|L\|$;
(ii) $\|B(x, y_j)\| < \varepsilon$ for $j = 1, \ldots, n$;
(iii) $\|B(x_i, y)\| < \varepsilon$ for $i = 1, \ldots, n$;
(iv) $\|x\| \leq C_1\|L\|^{1/2}$ and $\|y\| \leq C_2\|L\|^{1/2}$.

Lemma 3.5.2 *Suppose that B has the LRVAP. Let N be a fixed positive integer, $\gamma \in (c, 1)$, $x_1, \ldots, x_N \in \mathcal{X}$, $y_1, \ldots, y_N \in \mathcal{Y}$, $(L_{i,j})_{1\leq i,j\leq N} \subset \mathcal{Z}$ and $\delta_{i,j} > 0$ such that*

$$\|L_{i,j} - B(x_i, y_j)\| < \delta_{i,j}, \quad \textit{for all} \quad i, j = 1 \ldots, N.$$

Then there exist $u_1, \ldots, u_N \in \mathcal{X}$ and $v_1, \ldots, v_N \in \mathcal{Y}$ such that

(i) $\|B(u_i, v_j) - L_{i,j}\| \leq \gamma\delta_{i,j}$;
(ii) $\|x_i - u_i\| \leq C_1 \sum_{k=1}^N \delta_{i,k}^{1/2}$ for all $i = 1, \ldots, N$;
(iii) $\|y_j - v_j\| \leq C_2 \sum_{k=1}^N \delta_{k,j}^{1/2}$ for all $j - 1, \ldots, N$.

Proof Let $\varepsilon > 0$ be fixed later (depending on N, c, γ and $\delta_{i,j}$). Set $A_{i,j} = L_{i,j} - B(x_i, y_j)$. We will construct the vectors associated with $A_{i,j}$ in the LRVAP following the lexicographic order. More precisely, let $u_{1,1} \in \mathcal{X}$ and $v_{1,1} \in \mathcal{Y}$ such that

(i) $\|A_{1,1} - B(u_{1,1}, v_{1,1})\| \leq c\|A_{1,1}\|$;
(ii) $\|B(u_{1,1}, y_j)\| < \varepsilon$ for $j = 1, \ldots, N$;
(iii) $\|B(x_i, v_{1,1})\| < \varepsilon$ for $i = 1, \ldots, N$;
(iv) $\|u_{1,1}\| \leq C_1\|A_{1,1}\|^{1/2}$ and $\|v_{1,1}\| \leq C_2\|A_{1,1}\|^{1/2}$.

Suppose that we have constructed the vectors $u_{k',l'}$ and $v_{k',l'}$ for $(k', l') \prec (k, l)$, i.e., for $l' = 1, \ldots, N$ if $k' < k$ and $l' = 1, \ldots, l - 1$ if $k' = k$. Then using the LRVAP, there exist $u_{k,l}$ and $v_{k,l}$ such that

(i) $\|A_{k,l} - B(u_{k,l}, v_{k,l})\| \leq c\|A_{k,l}\|$;
(ii) $\|B(u_{k,l}, y_j)\| < \varepsilon$ for $j = 1, \ldots, N$ and $\|B(u_{k,l}, v_{k',l'})\| < \varepsilon$ for $(k', l') \prec (k, l)$;
(iii) $\|B(x_i, v_{k,l})\| < \varepsilon$ for $i = 1, \ldots, N$ and $\|B(u_{k',l'}, v_{k,l})\| < \varepsilon$ for $(k', l') \prec (k, l)$;
(iv) $\|u_{k,l}\| \leq C_1\|A_{k,l}\|^{1/2}$ and $\|v_{k,l}\| \leq C_2\|A_{k,l}\|^{1/2}$.

For $i, j = 1, \ldots, N$, set $u_i = \sum_{p=1}^N u_{i,p} + x_i$ and $v_j = \sum_{p=1}^N v_{p,j} + y_j$. Then we have $\|x_i - u_i\| \leq C_1 \sum_{k=1}^N \delta_{i,k}^{1/2}$ and $\|y_j - v_j\| \leq C_2 \sum_{k=1}^N \delta_{k,j}^{1/2}$, for $i, j = 1, \ldots, N$. Moreover, note that

$$\begin{aligned}\|L_{i,j} - B(u_i, v_j)\| \le \|A_{i,j} - B(u_{i,j}, v_{i,j})\| \\ + \sum_{p=1}^{N} (\|B(u_{i,p}, y_j)\| + \|B(x_i, v_{p,j})\|) \\ + \sum_{(k,l)\neq(j,i)} \|B(u_{i,k}, v_{l,j})\|.\end{aligned}$$

Note that $(k, l) \neq (j, i)$ implies $(k, l) \prec (j, i)$ or $(j, i) \prec (k, l)$. In both cases we have $\|B(u_{i,k}, v_{l,j})\| < \varepsilon$. It follows that

$$\|L_{i,j} - B(u_i, v_j)\| \le c\delta_{i,j} + 2N\varepsilon + (N^2 - 1)\varepsilon.$$

Since $\gamma > c$, there exists $\varepsilon > 0$ sufficiently small such that $c\delta_{i,j} + 2N\varepsilon + (N^2 - 1)\varepsilon < \gamma\delta_{i,j}$ (take $\varepsilon < (\gamma - c)/(N^2 + 2N - 1) \min_{i,j} \delta_{i,j}$). The proof of the lemma is now complete. □

Lemma 3.5.2 provided approximate solutions to $B(x_i, y_j) = L_{i,j}$ under the LRVAP assumption. We are now in a position to solve the equations exactly.

Proposition 3.5.3 *Suppose that B has the LRVAP. Let $x_1, \dots, x_n \in \mathcal{X}$, $y_1, \dots, y_n \in \mathcal{Y}$, $(L_{i,j})_{1\le i,j\le n} \subset \mathcal{Z}$ and $\delta_{i,j} > 0$ such that*

$$\|L_{i,j} - B(x_i, y_j)\| < \delta_{i,j}, \quad \textit{for all} \quad i, j = 1, \dots, n.$$

Then, for all $\gamma \in (c, 1)$, there exist $u_1, \dots, u_n \in \mathcal{X}$ and $v_1, \dots, v_n \in \mathcal{Y}$ such that

(i) $B(u_i, v_j) = L_{i,j}$;
(ii) $\|x_i - u_i\| \le \frac{C_1}{1-\sqrt{\gamma}} \sum_{k=1}^{n} \delta_{i,k}^{1/2}$ *for all* $i = 1, \dots, n$;
(iii) $\|y_j - v_j\| \le \frac{C_2}{1-\sqrt{\gamma}} \sum_{k=1}^{n} \delta_{k,j}^{1/2}$ *for all* $j = 1, \dots, n$.

Proof For all $i, j = 1, \dots, N$, set $x_0^i := x_i$ and $y_0^j := y_j$. By Lemma 3.5.2, there exist $x_1^i \subset \mathcal{X}$ and y_1^j for all $i, j = 1, \dots, N$ such that

(i) $\left\|B\left(x_1^i, y_1^j\right) - L_{i,j}\right\| \le \gamma\delta_{i,j}$;
(ii) $\|x_0^i - x_1^i\| \le C_1 \sum_{k=1}^{N} \delta_{i,k}^{1/2}$;
(iii) $\left\|y_0^j - y_1^j\right\| \le C_2 \sum_{k=1}^{N} \delta_{k,j}^{1/2}$.

Suppose that, for all $i, j = 1, \dots, N$, there exist $x_p^i \in \mathcal{X}$ and $y_p^j \in \mathcal{Y}$ such that

(i) $\left\| B\left(x_p^i, y_p^j\right) - L_{i,j} \right\| \le \gamma^p \delta_{i,j}$;

(ii) $\left\| x_p^i - x_{p-1}^i \right\| \le C_1 \gamma^{(p-1)/2} \sum_{k=1}^N \delta_{i,k}^{1/2}$;

(iii) $\left\| y_p^j - y_{p-1}^j \right\| \le C_2 \gamma^{(p-1)/2} \sum_{k=1}^N \delta_{k,j}^{1/2}$.

Applying Lemma 3.5.2 once more, there exist $x_{p+1}^i \in \mathcal{X}$ and $y_{p+1}^j \in \mathcal{Y}$ for all $i, j = 1, \ldots, N$ such that

(i) $\left\| B\left(x_{p+1}^i, y_{p+1}^j\right) - L_{i,j} \right\| \le \gamma^{p+1} \delta_{i,j}$;

(ii) $\left\| x_{p+1}^i - x_p^i \right\| \le C_1 \gamma^{p/2} \sum_{k=1}^N \delta_{i,k}^{1/2}$;

(iii) $\left\| y_{p+1}^j - y_p^j \right\| \le C_2 \gamma^{p/2} \sum_{k=1}^N \delta_{k,j}^{1/2}$.

It follows that, for all $i, j = 1, \ldots, N$, the sequences $\left(x_p^i\right)_p$ and $\left(y_p^j\right)_p$ are Cauchy sequences in the complete spaces $\mathcal{X}$ and $\mathcal{Y}$, respectively. Let u_i be the limit of $\left(x_p^i\right)_p$ and v_j the limit of $\left(y_p^j\right)_p$; then the assertion of the proposition is now clear. □

We have now reached the result that obtains exact solutions from approximate solutions.

Theorem 3.5.4 *If B has the LRVAP, then B has the property* $(\mathbb{A}_{\aleph_0})$.

Proof Let $\gamma \in (c, 1)$, where c is the positive constant in the LRVAP condition. Suppose first that for all i, j we have $\|L_{i,j}\| < \frac{1}{i^4 j^4}$. By Proposition 3.5.3, there exist $x_1^{(1)}$ and $y_1^{(1)}$ such that

$$L_{1,1} = B\left(x_1^{(1)}, y_1^{(1)}\right), \quad \left\| x_1^{(1)} \right\| \le \frac{C_1}{1-\sqrt{\gamma}} \quad \left\| y_1^{(1)} \right\| \le \frac{C_2}{1-\sqrt{\gamma}}.$$

Suppose that for all $i, j = 1, \ldots, n-1$, there exist $x_i^{n-1} \in \mathcal{X}$ and $y_j^{n-1} \in \mathcal{Y}$ such that

$$L_{i,j} = B\left(x_i^{n-1}, y_j^{n-1}\right).$$

Set $\left(\tilde{x}_1^{(n)}, \ldots, \tilde{x}_{n-1}^{(n)}, \tilde{x}_n^{(n)}\right) = \left(x_1^{(n-1)}, \ldots, x_{n-1}^{(n-1)}, 0\right)$ and $\left(\tilde{y}_1^{(n)}, \ldots, \tilde{y}_{n-1}^{(n)}, \tilde{y}_n^{(n)}\right) = \left(y_1^{(n-1)}, \ldots, y_{n-1}^{(n-1)}, 0\right)$. It follows that

$$\left\| L_{i,j} - B\left(\tilde{x}_i^{(n)}, \tilde{y}_j^{(n)}\right) \right\| = \begin{cases} 0 & \text{if} \quad i < n \quad \text{and} \quad j < n, \\ \|L_{i,j}\| & \text{if} \quad i = n \quad \text{or} \quad j = n. \end{cases}$$

Now, in order to apply Proposition 3.5.3, let $\delta_{i,j} > 0$ for $i < n$ and $j < n$ so that $\sum_{1\le i,j<n} \delta_{i,j}^{1/2} \le \varepsilon_n$, where $\varepsilon_n > 0$ satisfies $\sum_{n\ge 1} \varepsilon_n < \infty$. Note that $\left\| L_{i,j} - B\left(\tilde{x}_i^{(n)}, \tilde{y}_j^{(n)}\right)\right\| = 0 < \delta_{i,j}$ for all $i, j < n$. By Proposition 3.5.3, there exist $x_1^{(n)}, \ldots, x_n^{(n)} \in \mathcal{X}$ and $y_1^{(n)}, \ldots, y_n^{(n)} \in \mathcal{Y}$ such that

$$L_{i,j} = B\left(x_i^{(n)}, y_j^{(n)}\right), \quad i, j = 1, \ldots, n,$$

$$\left\| x_i^n - x_i^{n-1}\right\| \le \frac{C_1}{1-\sqrt{\gamma}}(\varepsilon_n + \|L_{i,n}\|^{1/2}) \le \frac{C_1}{1-\sqrt{\gamma}}\left(\varepsilon_n + \frac{1}{i^2 n^2}\right), \quad i < n,$$

$$\left\| y_j^n - y_j^{n-1}\right\| \le \frac{C_2}{1-\sqrt{\gamma}}(\varepsilon_n + \|L_{n,j}\|^{1/2}) \le \frac{C_2}{1-\sqrt{\gamma}}\left(\varepsilon_n + \frac{1}{j^2 n^2}\right), \quad j < n.$$

Now, let (i, j) be a fixed pair of positive integers. The above inequalities imply that $\left(x_i^{(n)}\right)_{n>i}$ (resp. $\left(y_j^{(n)}\right)_{n>j}$) is a Cauchy sequence in the Banach space $\mathcal{X}$ (resp. $\mathcal{Y}$). Set $u_i = \lim_{n\to\infty} x_i^{(n)}$ and $v_j = \lim_{n\to\infty} y_j^{(n)}$. Obviously we have $L_{i,j} = B(u_i, v_j)$.

Now we consider the general case. First take $a_1 > 0$ and $b_1 > 0$ such that

$$\frac{\|L_{1,1}\|}{a_1 b_1} < 1.$$

Suppose that there exist $a_1, \ldots, a_{n-1}, b_1, \ldots, b_{n-1} > 0$ such that

$$\frac{\|L_{i,j}\|}{a_i b_j} < \frac{1}{i^4 j^4} \quad \text{for all} \quad i, j < n.$$

Then one takes $a_n, b_n > 0$ such that

$$\frac{\|L_{n,n}\|}{a_n b_n} < \frac{1}{n^8}, \quad \frac{\|L_{i,n}\|}{a_i b_n} < \frac{1}{i^4 n^4}, \quad \frac{\|L_{n,j}\|}{a_n b_j} < \frac{1}{j^4 n^4} \quad \text{for all} \quad i, j < n.$$

By induction, one constructs sequences $(a_n)_n$ and $(b_n)_n$ of positive reals such that $L'_{i,j} := L_{i,j}/(a_i b_j)$ satisfies $\left\| L'_{i,j}\right\| < \frac{1}{i^4 j^4}$. Therefore there exist u'_i and v'_j such that $L'_{i,j} = B\left(u'_i, v'_j\right)$, and then $L_{i,j} = B\left(a_i u'_i, b_j v'_j\right)$. □

3.5.2 A sufficient condition for property $(\mathbb{A}_{1,\aleph_0})$

Let c be a constant in $(0, 1)$, $C_1, C_2 > 0$, and N a positive integer.

We say that B has the *N-right vanishing approximation property (N-RVAP)* if, for every arbitrary finite set $w_1, \ldots, w_p \in \mathcal{Y}$, $z \in \mathcal{X}$, $\delta > 0$ and $L_1, \ldots, L_N$ in $\mathcal{Z}$, there exist $u \in \mathcal{X}$ and $v_1, \ldots, v_N \in \mathcal{Y}$ such that

(i) $\|B(u+z, v_j) - L_j\| \leq c\|L_j\|$, for $j = 1, \ldots, N$;
(ii) $\|B(u, w_k)\| < \delta$ for $k = 1, \ldots, p$;
(iii) $\|u\| \leq C_1 \sum_{k=1}^{N} \|L_k\|^{1/2}$ and $\|v_j\| \leq C\|L_j\|^{1/2}$, for $j = 1, \ldots, N$.

As previously, the aim is to deduce from N-RVAP a factorization result. The next lemma is the key to the initialization of the approximation process.

Lemma 3.5.5 *Let $\gamma \in (c, 1)$ and suppose that B has the N-RVAP. Let $L_1, \ldots, L_N \in \mathcal{Z}$, $x \in \mathcal{X}$, $y_1, \ldots, y_N \in \mathcal{Y}$ and $\rho_j > 0$ (for $j = 1, \ldots, N$) such that*

$$\|B(x, y_j) - L_j\| < \rho_j, \quad j = 1, \ldots, N.$$

Then there exists $\xi \in \mathcal{X}$ and $\nu_1, \ldots, \nu_N \in \mathcal{Y}$ such that

$$\|B(\xi, \nu_j) - L_j\| < \gamma\rho_j, \quad j = 1, \ldots, N,$$

and

$$\|\xi - x\| \leq C_1 \sum_{k=1}^{N} \sqrt{\rho_k}, \quad \|\nu_j - y_j\| \leq C_2\sqrt{\rho_j}.$$

Proof Set $A_j = L_j - B(x, y_j)$ for $j = 1, \ldots, N$. Let $\delta > 0$ such that $c\|A_j\| + \delta < \gamma\rho_j$ for all $j = 1, \ldots, N$. Applying N-RVAP with $x, y_1, \ldots, y_N$ and $A_1, \ldots, A_N$, there exist $u \in \mathcal{X}$ and $v_1, \ldots, v_N \in \mathcal{Y}$ such that

(i) $\|L_j - B(x, y_j) - B(u+x, v_j)\| \leq c\|A_j\|$ for all $j = 1, \ldots, N$;
(ii) $\|B(u, y_j)\| < \delta$ for $j = 1, \ldots, N$;
(iii) $\|u\| \leq C_1 \sum_{k=1}^{N} \|A_k\|^{1/2}$ and $\|v_j\| \leq C_2\|A_j\|^{1/2}$ for $j = 1, \ldots, N$.

Now set $\xi = u + x$ and $\nu_j = y_j + v_j$. It is clear that $\|\xi - x\| \leq C_1 \sum_{k=1}^{N} \sqrt{\rho_k}$ and $\|\nu_j - y_j\| \leq C_2\sqrt{\rho_j}$. Moreover, we have

$$\begin{aligned}\|B(\xi, \nu_j) - L_j\| &\leq \|L_j - B(x, y_j) - B(u+x, v_j)\| + \|B(u, y_j)\| \\ &< c\|A_j\| + \delta < \gamma\rho_j.\end{aligned}$$ □

The next result provides a property of type $(\mathbb{A}_{1,N})$.

Theorem 3.5.6 *Let $\gamma \in (c, 1)$ and suppose that B has the N-RVAP. Let $L_1, \ldots, L_N \in \mathcal{Z}$, $x \in \mathcal{X}$, $y_1, \ldots, y_N \in \mathcal{Y}$ and $\rho_j > 0$ (for $j = 1, \ldots, N$) such that*

$$\|B(x, y_j) - L_j\| < \rho_j, \quad j = 1, \ldots, N.$$

Then there exist $u \in \mathcal{X}$, $v_1, \ldots, v_N \in \mathcal{Y}$ such that, for all $j = 1, \ldots, N$, we have

$$L_j = B(u, v_j)$$

and

$$\|u - x\| \leq \frac{C_1}{1-\sqrt{\gamma}} \sum_{k=1}^{N} \sqrt{\rho_k}, \quad \|y_j - v_j\| \leq \frac{C_2}{1-\sqrt{\gamma}} \sqrt{\rho_j}.$$

Proof The proof is by induction. By Lemma 3.5.5, for all $j = 1, \ldots, N$, there exist $x_0 \in \mathcal{X}$ and $y_{0,j} \in \mathcal{Y}$ such that

$$\|B(x_0, y_{0,j}) - L_j\| < \gamma \rho_j,$$

with

$$\|x_0 - x\| \leq C_1 \sum_{k=1}^{N} \sqrt{\rho_k} \qquad \text{and} \qquad \|y_{0,j} - y_j\| \leq C_2 \sqrt{\rho_j}.$$

Suppose that for a fixed $n \geq 1$ and for all $j = 1, \ldots, N$, we have constructed $x_l \in \mathcal{X}$ and $y_{l,j} \in \mathcal{Y}$ for $l = n-1, n$, such that

$$\|B(x_n, y_{n,j}) - L_j\| < \gamma^n \rho^j$$

and

$$\|x_n - x_{n-1}\| \leq C_1 \sum_{k=1}^{N} \sqrt{\rho_k} \gamma^{(n-1)/2}, \qquad \|y_{n,j} - y_{n-1,j}\| \leq C_2 \sqrt{\rho_j} \gamma^{(n-1)/2}.$$

By Lemma 3.5.5 applied with x_n and $y_{n,j}$, for all $j = 1, \ldots, N$, there exist x_{n+1} in $\mathcal{X}$ and $y_{n+1,j} \in \mathcal{Y}$ such that

$$\|B(x_{n+1}, y_{n+1,j}) - L_j\| < \gamma^{n+1} \rho^j$$

and

$$\|x_{n+1} - x_n\| \leq C_1 \sum_{k=1}^{N} \sqrt{\rho_k} \gamma^{n/2}, \quad \|y_{n+1,j} - y_{n,j}\| \leq C_2 \sqrt{\rho_j} \gamma^{n/2}.$$

It follows that $(x_n)_n$ and $(y_{n,j})_n$ are Cauchy sequences for all $j = 1, \ldots, N$, and thus converge to, say, u and v_j, satisfying the hypothesis of the theorem. □

Theorem 3.5.7 *If, for all positive integers N, B has the N-RVAP, then B has the property $(\mathbb{A}_{1,\aleph_0})$.*

Proof Let $(L_n)_n$ be a sequence in $\mathcal{Z}$. Suppose first that $\|L_j\| < \frac{1}{j^4}$ for all $j \geq 1$. By Theorem 3.5.6 with $x = 0$, $y_1 = 0$ and $\rho_1 = 1$, there exist $x^{(1)} \in \mathcal{X}$ and $y_1^{(1)} \in \mathcal{Y}$ such that

$$L_1 = B\left(x^{(1)}, y_1^{(1)}\right), \quad \|(x^{(1)}\| \leq \frac{C_1}{1-\sqrt{\gamma}} \quad \text{and} \quad \left\|y_1^{(1)}\right\| \leq \frac{C_2}{1-\sqrt{\gamma}}.$$

Suppose that, for all $j = 1, \dots, n-1$, we have

$$L_j = B\left(x^{(n-1)}, y_j^{(n-1)}\right).$$

Set $\tilde{x}^{(n)} = x^{(n-1)}$ and

$$\left(\tilde{y}_1^{(n)}, \dots, \tilde{y}_{n-1}^{(n)}, \tilde{y}_n^{(n)}\right) = \left(y_1^{(n)}, \dots, y_{n-1}^{(n)}, 0\right).$$

It follows that

$$\left\| L_j - B\left(\tilde{x}^{(n)}, \tilde{y}_j^{(n)}\right)\right\| = \begin{cases} 0 < \frac{\|L_n\|}{n^2} & \text{if} \quad j < n, \\ \|L_n\| < 2\|L_n\| & \text{if} \quad j = n. \end{cases}$$

By Theorem 3.5.6, there exist $x^{(n)} \in \mathcal{X}$ and $y_1^{(n)}, \dots, y_n^{(n)} \in \mathcal{Y}$ such that

$$L_j = B\left(x^{(n)}, y_j^{(n)}\right), \quad j = 1, \dots, n,$$

$$\|x^n - x^{n-1}\| \le \frac{C_1}{1-\sqrt{\gamma}}\left((n-1)/n + \sqrt{2}\right)\|L_n\|^{1/2} \le \frac{3C_1}{1-\sqrt{\gamma}}\|L_n\|^{1/2},$$

$$\left\|y_j^n - y_j^{n-1}\right\| \le \frac{C_2}{1-\sqrt{\gamma}}\sqrt{2}\|L_n\|^{1/2}, \quad j < n.$$

Since $\|L_n\| < 1/n^4$, the above inequalities imply that $(x^{(n)})_n$ (resp. $\left(y_j^{(n)}\right)_n$) is a Cauchy sequence in the Banach space $\mathcal{X}$ (resp. $\mathcal{Y}$). Set $u = \lim_{n\to\infty} x^{(n)}$ and $v_j = \lim_{n\to\infty} y_j^{(n)}$. Obviously we have $L_{i,j} = B(u, v_j)$.

In order to consider the general case, let $(b_n)_n$ be a sequence of positive reals such that $\frac{\|L_n\|}{b_n} < \frac{1}{n^4}$. Then there exist u' and v_j' such that $\frac{L_j}{b_j} =: L_j' = B\left(u_i', v_j'\right)$, and then $L_j = B\left(u', b_j v_j'\right)$. □

Exercises

1. Complete the proof of Proposition 3.1.5(i), by showing explicitly that $[P_\lambda] = [\tilde{x} \overset{T}{\cdot} \tilde{x}]$.
2. Show that, if $f \in L^\infty(\mathbb{T})$ with $|f| = 1$ a.e. and $f = g\overline{h}$ with $g, h \in H^\infty$ and $\|g\|_\infty \|h\|_\infty = 1$, then for some $\lambda > 0$ we have that λg and $\lambda^{-1}h$ are inner functions.
3. Suppose that $f \in L^\infty(\mathbb{T})$, with $|f| = 1$ a.e., and that f is not a.e. constant but satisfies $f(e^{it}) = 1$ on a set of positive measure. By considering the function $g - h$, show that f cannot be written as $f = g\overline{h}$ with g, h inner. (Taken with Exercise 2, this shows that the ε in Corollary 3.2.9 is necessary in general.)

4. Fill in the details of the case $q = \infty$ and $r < \infty$ in the proof of Corollary 3.2.10.
5. Prove that, if T is the bilateral shift on $L^2(\mathbb{T})$, then T factorizes every function in $L^1(\mathbb{T})$.
6. Let b be a finite Blaschke product of degree d, regarded as a mapping from $\mathbb{T}$ to itself. Show that $\mathbb{T}$ may be decomposed into d closed arcs, each of which is mapped bijectively onto $\mathbb{T}$.
7. Let θ be an arbitrary inner function. Prove that for almost all $\alpha \in \mathbb{D}$, the set of zeroes of θ' is contained in the closed convex hull of $\theta^{-1}(\{\alpha\}) \cup \{0\}$. (Hint: use Frostman's Theorem 1.4.18.)
8. Prove that a normal extension $N \in \mathcal{L}(\mathcal{K})$ of a subnormal operator $S \in \mathcal{L}(\mathcal{H})$ is minimal if and only if $\mathcal{K}$ is the closed linear span of $\{N^{*k}x : x \in \mathcal{H}, k \geq 0\}$.
9. Prove that two minimal normal extensions of a subnormal operator are unitarily equivalent.
10. Let $N \in \mathcal{L}(\mathcal{K})$ be the minimal normal extension of a subnormal operator $S \in \mathcal{L}(\mathcal{H})$. Show that $\partial\sigma(S) \subset \sigma(N) \subset \sigma(S)$, where $\partial\sigma(S)$ is the boundary of $\sigma(S)$. Prove also that a bounded connected component of $\mathbb{C} \setminus \sigma(N)$ either is contained in $\sigma(S)$ or is disjoint to $\sigma(S)$.

Comments

The developments in Section 3.1.1 arising from S. Brown's starting point can be found in Chalendar, Partington and Smith [77].

Chevreau [78] and Bercovici [31] independently proved the following seminal result: the class $\mathbb{A}$ coincides with the class $\mathbb{A}_1$. Moreover, Brown and Chevreau [52] proved that a contraction in the class $\mathbb{A}$ is even *reflexive*, a property that implies a rich lattice of invariant subspaces.

One can find characterizations and information about the classes $\mathbb{A}_{n,m}$ in Bercovici, Foias and Pearcy [35], Cassier, Chalendar and Chevreau [58, 59] and Chalendar and Jaeck [66].

Theorem 3.1.15 is theorem 5.2 in Chalendar and Partington [71], establishing conjecture 3.5 of Exner and Jung [103] by providing a constructive proof of the fact that the classes $\mathbb{A}_{n,m}$ are distinct. A non-constructive proof was given by Jung [127].

The results in Section 3.2 are based on a variety of sources. The presentation of the approximation (or 'correction') results as far as Theorem 3.2.3 is based on Chalendar and Esterle [62] (see also Kisliakov [133, 134]). The remainder of this section is based on the celebrated solution of the Douglas–Rudin problem sketched with a minimum of detail by Bourgain [48]. For many of the missing details and some improvements in the constants involved, we have drawn heavily on the work of Barclay [25, 24], whose results also include the solution of the even deeper matrix-valued (non-commutative) version of this problem. Thus, Theorem 3.2.8 and Corollaries 3.2.9 and 3.2.10 as given here are stronger forms of results originally due to Bourgain.

Most of the results of Section 3.3 come from Chalendar and Partington [71, 68]. Theorem 3.3.1 on the location of the zeroes of the derivative of a Blaschke product can be found in Cassier and Chalendar [57]. The case of a finite Blaschke product is due to Walsh [200].

The spectral inclusion linking a subnormal operator and its minimal normal extension, i.e., Lemma 3.4.4, is due to Halmos [116, 118]. An indispensable reference for subnormal operators is Conway [83].

S. Brown [50] first proved the existence of non-trivial invariant subspaces for subnormal operators. This seminal paper is very important since it already contains an approximation scheme which provided many important invariant-subspace results, such as the existence of non-trivial invariant subspaces for contractions whose spectrum is rich enough (see Chapter 3). Later, Thomson [193] gave another, simpler proof; this is the one presented here. Using methods similar to those used by Thomson, Yan [203] proved that each commuting subnormal tuple has a non-trivial joint invariant subspace.

The results of Section 3.5 can be found in Bercovici, Foias and Pearcy [36] and Rejasse [173]. It follows ideas introduced for particular bilinear mappings on Hilbert space by Chevreau and Pearcy [80]. The generalization to general sesquilinear or bilinear maps is due to Bercovici, Foias and Pearcy [36].

4

Polynomially bounded operators with rich spectrum

Much of the work in the previous two chapters has been preparatory, and in this chapter we arrive at some recent deep results, due to Ambrozie and Müller [8], concerning the existence of invariant subspaces for polynomially bounded operators on a complex Banach space with spectrum containing the unit circle. These results provide a powerful generalization of the celebrated results of Brown, Chevreau and Pearcy [53, 54], who proved the existence of invariant subspaces for contractions on a Hilbert space with spectrum containing the unit circle.

To follow this programme, it will be necessary to introduce a variety of themes that are of interest in their own right: Apostol sets, geometry of Banach spaces (in the form of Zenger's theorem) and Carleson interpolation. This chapter relies also on tools introduced in Chapters 2 and 3, namely, surjectivity of bilinear mappings and spectral measures.

We also emphasize the usefulness of a variety of functional calculi, such as those for holomorphic, $C^2(\mathbb{T})$ and H^∞ functions. Further ways of using the functional calculus are presented in Chapter 5.

4.1 Apostol's theorem

We begin with a nice application of the holomorphic functional calculus, and a classical technique of integration through the spectrum.

Definition 4.1.1 A subset $\Lambda \subset \mathbb{D}$ is called an *Apostol set* if the set of angles $\theta \in (-\pi, \pi]$ such that

$$\sup\{r \in [0, 1) : re^{i\theta} \in \Lambda\} < 1$$

is at most countable.

For $T \in \mathcal{L}(\mathcal{X})$, $\varepsilon > 0$ and k a positive integer, denote by $\Lambda_{\varepsilon,k}(T)$ the set

$$\{\lambda \in \mathbb{D} : \exists u \in \mathcal{X} \quad \text{with} \quad \|u\| = 1 \quad \text{and} \quad \|(T - \lambda \mathrm{I})u\| < \varepsilon(1 - |\lambda|)^k\}.$$

The points of $\Lambda_{\varepsilon,k}(T)$ can be seen as approximate eigenvalues of T, whose degree of approximation is quantified by ε and k.

Definition 4.1.2 An operator $T \in \mathcal{L}(\mathcal{X})$ satisfies the *Kreiss condition* if $\sigma(T) \subset \overline{\mathbb{D}}$ and if there exists a positive constant C such that

$$\|(T - \lambda \mathrm{I})^{-1}\| \leq \frac{C}{|\lambda| - 1} \quad \text{for all} \quad |\lambda| > 1.$$

If an operator T is power-bounded (i.e., $\sup_{n \geq 0} \|T^n\| < \infty$), then it satisfies the Kreiss condition. We leave this as an exercise.

Theorem 4.1.3 *Let $T \in \mathcal{L}(\mathcal{X})$ be an operator satisfying the Kreiss condition, and such that $\mathbb{T} \subset \sigma(T)$. If there exist $\varepsilon > 0$ and k a positive integer such that $\Lambda_{\varepsilon,k}(T)$ is not an Apostol set, then T has a non-trivial hyperinvariant subspace.*

Proof Suppose first that there exists $\lambda \in \mathbb{D} \cap \sigma(T)$ with $\lambda \notin \Lambda_{\varepsilon,k}(T)$. Therefore, we have

$$\|(T - \lambda \mathrm{I})x\| \geq \varepsilon(1 - |\lambda|)^k \|x\|,$$

for all $x \in \mathcal{X}$. This implies that $\mathrm{Ker}(T - \lambda \mathrm{I}) = \{0\}$ and the range of $T - \lambda \mathrm{I}$ is closed. Indeed, if $(T - \lambda \mathrm{I})x_n$ converges to y, then $(x_n)_n$ is a Cauchy sequence since $\|(T - \lambda \mathrm{I})x_n\| \geq \varepsilon(1 - |\lambda|)^k \|x_n\|$. Since $\mathcal{X}$ is a Banach space, (x_n) converges to x, which satisfies $(T - \lambda \mathrm{I})x = y$, since $T - \lambda \mathrm{I}$ is continuous. Now since $\lambda \in \sigma(T)$, it follows that the range of $T - \lambda \mathrm{I}$ is a non-trivial hyperinvariant subspace.

We may now assume that $\sigma(T) \cap \mathbb{D} \subset \Lambda_{\varepsilon,k}(T)$. Since $\Lambda_{\varepsilon,k}(T)$ is not an Apostol set, there exists an uncountable set $E \subset \mathbb{T}$ consisting of points that are not the radial limits of points in $\Lambda_{\varepsilon,k}(T)$. For $w \in E$, denote by s_w a segment $\{rw : r \in (r_w, 1)\}$ such that $s_w \cap \Lambda_{\varepsilon,k}(T) = \emptyset$. Since there are at most a countable number of connected components of the open set $\mathbb{D} \setminus \sigma(T)$, and since E is uncountable, there exists an (open) connected component of $\mathbb{D} \setminus \sigma(T)$ containing four distinct segments $s_{w_1}, \ldots, s_{w_4}$ associated with four points $w_1 = e^{i\theta_1}, \ldots, w_4 = e^{i\theta_4}$ in E such that $0 \leq \theta_1 < \cdots < \theta_4 < 2\pi$. Then we can find two simple rectifiable closed curves $\Gamma_{1,2}$ and $\Gamma_{3,4}$ having the following properties:

1. $\Gamma_{1,2} \cap \Gamma_{3,4} = \emptyset$.
2. $\Gamma_{1,2} \cap \sigma(T) = \{w_1, w_2\}$ and $\Gamma_{3,4} \cap \sigma(T) = \{w_3, w_4\}$.
3. $\Gamma_{1,2}$ and $\Gamma_{3,4}$ cross $\mathbb{T}$ radially.
4. For some $c > 0$, we have $\|(T - \lambda \mathrm{I})^{-1}\| \leq \frac{c}{(1-|\lambda|)^k}$ for all the points $\lambda \in \mathbb{D}$ lying on $\Gamma_{1,2}$ and $\Gamma_{3,4}$.

Indeed, in order to get properties 2 and 3, we construct $\Gamma_{1,2}$ (resp. $\Gamma_{3,4}$) using the segments s_{w_1}, s_{w_2} (resp. s_{w_3}, s_{w_4}), extending it radially through $\mathbb{T}$, and finally we join the segments s_{w_1} and s_{w_2} (resp. s_{w_3} and s_{w_4}), staying in the same connected component (see Figure 4.1).

We now only need to check property 4. If $\lambda \in \Gamma_{1,2}$ and $|\lambda| > 1$, then we use the fact that T satisfies the Kreiss condition. Since $k \geq 1$, there exists $c > 0$ such that $\frac{C}{|\lambda|-1} \leq \frac{c}{(|\lambda|-1)^k}$, which implies that

$$\|(T - \lambda \mathrm{I})^{-1}\| \leq \frac{c}{(|\lambda| - 1)^k},$$

for $\lambda \in \Gamma_{1,2}$ and $|\lambda| > 1$. If $\lambda \in \Gamma_{1,2} \cap \mathbb{D}$, by hypothesis, for all $x \in \mathcal{X}$, we have $\|(T - \lambda \mathrm{I})x\| \geq \varepsilon(1 - |\lambda|)^k \|x\|$, and thus

$$\|(T - \lambda \mathrm{I})^{-1}\| \leq \frac{1}{\varepsilon(1 - |\lambda|)^k}.$$

Similarly, $\Gamma_{3,4}$ satisfies property 4. Properties 1–4 allow us to define the following elements of $\mathcal{L}(\mathcal{X})$:

$$A_{1,2} = \frac{1}{2i\pi} \int_{\Gamma_{1,2}} (z - w_1)^k (z - w_2)^k (z\mathrm{I} - T)^{-1} dz,$$

$$A_{3,4} = \frac{1}{2i\pi} \int_{\Gamma_{3,4}} (z - w_3)^k (z - w_4)^k (z\mathrm{I} - T)^{-1} dz,$$

where $A_{1,2}$ and $A_{3,4}$ commute with T.

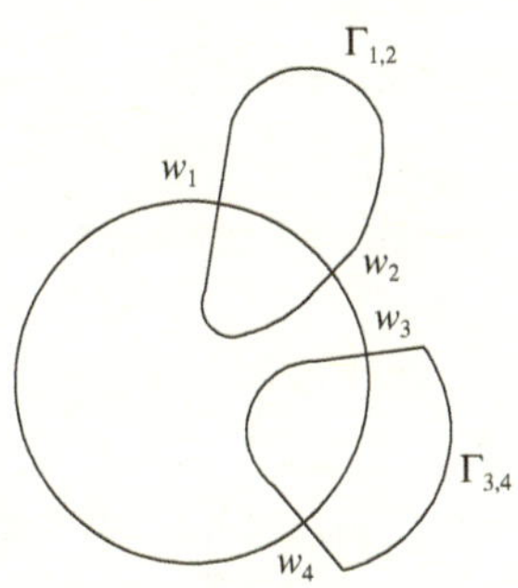

Figure 4.1 Examples of closed curves $\Gamma_{1,2}$ and $\Gamma_{3,4}$.

Moreover, $A_{1,2} \neq 0$, $A_{3,4} \neq 0$ and $A_{3,4}A_{1,2} = 0$. To prove this fact, let $\mathcal{B} \subset \mathcal{L}(\mathcal{H})$ be a commutative Banach algebra containing T, $A_{1,2}$ and $A_{3,4}$. By Zorn's lemma we may assume that $\mathcal{B}$ is maximal. Fix a point $w \in \mathbb{T}$ which is surrounded by $\Gamma_{1,2}$. Since $w \in \sigma(T)$, the Gelfand theory (Theorem 1.2.10) implies that there exists a continuous algebra homomorphism $\phi : \mathcal{B} \to \mathbb{C}$ such that $\phi(T) = w$.

Now, since $\mathcal{B}$ is maximal, we have $(\lambda I - T)^{-1} \in \mathcal{B}$ whenever $\lambda \notin \sigma(T)$, and hence we can calculate

$$\begin{aligned}
\phi(A_{1,2}) &= \frac{1}{2i\pi}\int_{\Gamma_{1,2}} (z-w_1)^k(z-w_2)^k\phi((zI-T)^{-1})\,dz \\
&= \frac{1}{2i\pi}\int_{\Gamma_{1,2}} (z-w_1)^k(z-w_2)^k(z-w)^{-1}\,dz \\
&= (w-w_1)^k(w-w_2)^k,
\end{aligned}$$

by the Cauchy theorem. Since $\phi(A_{1,2}) \neq 0$, we must have $A_{1,2} \neq 0$. The proof that $A_{3,4} \neq 0$ follows using similar arguments.

The proof that $A_{3,4}A_{1,2} = 0$ uses a standard technique from the Riesz–Dunford functional calculus (Section 1.2.3). In the following calculation we use the Cauchy theorem along with the *resolvent equation*,

$$\frac{1}{\mu-\lambda}((\lambda I - T)^{-1} - (\mu I - T)^{-1}) = (\lambda I - T)^{-1}(\mu I - T)^{-1}.$$

With the notation $p(\lambda) = (\lambda - w_1)(\lambda - w_2)$ and $q(\mu) = (\mu - w_3)(\mu - w_4)$, we have

$$\begin{aligned}
A_{3,4}A_{1,2} &= \frac{1}{(2i\pi)^2}\int_{\Gamma_{3,4}}\int_{\Gamma_{1,2}} p(\lambda)q(\mu)(\lambda I - T)^{-1}(\mu I - T)^{-1}\,d\lambda\,d\mu \\
&= \frac{1}{(2i\pi)^2}\int_{\Gamma_{3,4}}\int_{\Gamma_{1,2}} \frac{p(\lambda)q(\mu)}{\mu-\lambda}((\lambda I - T)^{-1} - (\mu I - T)^{-1})\,d\lambda\,d\mu \\
&= \frac{1}{(2i\pi)^2}\int_{\Gamma_{1,2}} p(\lambda)(\lambda I - T)^{-1}\left[\int_{\Gamma_{3,4}} \frac{q(\mu)}{\mu-\lambda}d\mu\right]d\lambda \\
&\quad -\frac{1}{(2i\pi)^2}\int_{\Gamma_{3,4}} q(\mu)(\mu I - T)^{-1}\left[\int_{\Gamma_{1,2}} \frac{p(\lambda)}{\mu-\lambda}d\lambda\right]d\mu \\
&= 0,
\end{aligned}$$

since the integrals in square brackets are zero by the Cauchy theorem.

Since $A_{1,2}$ commutes with T, the subspace $\mathcal{M}$ defined to be the closure of $A_{1,2}\mathcal{X}$ is a hyperinvariant subspace for T. Since $A_{1,2} \neq 0$, $\mathcal{M} \neq \{0\}$, and since $A_{3,4}A_{1,2} = 0$ with $A_{3,4} \neq 0$, it follows that $\mathcal{M} \neq \mathcal{X}$. □

4.2 $C^2(\mathbb{T})$ functional calculus and the Colojoară–Foiaş theorem

The Colojoară–Foiaş theorem provides an important intermediate result enabling us to reduce the study of polynomially bounded (even power-bounded) operators to the case when $\|T^n x\| \to 0$ or $\|T^{*n}x\| \to 0$ for each x. The main tool of the proof is the study of operators with a $C^2(\mathbb{T})$ functional calculus.

4.2.1 Operators with a $C^2(\mathbb{T})$ functional calculus

Definition 4.2.1 Let $\mathcal{X}$ be a complex Banach space and let $T \in \mathcal{L}(\mathcal{X})$. We say that T has a $C^2(\mathbb{T})$ *functional calculus* if there exists a continuous algebra homomorphism Φ_T from $C^2(\mathbb{T})$ into $\mathcal{L}(\mathcal{X})$ such that $\Phi_T(1) = \mathrm{I}$ and $\Phi_T(Z) = T$, where $1 : \xi \mapsto 1$ and $Z : \xi \mapsto \xi$ on $\mathbb{T}$.

We now present a sufficient condition which implies that T has a $C^2(\mathbb{T})$ functional calculus.

Proposition 4.2.2 *Let $T \in \mathcal{L}(\mathcal{X})$ such that* $\sup_{n\in\mathbb{Z}} \|T^n\| < \infty$. *Then T has a $C^2(\mathbb{T})$ functional calculus.*

Proof Let $f \in C^2(\mathbb{T})$ and note that, for all $n \in \mathbb{Z}$,

$$\hat{f}(n) := \frac{1}{2\pi}\int_0^{2\pi} f(e^{i\theta})e^{-in\theta}\,d\theta$$

is equal (via straightforward repeated integration by parts) to

$$\frac{1}{(in)^2}\frac{1}{2\pi}\int_0^{2\pi} g^{(2)}(\theta)e^{-in\theta}\,d\theta,$$

where $g(\theta) = f(e^{i\theta})$. We obtain

$$|\hat{f}(n)| \le \frac{1}{|n|^2}\|g^{(2)}\|_\infty.$$

Since $\sup_{n\in\mathbb{Z}} \|T^n\| < \infty$, the series $\Phi_T(f) := \sum_{n\in\mathbb{Z}} \hat{f}(n)T^n$ converges in the operator norm of $\mathcal{L}(\mathcal{X})$ and defines a continuous algebra homomorphism Φ_T, which is a $C^2(\mathbb{T})$ functional calculus. □

The next proposition provides useful information about the invariant subspaces of an operator with a $C^2(\mathbb{T})$ functional calculus, associated with any closed subset of $\mathbb{T}$.

The *support* of a function g, that is, the closure of the set of points where g is not zero, is denoted by $\operatorname{supp} g$.

Proposition 4.2.3 *Suppose that $T \in \mathcal{L}(\mathcal{X})$ has a $C^2(\mathbb{T})$ functional calculus. For any closed set $F \subset \mathbb{T}$, the closed subspace of $\mathcal{X}$ denoted by $E_T(F)$ and defined by*

$$E_T(F) = \bigcap \{\operatorname{Ker} \Phi_T(g) : g \in C^2(\mathbb{T}) \quad \text{and} \quad \operatorname{supp}(g) \cap F = \emptyset\}$$

is invariant for T and even under $\Phi_T(h)$ for all $h \in C^2(\mathbb{T})$. Moreover, we have the following properties:

1. *$\sigma\left(T_{|E_T(F)}\right) \subset F$.*
2. *If F_1 and F_2 are disjoint closed subsets of $\mathbb{T}$, then we have $E_T(F_1) \cap E_T(F_2) = \{0\}$.*
3. *If F_1 and F_2 are closed subsets of $\mathbb{T}$ such that $F_1 \cup F_2 = \mathbb{T}$, then $E_T(F_1) + E_T(F_2) = \mathcal{X}$, and if, moreover, $\emptyset \neq F_1 \cap \sigma(T) \neq \sigma(T)$, then $E_T(F_1)$ and $E_T(F_2)$ are non-trivial.*

Proof If $\operatorname{supp}(g) \cap F = \emptyset$, for all $h \in C^2(\mathbb{T})$, we have $\operatorname{supp}(gh) \cap F = \emptyset$. Therefore the closed subspace $E_T(F)$ is invariant under $\Phi_T(h)$ for all $h \in C^2(\mathbb{T})$.

1. Let $\lambda \in \mathbb{C} \setminus F$ and choose an open neighbourhood U of F for which $\lambda \notin \overline{U}$ and for which there exists a function $h \in C^2(\mathbb{T})$ such that h is equal to 1 on F and h is equal to 0 on $\mathbb{C} \setminus \overline{U}$. Note that there exists $\ell \in C^2(\mathbb{T})$ such that $h(\xi) = (\xi - \lambda)\ell(\xi)$, $\xi \in \mathbb{T}$. Since $\operatorname{supp}(1 - h) \cap F = \emptyset$, it follows that $x = \Phi_T(h)x$ for all $x \in E_T(F)$. Therefore we have

$$x = \Phi_T(h)x = (T - \lambda \mathrm{I})\Phi_T(\ell)x,$$

with $\Phi_T(\ell)x \in E_T(F)$, since we have proved that $E_T(F)$ is invariant under $\Phi_T(\ell)$. Thus $T - \lambda\mathrm{I}$ is invertible on $E_T(F)$, which proves the inclusion $\sigma\left(T_{|E_T(F)}\right) \subset F$.

2. Now, consider F_1 and F_2, two disjoint closed subsets of $\mathbb{T}$, and take U_1 and U_2, two neighbourhoods of F_1 and F_2, respectively, such that $U_1 \cap U_2 = \emptyset$. Then take a function $f \in C^2(\mathbb{T})$ such that f is equal to 1 on U_1 and f is zero on U_2. The properties of f imply immediately that, for $x \in E_T(F_2)$, we have $\Phi_T(f)x = 0$ and, for $x \in E_T(F_1)$, we have $\Phi_T(1 - f)x = 0$. It follows that if $x \in E_T(F_1) \cap E_T(F_2)$ then $\Phi_T(1)x = 0$, which gives $x = 0$.

3. Let $f \in C^2(\mathbb{T})$ satisfying $f = 0$ on $\mathbb{T} \setminus F_1$ and $f = 1$ on $\mathbb{T} \setminus F_2$. Note that for all $x \in \mathcal{X}$, $\Phi_T(f)x \in E_T(F_1)$. Indeed, if $g \in C^2(\mathbb{T})$ satisfies $\operatorname{supp}(g) \cap F_1 = \emptyset$, then $fg = 0$ and $\Phi_T(f)\Phi_T(g)x = 0$. Similarly, we have $\Phi_T(1 - f)x \in E_T(F_2)$. Now, since for all $x \in \mathcal{X}$ we have

$x = (\Phi_T(f) + \Phi_T(1-f))x$, it follows that $E_T(F_1)+E_T(F_2) = \mathcal{X}$. Moreover, since by 1. we have, for all $i \in \{1, 2\}$, $\sigma(T_{|E_T(F_i)}) \subset F_i$, it is clear that $E_T(F_i)$ is not equal to the whole space $\mathcal{X}$, and thus cannot be trivial. □

We now require the definition of a quasiaffine transformation.

Definition 4.2.4 Let $T_1 \in \mathcal{L}(\mathcal{X}_1)$, $T_2 \in \mathcal{L}(\mathcal{X}_2)$ and $A \in \mathcal{L}(\mathcal{X}_1, \mathcal{X}_2)$, where A is injective with dense range and where $\mathcal{X}_1, \mathcal{X}_2$ are complex Banach spaces. If $AT_1 = T_2 A$, we say that T_2 is a *quasiaffine transformation* of T_1, denoted by $T_1 \prec T_2$.

Let $\mathcal{X}_1, \mathcal{X}_2$ and $\mathcal{X}$ be complex Banach spaces.

Proposition 4.2.5 *Suppose that $T_1 \prec T \prec T_2$, where $T_1 \in \mathcal{L}(\mathcal{X}_1)$ and $T_2 \in \mathcal{L}(\mathcal{X}_2)$ both have a $C^2(\mathbb{T})$ functional calculus. If $\sigma(T_1)$ is not just a singleton, then $T \in \mathcal{L}(\mathcal{X})$ has non-trivial hyperinvariant subspaces.*

Proof Let $A \in \mathcal{L}(\mathcal{X}_1, \mathcal{X})$ and $B \in \mathcal{L}(\mathcal{X}, \mathcal{X}_2)$ be some injective and dense range operators such that $TA = AT_1$ and $T_2 B = BT$. Let F be a closed subset of $\mathbb{T}$ such that $\emptyset \neq F \cap \sigma(T_1) \neq \sigma(T_1)$. By 3. of Proposition 4.2.3 the subspace $E_{T_1}(F)$ is not trivial. Let

$$E := \text{span}\{WAE_{T_1}(F) : W \in \{T\}'\},$$

where as usual $\{T\}'$ denotes the set of operators commuting with T. Since $E_{T_1}(F) \neq \{0\}$ and since $AE_{T_1}(F) \subset E$, the injectivity of A implies $E \neq \{0\}$.

Note that $BWAT_1 = BWTA = T_2 BWA$, which implies that, for $x \in E_{T_1}(F)$ and $g \in C^2(\mathbb{T})$, we have $\Phi_{T_2}(g)BWAx = BWA\Phi_{T_1}(g)x$. Therefore, we get $BE \subset E_{T_2}(F)$. Since B has dense range, in order to prove that $E \neq \mathcal{X}$ it is sufficient to prove that $E_{T_2}(F) \neq \mathcal{X}_2$. Let $F' \neq \emptyset$ be a closed subset of in $\sigma(T_1) \setminus F$. Let

$$E' := \text{span}\{WAE_{T_1}(F') : W \in \{T\}'\}.$$

As previously, we have $BE' \subset E_{T_2}(F')$ and $E' \neq \{0\}$. Since $F \cap F' = \emptyset$, using 2. of Proposition 4.2.3, we have $E_{T_2}(F) \cap E_{T_2}(F') = \{0\}$. It follows that $E_{T_2}(F)$ and $E_{T_2}(F')$ cannot both be equal to $\mathcal{X}_2$, which ends the proof. □

We now prove a nice property of an operator with a $C^2(\mathbb{T})$ functional calculus whose spectrum is just a single point.

Proposition 4.2.6 *Suppose that $T \in \mathcal{L}(\mathcal{X})$ has a $C^2(\mathbb{T})$ functional calculus and that its spectrum is just a single point $\{\lambda\}$. Then $T - \lambda I$ is a nilpotent operator of order at most* 3.

Proof Let $f \in C^2(\mathbb{T})$ such that $f = 0$ in a neighbourhood of $\sigma(T) = \{\lambda\}$. We first check that $\Phi_T(f) = 0$. Let $\Psi : \mathbb{C} \to \mathcal{L}(\mathcal{X})$ defined by $\Psi(\xi) = \Phi_T(f_\xi)$ where $f_\xi(\xi) = 0$ and $f_\xi(z) = \frac{f(z)}{\xi - z}$ for $z \in \mathbb{T} \setminus \{\xi\}$. We have

$$\Phi_T(f_\xi) = (\xi \mathrm{I} - T)^{-1}\Phi_T(f)$$

for all $\xi \in \mathbb{C} \setminus (\mathrm{supp}(f) \cup \{\lambda\})$. Now we define h on $\mathbb{C}$ by

$$h(\xi) = \begin{cases} (\xi \mathrm{I} - T)^{-1}\Phi_T(f) & \text{if } \xi \neq \lambda, \\ \Phi_T(f_\xi) & \text{if } \xi \notin \mathrm{supp}(f). \end{cases}$$

Since f satisfies $\mathrm{supp}(f) \cap \{\lambda\} = \emptyset$, h is an entire function and it satisfies $\lim_{|\xi| \to \infty} \|h(\xi)\| = 0$: using Liouville's theorem h is identically equal to 0. In particular we have $\Phi_T(f_\xi) = 0$ on $\mathbb{C} \setminus \mathrm{supp}(f)$, which implies that $\Phi_T(f) = 0$.

Now, note that for $x \in \mathcal{X}$ and $x^* \in \mathcal{X}^*$, $A_{x,x^*} : g \mapsto \langle \Phi_T(g)x, x^* \rangle$ is in $C^2(\mathbb{T})^*$. Since for all g equal to 0 in a neighbourhood of λ we have $\Phi_T(g) = 0$, A_{x,x^*} is associated with a numerical distribution of the form $\sum_{k=0}^{2} a_k \delta_\lambda^{(k)}$, where δ_λ is the Dirac distribution at λ. Consequently, there exists a positive constant M_{x,x^*} (depending only on x and x^*) such that

$$|\langle \Phi_T(g)x, x^* \rangle| \leq M_{x,x^*} \max_{k=0,1,2} |g^{(k)}(\lambda)|.$$

By the Banach–Steinhaus theorem (Theorem 1.1.7), there exists a positive constant M such that $\|\Phi_T(g)\| \leq M \max_{k=0,1,2} |g^{(k)}(\lambda)|$ for all $g \in C^2(\mathbb{T})$. In particular, for $g(\xi) = (\xi - \lambda)^3$, we get $(T - \lambda \mathrm{I})^3 = 0$. □

4.2.2 The Colojoară–Foiaş theorem

Let $T \in \mathcal{L}(\mathcal{X})$, where $\mathcal{X}$ is a complex Banach space. We write

$$\mathcal{X}_1 = \{x \in \mathcal{X} : \lim_{n \to \infty} \|T^n x\| = 0\}$$

and

$$\mathcal{Y}_1 = \{x^* \in \mathcal{X}^* : \lim_{n \to \infty} \|T^{*n} x^*\| = 0\}.$$

The operator T is said to be of *class* C_{11} if $\mathcal{X}_1 = \{0\}$ and $\mathcal{Y}_1 = \{0\}$.

Lemma 4.2.7 *Let $T \in \mathcal{L}(\mathcal{X})$. If $\mathcal{X}_1$ is non-trivial then T has a non-trivial hyperinvariant subspace. If $\mathcal{Y}_1$ is non-trivial then T^* has a non-trivial hyperinvariant subspace.*

Proof It is clear that $\mathcal{X}_1$ is a closed hyperinvariant subspace for T, whereas $\mathcal{Y}_1$ is a closed hyperinvariant subspace for T^*. □

Note that if $\mathcal{X}_1$ is non-trivial, it follows that T^* has also a non-trivial hyperinvariant subspace, namely $\mathcal{X}_1^{\perp}$. We recall a notion mentioned briefly in the Exercises to Chapter 2.

Definition 4.2.8 Let $T \in \mathcal{L}(\mathcal{X})$, where $\mathcal{X}$ is a complex Banach space. Then T is said to be *power-bounded* if $\sup_{n\in\mathbb{N}} \|T^n\| < \infty$. In particular this class contains all the polynomially bounded operators.

Theorem 4.2.9 *Let $T \in \mathcal{L}(\mathcal{X})$, where $\mathcal{X}$ is a complex Banach space. If T is power-bounded and of class C_{11}, then either T^* has a non-trivial hyperinvariant subspace or T is a scalar multiple of the identity.*

Proof For $x \in \mathcal{X}$ define $|||x||| = \limsup_{n\to\infty} \|T^n x\|$. Since T is power-bounded, there exists $C > 0$ such that $|||x||| \le C\|x\|$. Since $\mathcal{X}_1 = \{0\}$, $|||\cdot|||$ is a norm. Let $\tilde{\mathcal{X}}$ be the completion of $\mathcal{X}$ with respect to the norm $|||\cdot|||$. Let $A : \mathcal{X} \to \tilde{\mathcal{X}}$ be the natural embedding of $\mathcal{X}$ into $\tilde{\mathcal{X}}$. Then A is a quasi-affinity, i.e., it is a bounded injective linear operator with dense range. Since $|||Tx||| = |||x|||$ for all $x \in \mathcal{X}$, the operator T extends continuously to an isometry $\tilde{T}$ from $\tilde{\mathcal{X}}$ to $\tilde{\mathcal{X}}$. We can assume that $T\mathcal{X}$ is dense in $\mathcal{X}$, since otherwise $(T\mathcal{X})^{\perp}$ is a non-trivial hyperinvariant subspace with respect to T^*. It follows that $\tilde{T}\tilde{\mathcal{X}}$ is dense in $\tilde{\mathcal{X}}$. Therefore $\tilde{T}$ is an invertible isometric operator. By Proposition 4.2.2, $\tilde{T}$ has a $C^2(\mathbb{T})$ functional calculus.

It is easy to see that $AT = \tilde{T}A$, and thus $T^*A^* = A^*\tilde{T}^*$, where A^* is still a quasiaffinity. In other words we have $\tilde{T}^* \prec T^*$. Applying the same argument to T^* instead of T, we get the existence of an operator $T^\sharp$ admitting a $C^2(\mathbb{T})$ functional calculus and satisfying $T^* \prec T^\sharp$. It follows that $\tilde{T}^* \prec T^* \prec T^\sharp$, where both $\tilde{T}^*$ and $T^\sharp$ have a $C^2(\mathbb{T})$ functional calculus.

By Proposition 4.2.5, if the spectrum of $\tilde{T}^*$ is not just a singleton, then T^* has a non-trivial hyperinvariant subspace. In the case where $\sigma(\tilde{T}^*) = \{\lambda\}$, using Proposition 4.2.6, $\tilde{T}^* - \lambda I$ is a nilpotent operator of order at most 3. Since $(T^* - \lambda I)A^* = A^*(\tilde{T}^* - \lambda I)$, it follows that $(T^* - \lambda I)^3 A^* = 0$. Since A^* has dense range, $(T^* - \lambda I)^3 = 0$. Therefore $T^* = \lambda I = T$, or T^* has a non-trivial hyperinvariant subspace. Indeed, if $T^* \ne \lambda I$, then the closure of the range of $T^* - \lambda I$ is a non-trivial hyperinvariant subspace for T^*. □

4.3 Zenger's theorem

Our theme now is the construction of certain key functionals on Banach spaces, by largely geometrical techniques. As an application, in Theorem 4.6.3, we obtain an approximation result for polynomially bounded operators with a weak*-continuous H^∞ functional calculus.

4.3.1 Zenger's theorem and a factorization result

Theorem 4.3.1 *Let $\mathcal{X}$ be a complex Banach space, and let $u_1, \ldots, u_n \in \mathcal{X}$ be linearly independent. Let $\alpha_1, \ldots, \alpha_n$ be positive numbers such that $\sum_{j=1}^n \alpha_j = 1$. Then there exist complex numbers $w_1, \ldots, w_n$ and $\varphi \in \mathcal{X}^*$ such that $\left\| \sum_{j=1}^n w_j u_j \right\| \le 1$, $\|\varphi\| = 1$ and $\varphi(w_j u_j) = \alpha_j$ for all $j = 1, \ldots, n$.*

The first step in the proof of Theorem 4.3.1 is the following lemma.

Lemma 4.3.2 *Let $\| \ldots \|$ be a norm on $\mathbb{C}^n$ let $\{e_1, \ldots, e_n\}$ be a basis of $\mathbb{C}^n$, and let $\alpha_1, \cdots, \alpha_n$ be positive numbers such that $\sum_{j=1}^n \alpha_j = 1$. Then there exists $w = \sum_{j=1}^n w_j e_j \in \mathbb{C}^n$ with $\|w\| = 1$ and $w_1, \ldots, w_n \neq 0$, and a functional Φ on $\mathbb{C}^n$ such that $\Phi(w_j e_j) = \alpha_j$ for $j = 1, \ldots, n$ and $\|\Phi\| = 1$.*

Proof Consider the function F defined on $\mathbb{C}^n$ by $F(z) = \prod_{k=1}^n |z_k|^{\alpha_k}$ for $z = \sum_{j=1}^n z_k e_k$. Let $c = \sup\{F(z) : z \in \mathbb{C}^n, \|z\| \le 1\}$. The compactness of the closed unit ball of $\mathbb{C}^n$ implies that c is attained at a point $w \in \mathbb{C}^n$ with $\|w\| \le 1$. Moreover, since $F(\lambda z) = \lambda F(z)$ for all $z \in \mathbb{C}^n$ and $\lambda > 0$, we have $\|w\| = 1$. Since $\alpha_k > 0$, it is clear that $w_k > 0$, for otherwise $F(w) = 0$, and then F is identically equal to 0. Let $w_k = u_k + i v_k$ with $u_k, v_k \in \mathbb{R}$, and regard F as a function of $2n$ real variables x_k, y_k (for $k = 1, \ldots, n$):

$$F(z) = \prod_{k=1}^n \left(x_k^2 + y_k^2\right)^{\alpha_k/2}, \qquad z = \sum_{k=1}^n (x_k + i y_k) e_k.$$

Then F is differentiable at w with

$$\frac{\partial F}{\partial x_k}(w) = c\alpha_k u_k |w_k|^{-2} \qquad \text{and} \qquad \frac{\partial F}{\partial y_k}(w) = c\alpha_k v_k |w_k|^{-2}.$$

Define a real linear functional Ψ on $\mathbb{C}^n$ by taking

$$\Psi(z) = \sum_{k=1}^n \alpha_k |w_k|^{-2}(u_k x_k + v_k y_k), \qquad z = \sum_{k=1}^n (x_k + i y_k) e_k.$$

Clearly we have $\Psi(w) = 1$, $F'(w) = c\Psi = F(w)\Psi$. The definition of $F'(w)$ also gives, for $z \to w$,

$$F(z) - F(w) = c\Psi(z - w) + o(\|z - w\|).$$

Thus $F(z) = c\Psi(z) + o(\|z - w\|)$. Suppose that there exists $a \in \mathbb{C}^n$ such that $\|a\| \le 1$ and $\Psi(a) = 1 + \delta$ with $\delta > 0$. Then, for all $0 < t \le 1$, the convexity of $B := \{z \in \mathbb{C}^n : \|z\| \le 1\}$ implies that $ta + (1 - t)w \in B$ and

$$F(ta + (1-t)w) = ct(1+\delta) + (1-t)c + o(t) = c + ct\delta + o(t) > c,$$

for t sufficiently small. This contradicts the definition of c and implies that $\|\psi\| \leq 1$. Now consider the functional Φ defined by

$$\Phi(z) = \sum_{k=1}^{n} \alpha_k w_k^{-1} z_k, \qquad z = \sum_{k=1}^{n} z_k e_k.$$

We have

$$\begin{aligned}\Phi(z) &= \sum_{k=1}^{n} \alpha_k |w_k|^{-2} \overline{w_k} z_k \\ &= \sum_{k=1}^{n} \alpha_k |w_k|^{-2}(u_k x_k + v_k y_k) - i \sum_{k=1}^{n} \alpha_k |w_k|^{-2}(v_k x_k - u_k y_k) \\ &= \Psi(z) - i\Psi(iz).\end{aligned}$$

Then $\mathrm{Re}(\Phi) = \Psi$. Obviously we have $\|\Phi\| \geq |\Psi\|$ and the reverse inequality is also true since, for all $z \in \mathbb{C}^n$, there exists $\xi \in \mathbb{C}$ of modulus 1 such that $\mathrm{Re}(\xi\Phi(z)) = |\Phi(z)|$, and thus $\Psi(\xi z) = |\Phi(z)|$ with $|\xi| = 1$. Then we have $\|\Phi\| = 1$, and the lemma is proved. □

Proof of Theorem 4.3.1 Let $\mathcal{X}_0$ be the subspace of $\mathcal{X}$ generated by the vectors $u_1, \ldots, u_n$. We can identify $\mathcal{X}_0$ with $\mathbb{C}^n$, using the norm

$$\|(w_1, \ldots, w_n)\| = \left\| \sum_{j=1}^{n} w_j u_j \right\|_{\mathcal{X}}.$$

By Lemma 4.3.2, there exist $w_1, \ldots, w_n \in \mathbb{C}$ and $\Phi \in \mathcal{X}_0^*$ of norm 1 such that $\Phi(w_j u_j) = \alpha_j$ for $j = 1, \ldots, n$. By the Hahn–Banach theorem, there exists $\varphi \in \mathcal{X}^*$ with $\varphi_{|\mathcal{X}_0} = \Phi$ and $\|\varphi\| = \|\Phi\|$. □

As an application of Theorem 4.3.1, we get the following factorization result for polynomially bounded operators. Recall that the notation $[x \overset{T}{\cdot} x^*]$ was introduced after Definition 2.3.4.

Corollary 4.3.3 *Let T be a polynomially bounded operator on a complex Banach space $\mathcal{X}$. Suppose that T has a weak*-continuous H^∞ functional calculus. Let $\lambda_1, \ldots, \lambda_n$ be distinct eigenvalues of T. Then there exist positive reals $\alpha_1, \ldots, \alpha_n$ such that $\sum_{j=1}^{n} \alpha_j = 1$, and $x \in \mathcal{X}, x^* \in \mathcal{X}^*$ with $\|x\|, \|x^*\| \leq 1$ and such that $[x \overset{T}{\cdot} x^*] = \left[\sum_{i=1}^{n} \alpha_i P_{\lambda_i}\right]$, where P_λ denotes the Poisson kernel at λ.*

Proof Let $u_1, \ldots, u_n$ be linearly independent eigenvectors of T associated with $\lambda_1, \ldots, \lambda_n$, respectively. By Theorem 4.3.1, there exist $w_1, \ldots, w_n \in \mathbb{C}$ and $x^* \in \mathcal{X}^*$ such that $\|x^*\| = 1$ and $x^*(w_j u_j) = \alpha_j$ for each j. Therefore, taking $x = \sum_{j=1}^n w_j u_j$, we get, for all $f \in H^\infty$,

$$\langle [x \overset{T}{\cdot} x^*], f\rangle = \left\langle \sum_{j=1}^n w_j f(T) u_j, x^* \right\rangle = \sum_{j=1}^n f(\lambda_j) \left\langle \sum_{j=1}^n w_j u_j, x^* \right\rangle$$

$$= \sum_{j=1}^n \alpha_j f(\lambda_j) = \left\langle \left[\sum_{j=1}^n \alpha_j P_{\lambda_j} \right], f \right\rangle.$$

□

4.3.2 A stronger version of Zenger's theorem

In the following discussion we will need a stronger version of Theorem 4.3.1. Roughly speaking, we would like to find the functional φ in a ball centred at some given point, and not necessarily at the origin.

The next result is the real version of the required generalization, formulated dually.

Proposition 4.3.4 *Let $\| \cdot \|$ be a norm on $\mathbb{R}^n$, $\alpha_1, \ldots, \alpha_n$ positive numbers such that $\sum_{j=1}^n \alpha_j = 1$, $\{e_1, \ldots, e_n\}$ a basis in $\mathbb{R}^n$, and $s = \sum_{i=1}^n s_i e_i \in \mathbb{R}^n$. Then there exists $w = \sum_{i=1}^n w_j e_j \in \mathbb{R}^n$ and $\psi \in (\mathbb{R}^n, \| \cdot \|)^*$ such that $\|\psi\| = 1$, $\|w - s\| \leq 1$ and $\psi(w_j e_j) = \alpha_j$ for $j = 1, \ldots, n$.*

Proof Let $B = \{x \in \mathbb{R}^n : \|x - s\| \leq 1\}$ and $B_+ = \{x \in B : x_j s_j \geq 0 \text{ for all } j\}$. Let $F : \mathbb{R}^n \to [0, \infty)$ be defined by

$$F(x) = \prod_{j=1}^n |x_j|^{\alpha_j}, \qquad \text{where} \quad x = \sum_{j=1}^n x_j e_j.$$

Since B_+ is compact, there exists $w \in B_+$ such that

$$F(w) = \sup\{F(x) : x \in B_+\} := m.$$

Since F is not identically equal to 0 on B_+, $w_j \neq 0$ for all j. Let $\Psi : \mathbb{R}^n \to \mathbb{R}$ be defined by

$$\Psi(x) = \sum_{j=1}^n \alpha_j x_j w_j^{-1}, \qquad \text{where} \quad x = \sum_{j=1}^n x_j e_j.$$

Obviously $\Psi(w_j e_j) = \alpha_j$ for $j = 1, \ldots, n$. It remains to prove that $\|\Psi\| = 1$. Note that $\Psi(w) = 1$ and in a neighbourhood of w we have

$$\frac{\partial F}{\partial x_j}(x) = \alpha_j |x_j|^{\alpha_j - 1} \operatorname{sign}(x_j) \prod_{k \neq j} |x_k|^{\alpha_k} = \frac{\alpha_j F(x)}{x_j},$$

for all $j = 1, \ldots, n$. In particular, we get $\frac{\partial F}{\partial x_j}(w) = \frac{m\alpha_j}{w_j}$, and thus

$$F'(w) = m\Psi = F(w)\Psi.$$

The definition of $F'(w)$ also gives, for $x \to w$,

$$F(x) - F(w) = F'(w)(x - w) + o(\|x - w\|) = F(w)\Psi(x - w) + o(\|x - w\|),$$

and thus $F(x) = m\Psi(x) + o(\|x - w\|)$. In order to prove that $\Psi(x) \leq 1$ for all $x \in B$, suppose on the contrary that there is an $x \in B$ with $\Psi(x) > 1$. For $t \in (0, 1)$, let $y_t = (1 - t)w + tx = w + t(x - t)$. Since B is convex, $y_t \in B$ for all $t \in (0, 1)$. Furthermore, since $w_j \not\to$ for all j, $y_t \in B_+$ for all t small enough, and $F(y_t) = m + mt\Psi(x - w) + o(t)$. Since $\Psi(x) > 1 = \Psi(m)$, it follows that $F(y_t) > m$ for t small enough, which contradicts the fact that $y_t \in B_+$. Since Ψ is a linear form on a finite-dimensional space,

$$\|\Psi\| = \max\{\Psi(x) - \Psi(s) : x \in B\}.$$

Since $w \in B_+$, $\Psi(s) \geq 0$ and thus $\|\Psi\| \leq 1$. □

The complex version of Proposition 4.3.4 will be given under an additional assumption that the norm is rather regular. We now introduce a definition to make the condition on the regularity of the norm more explicit.

Definition 4.3.5 Let $\mathcal{X}$ be a complex Banach space, and let $u_1, \ldots, u_n \in \mathcal{X}$ be non-zero vectors and $L > 0$. We say that the vectors $u_1, \ldots, u_n$ are *L-unconditional* if

$$\left\| \sum_{j=1}^{n} \beta_j u_j \right\| \leq L \left\| \sum_{j=1}^{n} \gamma_j u_j \right\|$$

whenever $\beta_j, \gamma_j \in \mathbb{C}$ and $|\beta_j| \leq |\gamma_j|$ for $j = 1, \ldots, n$.

It is easy to see that L-unconditional vectors are linearly independent.

Lemma 4.3.6 *Let $L > 0$, and let $\{e_1, \ldots, e_n\}$ be a basis for $\mathbb{C}^n$ equipped with a norm $\|.\|$ such that $e_1, \ldots, e_n$ is L-unconditional. Let $\alpha_1, \ldots, \alpha_n$ be positive numbers such that $\sum_{j=1}^{n} \alpha_j = 1$, and let $s = \sum_{j=1}^{n} s_j e_j \in \mathbb{C}^n$. Then there exist $w = \sum_{j=1}^{n} w_j e_j \in \mathbb{C}^n$ and a complex-linear functional*

$\Psi \in (\mathbb{C}^n, \|\cdot\|)^*$ such that $\|\Psi\| \le L\sqrt{2}$, $\|w - s\| \le 1$ and $\Psi(w_j e_j) = \alpha_j$ for $j = 1, \ldots, n$.

Proof For $j = 1, \ldots, n$, write $s_j = |s_j| e^{i\theta_j}$, where $\theta_j \in [0, 2\pi)$ (if $s_j = 0$, then choose any $\theta_j \in [0, 2\pi)$). Consider the real half-space X in $\mathbb{C}^n$ generated by the vectors $e^{i\theta_j} e_j$, $j = 1, \ldots, n$. Clearly $s \in X$ and $X + iX = \mathbb{C}^n$. By Proposition 4.3.4, there exist real numbers t_j, $j = 1, \ldots, n$, and a real linear functional $\Psi_0 : X \to \mathbb{R}$ such that $\Psi_0\left(t_j e^{i\theta_j} e_j\right) = \alpha_j$ for $j = 1, \ldots, n$, $\left\|\sum_{j=1}^n t_j e^{i\theta_j} e_j - s\right\| \le 1$ and $\|\Psi_0\| \le 1$. Set $w_j = t_j e^{i\theta_j}$ and extend Ψ_0 to $\mathbb{C}^n$ by

$$\Psi(x + iy) = \Psi_0(x) + i\Psi_0(y),$$

for all $x, y \in X$. It is easy to check that Ψ is a complex-linear functional and that $\Psi(w_j e_j) = \alpha_j$ for all $j = 1, \ldots, n$. Let $b_j, c_j \in \mathbb{R}$. Note that if $\left\|\sum_{j=1}^n (b_j + ic_j) e^{i\theta_j} e_j\right\| \le 1$, then

$$\left\|\sum_{j=1}^n b_j e^{i\theta_j} e_j\right\| \le 1 \qquad \text{and} \qquad \left\|\sum_{j=1}^n c_j e^{i\theta_j} e_j\right\| \le 1.$$

It follows that

$$\left|\Psi\left(\sum_{j=1}^n (b_j + ic_j) e^{i\theta_j} e_j\right)\right|^2 = \left|\Psi_0\left(\sum_{j=1}^n b_j e^{i\theta_j} e_j\right)\right|^2 + \left|\Psi_0\left(\sum_{j=1}^n c_j e^{i\theta_j} e_j\right)\right|^2 \le 2L^2,$$

and therefore $\|\Psi\| \le L\sqrt{2}$. □

The next result is a dual version of Lemma 4.3.6.

Lemma 4.3.7 *Let $L > 0$, and let $\{e_1, \ldots, e_n\}$ be a basis for $\mathbb{C}^n$ equipped with a norm $\|.\|$ such that $e_1, \ldots, e_n$ is L-unconditional. Let $\alpha_1, \ldots, \alpha_n$ be positive numbers such that $\sum_{j=1}^n \alpha_j = 1$, and let $\Phi \in (\mathbb{C}^n, \|\cdot\|)^*$. Then there exist $w = \sum_{j=1}^n w_j e_j \in \mathbb{C}^n$ and a complex-linear functional $\Psi \in (\mathbb{C}^n, \|\cdot\|)^*$ such that $\|\Psi - \Phi\| \le 1$, $\|w\| \le L\sqrt{2}$ and $\Psi(w_j e_j) = \alpha_j$ for $j = 1, \ldots, n$.*

Proof Let $f_1, \ldots, f_n \in (\mathbb{C}^n, \|\cdot\|)^*$ be defined by $\langle e_j, f_k \rangle = \delta_{j,k}$, where $\delta_{j,k}$ is the Kronecker symbol. We prove first that $f_1, \ldots, f_n$ is L-unconditional in $(\mathbb{C}^n, \|\cdot\|)^*$. Let $\beta_j, \gamma_j \in \mathbb{C}$ with $|\beta_j| \le |\gamma_j|$ for $j = 1, \ldots, n$ and let

$$F = \{j \in \{1, \ldots, n\} : \beta_j \ne 0\}.$$

Then we have

$$\begin{aligned}\left\|\sum_{j=1}^{n}\beta_j f_j\right\| &= \sup\left\{\left|\left\langle\sum_{j\in F}\beta_j f_j, \sum_{j=1}^{n} w_j e_j\right\rangle\right| : \left\|\sum_{j=1}^{n} w_j e_j\right\| \le 1\right\}\\ &= \sup\left\{\left|\sum_{j\in F}\beta_j w_j\right| : \left\|\sum_{j=1}^{n} w_j e_j\right\| \le 1\right\}\\ &= \sup\left\{\left|\left\langle\sum_{j=1}^{n}\gamma_j f_j, \sum_{j\in F} w_j\beta_j\gamma_j^{-1} e_j\right\rangle\right| : \left\|\sum_{j=1}^{n} w_j e_j\right\| \le 1\right\}\\ &\le \sup\left\{\left|\left\langle\sum_{j=1}^{n}\gamma_j f_j, \sum_{j=1}^{n} \mu_j e_j\right\rangle\right| : \left\|\sum_{j=1}^{n} \mu_j e_j\right\| \le L\right\}\\ &= L\left\|\sum_{j=1}^{n}\gamma_j f_j\right\|.\end{aligned}$$

By Lemma 4.3.6, there exist $\Psi \in (\mathbb{C}^n, \|\cdot\|)^*$, $\Psi = \sum_{j=1}^{n}\Psi_j f_j$ such that $\|\Psi - \Phi\| \le 1$ and $w = (w_1, \ldots, w_n) \in \mathbb{C}^n$ such that $\|w\| \le L\sqrt{2}$ and

$$\alpha_l = \left\langle\sum_{j=1}^{n} w_j e_j, \Psi_l f_l\right\rangle = w_l\Psi_l = \langle w_l e_l, \Psi\rangle,$$

for all $l = 1, \ldots, n$. □

We are now ready to give the required generalization of Zenger's theorem, formulated dually.

Theorem 4.3.8 *Let $u_1, \ldots, u_n$ be L-unconditional vectors in a complex Banach space $\mathcal{X}$. Let $\alpha_1, \ldots, \alpha_n$ be positive numbers such that $\sum_{j=1}^{n}\alpha_j = 1$, and let $\Phi \in \mathcal{X}^*$. Then there exist complex numbers $w_j \in \mathbb{C}$, $j = 1, \ldots, n$ and a complex-linear functional $\Psi \in \mathcal{X}^*$ such that $\|\Psi - \Phi\| \le 1$, $\left\|\sum_{j=1}^{n} w_j u_j\right\| \le L\sqrt{2}$ and $\Psi(w_j u_j) = \alpha_j$ for $j = 1, \ldots, n$.*

Proof Let $\mathcal{X}_0$ be the subspace of $\mathcal{X}$ generated by the vectors $u_1, \ldots, u_n$. We can identify $\mathcal{X}_0$ with $\mathbb{C}^n$ with the norm $\|(w_1, \ldots, w_n)\| = \left\|\sum_{j=1}^{n} w_j u_j\right\|_{\mathcal{X}}$. By Lemma 4.3.7, there exist $w_j \in \mathbb{C}$ and $\Psi \in \mathcal{X}_0^*$ such that $\left\|\sum_{j=1}^{n} w_j u_j\right\| \le L\sqrt{2}$, $\langle w_j u_j, \Psi\rangle = \alpha_j$ and $\|\Psi - \Phi_{|\mathcal{X}_0}\| \le 1$. By the Hahn–Banach theorem, we can extend Ψ to $\mathcal{X}$ such that $\|\Psi - \Phi\|_{\mathcal{X}^*} \le 1$. □

A technical problem in the following discussion is that the different versions of Zenger's theorem give no estimate of the coefficients w_j. Such estimates can be obtained using the classical Carleson interpolation theorem, discussed in the next section.

4.4 Carleson's interpolation theorem

We start with the notion of a separated set in $\mathbb{D}$, more restrictive than the usual notion associated with a Carleson interpolation set (which was discussed in Theorem 1.4.29).

Definition 4.4.1 For $\lambda = re^{i\theta}$, denote by I_λ the set $\{e^{it} : |t-\theta| < 2(1-r)\}$. A subset F of $\mathbb{D}$ is called *angularly separated* if, for all $\lambda, \mu \in F, \lambda \neq \mu$, we have $I_\lambda \cap I_\mu = \emptyset$.

Remark 4.4.2 Let F be a finite and angularly separated subset of $\mathbb{D}$ and suppose that $\bigcup_{\lambda \in F} I_\lambda \subset J$, where J is a sub-arc of $\mathbb{T}$. Then we have

$$4\sum_{\lambda\in F}(1-|\lambda|) \leq 2\pi\ell(J),$$

where $\ell(J)$ denotes the normalized length of the arc J.

We start with a geometric lemma.

Lemma 4.4.3 *For all $z, \lambda \in \mathbb{D} \setminus \frac{3}{4}\mathbb{D}$, if $I_\lambda \cap I_z = \emptyset$, then $\left|\frac{z-\lambda}{1-\bar{\lambda}z}\right| \geq \frac{3}{2\pi+3}$.*

Proof Without loss of generality we may assume that $\lambda = r \in [3/4, 1)$. Let $z = se^{i\theta}$ with $s \in [3/4, 1)$ and $\theta \in [-\pi, \pi)$. Then we have

$$|\lambda - z|^2 = (r-s)^2 + 2rs(1-\cos\theta) \geq 4rs\sin^2(\theta/2) \geq \frac{9}{4}\sin^2(\theta/2).$$

Using the inequality $\sin x \geq \frac{2}{\pi}x$ for all $x \in [0, \pi/2]$, we have $|\lambda - z| \geq \frac{3|\theta|}{2\pi}$. Since $e^{i\theta} \notin I_r$, $|\theta| \geq 2(1-r)$ and then $|\lambda - z| \geq \frac{3(1-r)}{\pi}$. Moreover, since

$$|1-\bar{\lambda}z| = |1-rz| \leq |1-r^2| + |r^2 - rz| \leq 2(1-r) + r|r-z|,$$

we have

$$\left|\frac{z-\lambda}{1-\bar{\lambda}z}\right| \geq \frac{|z-r|}{2(1-r)+r|r-z|} = \frac{1}{2\frac{1-r}{|z-\lambda|}+r} \geq \frac{1}{\frac{2\pi}{3}+1} = \frac{3}{2\pi+3}.$$

□

We shall now need the concept of a Carleson measure, as in Definition 1.4.32 and Theorem 1.4.33. We construct a particular example.

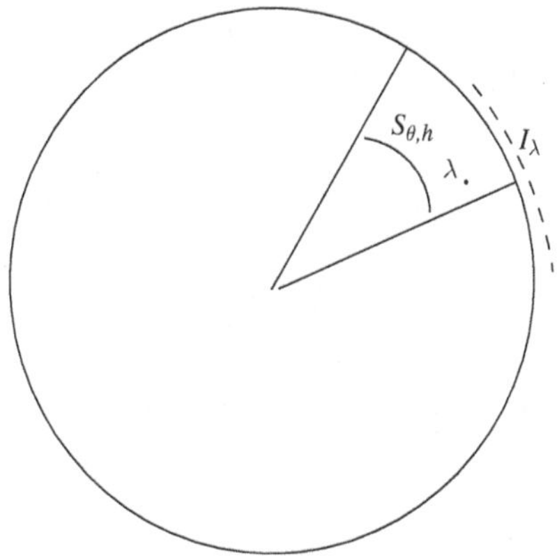

Figure 4.2 A sector $S_{\theta,h}$ and an arc I_λ with $\lambda \in S_{\theta,h}$.

Lemma 4.4.4 *Let $F \subset \mathbb{D}$ be a finite angularly separated set. Denote by δ_λ the Dirac measure at λ. Then $\mu := \sum_{\lambda \in F}(1 - |\lambda|)\delta_\lambda$ is a Carleson measure with $c_\mu \leq 1$.*

Proof Let $S_{\theta,h} = \{re^{it} : 1-h \leq r < 1, |t-\theta| \leq h\}$, let $\Gamma := \{e^{it} : |t-\theta| \leq h\}$ and let $\lambda \in F \cap S_{\theta,h}$ (see Figure 4.2). Then at least one of the arcs

$$\{e^{it} : t \in (s, s+2(1-|\lambda|))\} \qquad \text{or} \qquad \{e^{it} : t \in (s-2(1-|\lambda|), s)\}$$

is included in Γ. It follows that $m(I_\lambda \cap \Gamma) \geq 2(1-|\lambda|)$, where m denotes the Lebesgue measure on $\mathbb{T}$. Therefore we have

$$\mu(S_{\theta,h}) = \sum_{\lambda \in F \cap S_{\theta,h}} (1-|\lambda|) \leq \frac{1}{2} m\left(\Gamma \cap \bigcup_{\lambda \in S_{\theta,h} \cap \Gamma} I_\lambda\right) \leq \frac{1}{2} m(\Gamma) = h.$$

□

It follows from the Carleson interpolation theory, combined with Lemmas 4.4.3 and 4.4.4, that, given a finite angularly separated set $F \in \mathbb{D} \setminus \frac{3}{4}\mathbb{D}$ and numbers $c_\lambda \in \mathbb{C}$, there exists $f \in H^\infty(\mathbb{D})$ such that $f(\lambda) = c_\lambda$ for all $\lambda \in F$, with $\|f\|_\infty \leq b \sup_{\lambda \in F} |c_\lambda|$, where b is an absolute constant independent of F and c_λ.

In order to provide a self-contained proof, we now give all the details. The following characterization of Carleson measures will be useful.

Lemma 4.4.5 *A positive measure μ on $\mathbb{D}$ is a Carleson measure with constant c_μ if and only if*

$$\sup_{z_0 \in \mathbb{D}} \int_{\mathbb{D}} \frac{1-|z_0|^2}{|1-\overline{z_0}z|^2} d\mu(z) = M < \infty, \tag{4.1}$$

where $C_1 c_\mu \leq M \leq C_2 c_\mu$ for absolute constants C_1 and C_2.

Proof Suppose that (4.1) holds and let

$$S_{\theta,h} := \{re^{it} : 1-h \leq r < 1, |t-\theta| \leq h\}.$$

Since (4.1) with $z_0 = 0$ shows that $\mu(\mathbb{D}) \leq M$, we can suppose that $h < \frac{1}{4}$. Indeed, for $h \geq \frac{1}{4}$, $\mu(S_{\theta,h}) \leq \mu(\mathbb{D}) \leq M \leq 4Mh$.

Let $h < \frac{1}{4}$ and consider $z_0 = \left(1 - \frac{1}{2}h\right)e^{i\theta}$. Note that for $z \in S_{\theta,h}$, we have

$$\frac{1-|z_0|^2}{|1-\overline{z_0}z|^2} \geq \frac{C}{1-|z_0|^2}$$

for some absolute constant $C > 0$. Therefore we get

$$(1-|z_0|) \int_{S_{\theta,h}} \frac{1-|z_0|^2}{|1-\overline{z_0}z|^2} d\mu(z) \geq \frac{C}{1+|z_0|}\mu(S_{\theta,h}) \geq \frac{C}{2}\mu(S_{\theta,h}).$$

It follows that

$$\begin{aligned}\mu(S_{\theta,h}) &\leq \frac{2}{C}(1-|z_0|) \int_{S_{\theta,h}} \frac{1-|z_0|^2}{|1-\overline{z_0}z|^2} d\mu(z) \\ &\leq \frac{2}{C}((1-|z_0|)M = \frac{2}{C}M\frac{h}{2} \\ &= \frac{M}{C}h.\end{aligned}$$

Conversely, suppose that μ is a Carleson measure and let $z_0 \in \mathbb{D}$. If $|z_0| \leq \frac{3}{4}$, we have

$$\frac{1-|z_0|^2}{|1-\overline{z_0}z|^2} \leq \frac{1-|z_0|^2}{(1-|z_0|)^2} = \frac{1+|z_0|}{1-|z_0|} \leq 7.$$

Since $\mu(\mathbb{D}) \leq 4c_\mu$, it follows that $\int_{S_{\theta,h}} \frac{1-|z_0|^2}{|1-\overline{z_0}z|^2} d\mu(z) \leq 28c_\mu$. If $|z_0| > \frac{3}{4}$, for all positive integers n, we let

$$E_n = \left\{z \in \mathbb{D} : \left|z - \frac{z_0}{|z_0|}\right| < 2^n(1-|z_0|)\right\}.$$

For $z \in E_1$, note that

$$\frac{1-|z_0|^2}{|1-\overline{z_0}z|^2} \leq \frac{1+|z_0|}{1-|z_0|} \leq \frac{2}{1-|z_0|}.$$

Using the identity

$$|1 - \overline{z_0}z|^2 = |z - z_0|^2 + (1 - |z_0|^2)(1 - |z|^2),$$

for all $n \geq 2$ and $z \in E_n \setminus E_{n-1}$, we have

$$\begin{aligned} |1 - \overline{z_0}z|^2 \geq |z - z_0|^2 &\geq (2^{n-1} - 1)^2(1 - |z_0|)^2 \\ &\geq (2^{n-2})^2(1 - |z_0|)^2 = \frac{2^{2n}}{2^4}(1 - |z_0|)^2, \end{aligned}$$

and then

$$\frac{1 - |z_0|^2}{|1 - \overline{z_0}z|^2} \leq \frac{16}{2^{2n}(1 - |z_0|)^2}\frac{16}{2^{2n}(1 - |z_0|)}.$$

Moreover, since $E_n \subset S_{\theta,2^{n+1}(1-|z_0|)}$, it follows that $\mu(E_n) \leq 2^{n+1}(1-|z_0|)c_\mu$ for all $n \geq 1$. Then we obtain

$$\begin{aligned} \int_{\mathbb{D}} \frac{1 - |z_0|^2}{|1 - \overline{z_0}z|^2} d\mu(z) &= \int_{E_1} \frac{1 - |z_0|^2}{|1 - \overline{z_0}z|^2} d\mu(z) + \sum_{n\geq 2} \int_{E_n\setminus E_{n-1}} \frac{1 - |z_0|^2}{|1 - \overline{z_0}z|^2} d\mu(z) \\ &\leq 8c_\mu + 32 \sum_{n\geq 2} \frac{1}{2^n} c_\mu = 24c_\mu. \end{aligned}$$

□

Lemma 4.4.6 *There exists a numerical constant $\delta > 0$ such that for every angularly separated finite set F in $\mathbb{D} \setminus \frac{3}{4}\mathbb{D}$, and each $\lambda_0 \in F$, we have*

$$\prod_{\lambda \in F\setminus\{\lambda_0\}} \left| \frac{\lambda - \lambda_0}{1 - \overline{\lambda_0}\lambda} \right| \geq \delta.$$

Proof Using Lemma 4.4.4, $\mu := \sum_{\lambda\in F}(1 - |\lambda|)\delta_\lambda$ is a Carleson measure with $c_\mu \leq 1$. By Lemma 4.4.5, for each $\lambda_0 \in F$, we have

$$\int_{\mathbb{D}} \frac{1 - |z_0|^2}{|1 - \overline{z_0}z|^2} d\mu(z) \leq C_2,$$

where C_2 is an absolute constant. The definition of μ implies that, for each $\lambda_0 \in F$,

$$\sum_{\lambda\in F} \frac{(1 - |\lambda|)(1 - |\lambda_0|^2)}{|1 - \overline{\lambda_0}\lambda|^2} \leq C_2.$$

Let c be the constant from Lemma 4.4.3, i.e., $c = \frac{3}{2\pi+3}$. Since $c < 1$, using the concavity of $t \mapsto \log t$, for any $t \in (c^2, 1)$, we have

$$\log t \geq \frac{2 \log c}{1 - c^2}(1 - t).$$

For $\lambda \in F\backslash\{\lambda_0\}$, using the identity $|1-\overline{\lambda}\lambda_0|^2-|\lambda_0-\lambda|^2 = (1-|\lambda|^2)(1-|\lambda_0|^2)$ and Lemma 4.4.3, we have

$$\log\left|\frac{\lambda_0-\lambda}{1-\overline{\lambda_0}\lambda}\right|^2 \geq \frac{2\log c}{1-c^2}\left(1-\left|\frac{\lambda_0-\lambda}{1-\overline{\lambda_0}\lambda}\right|^2\right) = \frac{2\log c}{1-c^2}\frac{(1-|\lambda|^2)(1-|\lambda_0|^2)}{|1-\overline{\lambda}\lambda_0|^2}$$
$$\geq \frac{4\log c}{1-c^2}\frac{(1-|\lambda|)(1-|\lambda_0|^2)}{|1-\overline{\lambda}\lambda_0|^2}.$$

Set $B(z) = \prod_{\lambda\in F\backslash\{\lambda_0\}}\frac{z-\lambda}{1-\overline{\lambda}z}$. Then

$$\log|B(\lambda_0)|^2 \geq \frac{4\log c}{1-c^2}\sum_{\lambda\in F\backslash\{\lambda_0\}}\frac{(1-|\lambda|)(1-|\lambda_0|^2)}{|1-\overline{\lambda_0}\lambda|^2} \geq \frac{4\log c}{1-c^2}C_2.$$

Thus we get $|B(\lambda_0)| \geq \delta$, where $\delta = \exp\left(\frac{2\log c}{1-c^2}C_2\right)$. □

A finite version of the Carleson interpolation theorem was given earlier as Theorem 1.4.31. As an immediate corollary, we have the following interpolation result.

Corollary 4.4.7 *There exists an absolute constant $b > 0$ such that, for every angularly separated finite set F in $\mathbb{D}\backslash\frac{3}{4}\mathbb{D}$ and every sequence of complex numbers $(c_\lambda)_{\lambda\in F}$, there exists a rational function $f \in A(\mathbb{D})$ such that $f(\lambda)=c_\lambda$ and $\|f\|_\infty \leq b\max_{\lambda\in F}|c_\lambda|$.*

We now derive a technical result which provides a control of the coefficients of the approximate eigenvalues. This will be used in Section 4.5.

Theorem 4.4.8 *Let $T \in \mathcal{L}(\mathcal{X})$ be a polynomially bounded operator with polynomial bound M. Let b be the absolute constant from Corollary 4.4.7. Let $F \subset \mathbb{D}$ be a finite set with $(I_\lambda)_{\lambda\in F}$ pairwise disjoint and $|\lambda| \geq 3/4$ for all $\lambda \in F$. Suppose that there are vectors $u_\lambda \in \mathcal{X}$ and complex numbers μ_λ $(\lambda\in F)$ such that $\|u_\lambda\| \leq 1$, $\|(T-\lambda\mathrm{I})u_\lambda\| < \frac{1}{2Mb\pi}(1-|\lambda|)^2$ and $\|\sum_{\lambda\in F}\mu_\lambda u_\lambda\| = 1$. Then $|\mu_\lambda| \leq 2Mb$ for all $\lambda\in F$.*

Proof Let $\lambda_0 \in F$ satisfy $|\mu_{\lambda_0}| = \max_{\lambda\in F}|\mu_\lambda|$. By Corollary 4.4.7, there is a function $f \in A(\mathbb{D})$ such that $\|f\|_\infty \leq b$, $f(\lambda_0) = 1$ and $f(\lambda) = 0$ for all $\lambda \in F\setminus\{\lambda_0\}$. Then $f(T)$ is well-defined with $\|f(T)\| \leq Mb$, and $\|f(T)u\| \leq Mb$ with $u = \sum_{\lambda\in F}\mu_\lambda u_\lambda$.

For $\lambda \in F$ let g_λ denote the analytic function in a neighbourhood of $\overline{\mathbb{D}}$ defined by

$$g_\lambda(z) = \frac{f(z)-f(\lambda)}{z-\lambda}.$$

Then $\|g_\lambda\|_\infty \le 2\|f\|_\infty(1-|\lambda|)^{-1} \le 2b(1-|\lambda|)^{-1}$.

Therefore we get

$$\begin{aligned}
Mb &\ge \|f(T)u\| \\
&\ge \left\|\sum_{\lambda\in F} f(\lambda)\mu_\lambda u_\lambda\right\| - \left\|\sum_{\lambda\in F} \mu_\lambda(f(\lambda)-f(T))u_\lambda\right\| \\
&\ge \|\mu_{\lambda_0}u_{\lambda_0}\| - \sum_{\lambda\in F} |\mu_\lambda|\|g_\lambda(T)(T-\lambda \mathrm{I})u_\lambda\| \\
&\ge |\mu_{\lambda_0}| - |\mu_{\lambda_0}|\sum_{\lambda\in F} 2Mb(1-|\lambda|)^{-1}\frac{1}{2Mb\pi}(1-|\lambda|)^2 \\
&\ge |\mu_{\lambda_0}|\left(1-\sum_{\lambda\in F}\pi^{-1}(1-|\lambda|)\right) \ge \frac{|\mu_{\lambda_0}|}{2},
\end{aligned}$$

since $\sum_{\lambda\in F}(1-|\lambda|) \le \frac{1}{4}m(\bigcup_{\lambda\in F} I_\lambda) \le \frac{\pi}{2}$. Therefore we have $|\mu_\lambda| \le |\mu_{\lambda_0}| \le 2Mb$ for all $\lambda \in F$. □

Note that a possible interpolating function f in the proof of Theorem 4.4.8 is $f(z) = B(z)/B(\lambda_0)$, with

$$B(z) = \prod_{\lambda\in F,\lambda\neq\lambda_0} (z-\lambda)(1-\overline{\lambda}z)^{-1}$$

and $\|f\|_\infty \le 1/\delta$. However, this fact is not used in the subsequent development.

4.5 Approximation using Apostol sets

The main idea that we wish to present now is that every positive integrable function can be approximated in $L^1(\mathbb{T})$ norm by a linear combination of Poisson kernels P_λ with the λ angularly separated and lying in an Apostol set; the important fact here is that this can be done with a control on the coefficients. The application in Section 4.6 will be to the factorization of L^1 functions and hence the existence of invariant subspaces.

4.5.1 Approximation of integrable non-negative functions

Before giving an approximation result using Apostol sets, we first introduce a 'truncated' Poisson kernel, denoted by Q_λ, and we give some estimates of its integral. Recall that, for $\lambda \in \mathbb{D}$, the Poisson kernel at λ is defined by $P_\lambda(t) = \frac{1-|\lambda|^2}{|\lambda-e^{it}|^2}$, and it satisfies $\int_{-\pi}^{\pi} P_\lambda(t)\,dt = 2\pi$ and $\max_t P_\lambda(t) = \frac{1+|\lambda|}{1-|\lambda|}$.

The 2π-periodic function Q_λ associated with $\lambda \in \mathbb{D}$ is defined on $\mathbb{R}$ by $Q_\lambda(t) = P_\lambda(t)$ if $e^{it} \in I_\lambda$ and $Q_\lambda(t) = 0$ otherwise.

Lemma 4.5.1 *For $|\lambda| \geq 3/4$, we have*

$$\int_{-\pi}^{\pi} Q_\lambda(t)\, dt \geq \frac{7\pi}{6}, \tag{4.2}$$

and thus

$$\int_{-\pi}^{\pi} (P_\lambda(t) - Q_\lambda(t))\, dt \leq \frac{5}{7} \int_{-\pi}^{\pi} Q_\lambda(t)\, dt. \tag{4.3}$$

Proof Without loss of generality we may assume that $\lambda = r \geq 3/4$. Note that

$$\int_{-\pi}^{\pi} Q_\lambda(t)\, dt = \int_{-2(1-r)}^{2(1-r)} \frac{1-r^2}{|r-e^{it}|^2}\, dt = 2(1-r^2) \int_0^{2(1-r)} \frac{dt}{|r-e^{it}|^2}.$$

Since $\sin^2(1-r) \leq \sin(1-r) \leq (1-r)$, for all $e^{it} \in I_\lambda$, we have

$$\cos t \geq \cos 2(1-r) = 1 - 2\sin^2(1-r) \geq 2r - 1,$$

and thus $1 - r \geq \cos t - r \geq r - 1$. It follows that

$$|r - e^{it}|^2 = (r - \cos t)^2 + \sin^2 t \leq (1-r)^2 + t^2.$$

Therefore we get

$$\begin{aligned}
\int_{-\pi}^{\pi} Q_\lambda(t)\, dt &\geq 2(1-r^2) \int_0^{2(1-r)} \frac{dt}{(1-r)^2 + t^2} \\
&= 2(1-r^2) \left[\frac{1}{1-r} \tan^{-1} \frac{t}{1-r} \right]_0^{2(1-r)} \\
&= 2(1+r) \tan^{-1} 2 \geq \frac{7}{2} \tan^{-1} \sqrt{3} = \frac{7\pi}{6},
\end{aligned}$$

which proves (4.2). In order to prove (4.3), note that, using (4.2), we have the following estimate:

$$\frac{\int_{-\pi}^{\pi} (P_\lambda(t) - Q_\lambda(t))\, dt}{\int_{-\pi}^{\pi} Q_\lambda(t)\, dt} = \frac{\int_{-\pi}^{\pi} P_\lambda(t)\, dt}{\int_{-\pi}^{\pi} Q_\lambda(t)\, dt} - 1 \leq 2\pi \left(\frac{7\pi}{6} \right)^{-1} - 1 = \frac{5}{7}.$$

□

Theorem 4.5.2 *Let $\Lambda \subset \mathbb{D}$ be an Apostol set. Let t_1, t_2 be real numbers satisfying $-\pi \leq t_1 < t_2 \leq \pi$. Let $f(t) := 1$ if $t_1 \leq t \leq t_2$, and $f(t) := 0$ otherwise. Then there is $n_0 \geq 1$ such that for every $n \geq n_0$ there exist a* *finite set* $F \subset \Lambda$ *and positive real numbers* α_λ $(\lambda \in F)$ *with the following properties:*

(i) $I_\lambda \subset \{e^{it} : t_1 < t < t_2\}$ *for all* $\lambda \in F$.
(ii) *The sets* I_λ $(\lambda \in F)$ *are pairwise disjoint.*
(iii) $m\left(\bigcup_{\lambda \in F} I_\lambda\right) \geq \frac{1}{20\pi}(t_2 - t_1)$.
(iv) $|\lambda| \geq \frac{3}{4}$ *and* $|\lambda^n - 1| < \frac{1}{9}$ *for all* $\lambda \in F$.
(v) $\sum_{\lambda \in F} \alpha_\lambda \leq \frac{t_2 - t_1}{7}$.
(vi) $\int_{-\pi}^{\pi} \left|\sum_{\lambda \in F} \alpha_\lambda \lambda^n P_\lambda(t) - f(t)\right| dt \leq c_1(t_2 - t_1)$, *where* $c_1 = 1 - \frac{11}{8640}$.

Proof Let n be a positive integer such that the arc between e^{int_1} and e^{int_2} contains the unit circle (i.e., $n \geq \frac{2\pi}{t_2 - t_1}$). Set

$$M_n = \{t \in (t_1, t_2) : |e^{int} - 1| \leq 1/10\}$$

and

$$L_n = \{x \in (nt_1, nt_2) : |e^{ix} - 1| \leq 1/10\},$$

so that $L_n = nM_n$. Note that the number of complete revolutions between e^{int_1} and e^{int_2} is the integer part of $\frac{n(t_2 - t_1)}{2\pi}$; moreover,

$$m(\{t \in (-\pi, \pi] : |e^{it} - 1| \leq 1/10\}) := a > 1/5.$$

It follows that $m(L_n) > \frac{1}{5}\frac{n(t_2 - t_1)}{2\pi}$ for n sufficiently large, and thus

$$m(M_n) > \frac{t_2 - t_1}{10\pi}. \tag{4.4}$$

Fix n satisfying (4.4) and let $\varepsilon > 0$ such that $m(M_n) - \varepsilon > \frac{t_2 - t_1}{10\pi}$. Let $S \subset (t_1, t_2)$ be the exceptional set of the Apostol set Λ, i.e., $\sup\{0 \leq r < 1 : re^{i\theta} \in \Lambda\} = 1$ for all $\theta \in (t_1, t_2) \setminus S$. Since S is at most countable, it can be covered by a countable union U of open intervals with $m(U) < \frac{\varepsilon}{2}$. Then the set M' defined by

$$M' = (M_n \cap [t_1 + \varepsilon/4, t_2 - \varepsilon/4]) \setminus U$$

is compact with $m(M') > (t_2 - t_1)/10\pi$. For each $t \in M'$ take $r_t \geq 3/4$ such that $\lambda_t := r_t e^{it} \in \Lambda$, $\left|\lambda_t^n - 1\right| < 1/9$ and $I_{\lambda_t} \subset \{e^{is} : t_1 < s < t_2\}$. Note that $\{e^{is} : s \in M'\} \subset \bigcup_{t \in M'} I_{\lambda_t}$. Since $\{e^{is} : s \in M'\}$ is a compact subset of the one-dimensional set $\mathbb{T}$, there exists a finite subcover of $(I_{\lambda_t})_{t \in M'}$ such that any three of these subsets have empty intersection.

Considering a cover of the minimal cardinality with this property, it is easy to see that we can find numbers $\lambda_1, \ldots, \lambda_k \in \Lambda$ with $\lambda_j = |\lambda_j| e^{is_j}$ such that

$t_1 < s_1 < \cdots < s_k < t_2$, $\bigcup_{j=1}^{k} I_{\lambda_j} \supset \{e^{is} : s \in M'\}$ and $I_{\lambda_j} \cap I_{\lambda_{j'}} = \emptyset$ if $|j - j'| \geq 2$. Let $F_1 = \{\lambda_1, \lambda_3, \ldots\}$ and $F_2 = \{\lambda_2, \lambda_4, \ldots\}$. Then, let F be one of the sets F_1, F_2 such that

$$m\left(\bigcup_{\lambda \in F} I_\lambda\right) = \max\left\{m\left(\bigcup_{\lambda \in F_1} I_\lambda\right), m\left(\bigcup_{\lambda \in F_2} I_\lambda\right)\right\}.$$

It follows that $I_\lambda \cap I_{\lambda'} = \emptyset$ for all distinct λ, λ' in F, and

$$m\left(\bigcup_{\lambda \in F} I_\lambda\right) \geq m(M')/2 > (t_2 - t_1)/20\pi.$$

For any $\lambda \in F$, set $\alpha_\lambda = (1 - |\lambda|)(1 + |\lambda|)^{-1}$. Now, since $|\lambda| \geq 3/4$ and we have $I_\lambda \subset \{e^{is} : t_1 < s < t_2\}$, we get

$$\sum_{\lambda \in F} \alpha_\lambda \leq \frac{4}{7} \sum_{\lambda \in F} (1 - |\lambda|) = \frac{1}{7} \sum_{\lambda \in F} m(I_\lambda) \leq \frac{t_2 - t_1}{7}.$$

In order to check the last condition, we use the 2π-periodic function Q_λ defined previously. First note that $\int_{t_1}^{t_2} \left(1 - \sum_{\lambda \in F} \alpha_\lambda Q_\lambda(t)\right) dt \geq 0$. Indeed, we have

$$\begin{aligned}
\int_{t_1}^{t_2} \left(1 - \sum_{\lambda \in F} \alpha_\lambda Q_\lambda(t)\right) dt &= (t_2 - t_1) - \int_{t_1}^{t_2} \sum_{\lambda \in F} \alpha_\lambda Q_\lambda(t)\, dt \\
&\geq (t_2 - t_1) - 2\pi \sum_{\lambda \in F} \alpha_\lambda \\
&\geq (t_2 - t_1)(1 - 2\pi/7).
\end{aligned}$$

Let $A := \int_{-\pi}^{\pi} \left|\sum_{\lambda \in F} \alpha_\lambda \lambda^n P_\lambda(t) - f(t)\right| dt$. Using the triangle inequality, we have

$$\begin{aligned}
A \leq{}& \int_{-\pi}^{\pi} \left|\sum_{\lambda \in F} \alpha_\lambda \lambda^n (P_\lambda(t) - Q_\lambda(t))\right| dt + \int_{-\pi}^{\pi} \sum_{\lambda \in F} \alpha_\lambda |\lambda^n - 1| Q_\lambda(t)\, dt \\
&+ \int_{t_1}^{t_2} \left(1 - \sum_{\lambda \in F} \alpha_\lambda Q_\lambda(t)\right) dt \\
\leq{}& \sum_{\lambda \in F} \alpha_\lambda \int_{-\pi}^{\pi} (P_\lambda(t) - Q_\lambda(t))\, dt + \frac{1}{9} \int_{t_1}^{t_2} \sum_{\lambda \in F} \alpha_\lambda Q_\lambda(t)\, dt \\
&+ (t_2 - t_1) - \int_{t_1}^{t_2} \sum_{\lambda \in F} \alpha_\lambda Q_\lambda(t)\, dt,
\end{aligned}$$

by (iv). Now, using (4.3), we get

$$\begin{aligned} A &\le (t_2 - t_1) + \left(\frac{5}{7} + \frac{1}{9} - 1\right) \int_{t_1}^{t_2} \sum_{\lambda \in F} \alpha_\lambda Q_\lambda(t)\, dt \\ &= (t_2 - t_1) - \frac{11}{63} \int_{t_1}^{t_2} \sum_{\lambda \in F} \alpha_\lambda Q_\lambda(t)\, dt. \end{aligned}$$

Using (4.2), we have

$$\begin{aligned} A &\le (t_2 - t_1) - \frac{11}{63} \sum_{\lambda \in F} \frac{1 - |\lambda|}{1 + |\lambda|} \frac{7\pi}{6} \\ &\le (t_2 - t_1) - \frac{11\pi}{108} \sum_{\lambda \in F} (1 - |\lambda|) = (t_2 - t_1) - \frac{11\pi}{432} \sum_{\lambda \in F} m(I_\lambda). \end{aligned}$$

By (iii), it follows that $A \le c_1(t_2 - t_1)$ where $c_1 = 1 - 11/8640$. □

Corollary 4.5.3 *Let $c_2 \in (c_1, 1)$ where $c_1 = 1 - \frac{11}{8640}$. Let $f : (-\pi, \pi] \to [0, \infty)$ be an integrable function and let Λ be an Apostol set. Then, for any n sufficiently large there are a finite set $F \subset \Lambda$ and a sequence of positive numbers $(\alpha_\lambda)_{\lambda \in F}$ such that*

(i) The sets I_λ $(\lambda \in F)$ are pairwise disjoint.
(ii) $|\lambda| \ge \frac{3}{4}$ and $|\lambda^n - 1| < \frac{1}{9}$ for all $\lambda \in F$.
(iii) $\sum_{\lambda \in F} \alpha_\lambda \le \frac{1}{2\pi} \int_{-\pi}^{\pi} f(t)\, dt$.
(iv) $\int_{-\pi}^{\pi} \left|\sum_{\lambda \in F} \alpha_\lambda \lambda^n P_\lambda(t) - f(t)\right| dt \le c_2 \int_{-\pi}^{\pi} f(t)\, dt$.

Proof Take ε such that $0 < \varepsilon < \min\left\{\frac{c_2 - c_1}{2}, \frac{7}{2\pi} - 1\right\}$. Let $g : (-\pi, \pi] \to [0, \infty)$ be a step function such that

$$\int_{-\pi}^{\pi} |(f - g)(t)|\, dt \le \varepsilon \int_{-\pi}^{\pi} f(t)\, dt.$$

By Theorem 4.5.2 applied to each interval where g is constant, we can find a finite set $F \subset \Lambda$ and positive numbers α_λ, $\lambda \in F$, satisfying (i), (ii) and

$$\begin{aligned} \sum_{\lambda \in F} \alpha_\lambda &\le \frac{1}{7} \int_{-\pi}^{\pi} g(t)\, dt \\ &\le \frac{1}{7} \left(\int_{-\pi}^{\pi} f(t)\, dt + \int_{-\pi}^{\pi} |(f - g)(t)|\, dt \right) \\ &\le \frac{1}{7}(1 + \varepsilon) \int_{-\pi}^{\pi} f(t)\, dt \le \frac{1}{2\pi} \int_{-\pi}^{\pi} f(t)\, dt, \end{aligned}$$

since $\varepsilon < \frac{7}{2\pi} - 1$. In order to check the last assertion, note that, by Theorem 4.5.2,

$$\int_{-\pi}^{\pi} \left| \sum_{\lambda \in F} \alpha_\lambda \lambda^n P_\lambda(t) - g(t) \right| dt \leq c_1 \int_{-\pi}^{\pi} g(t)\, dt.$$

Then we have

$$\begin{aligned} \int_{-\pi}^{\pi} \left| \sum_{\lambda \in F} \alpha_\lambda \lambda^n P_\lambda(t) - f(t) \right| dt &\leq \int_{-\pi}^{\pi} \left| \sum_{\lambda \in F} \alpha_\lambda \lambda^n P_\lambda(t) - g(t) \right| dt \\ &\quad + \int_{-\pi}^{\pi} |f(t) - g(t)|\, dt \\ &\leq c_1 \int_{-\pi}^{\pi} g(t)\, dt + \varepsilon \int_{-\pi}^{\pi} f(t)\, dt \\ &\leq (c_1 + 2\varepsilon) \int_{-\pi}^{\pi} f(t)\, dt \leq c_2 \int_{-\pi}^{\pi} f(t)\, dt, \end{aligned}$$

since $\varepsilon < \frac{c_2 - c_1}{2}$. □

4.5.2 Approximate eigenvalues

Combining the results of Section 4.4 and the result of approximation of integrable non-negative functions, we get the following result.

Proposition 4.5.4 *Let $T \in \mathcal{L}(\mathcal{X})$ be a polynomially bounded operator with polynomial bound M. Let $c_2 \in (c_1, 1)$ where $c_1 = 1 - \frac{11}{8640}$. Let $f : (-\pi, \pi] \to [0, \infty)$ be an integrable function and let Λ be an Apostol set. Then, for any n sufficiently large there are a finite set $F \subset \Lambda$ and a sequence of positive numbers $(\alpha_\lambda)_{\lambda \in F}$ such that*

(i) The sets I_λ $(\lambda \in F)$ are pairwise disjoint.
(ii) $|\lambda| \geq \frac{3}{4}$ and $|\lambda^n - 1| < \frac{1}{9}$ for all $\lambda \in F$.
(iii) $\sum_{\lambda \in F} \alpha_\lambda \leq \frac{1}{2\pi} \int_{-\pi}^{\pi} f(t)\, dt$.
(iv) $\int_{-\pi}^{\pi} \left| \sum_{\lambda \in F} \alpha_\lambda \lambda^n P_\lambda(t) - f(t) \right| dt \leq c_2 \int_{-\pi}^{\pi} f(t)\, dt$.

Moreover, if $\Lambda = \Lambda_{\varepsilon,2}$, with $\varepsilon \in \left(0, \frac{1}{2Mb\pi}\right)$ and where b is the constant in Corollary 4.4.7, then there exists a finite sequence of vectors $(u_\lambda)_{\lambda \in F}$ in $\mathcal{X}$ which are $2Mb$-unconditional, $\|u_\lambda\| = 1$ and $\|(T - \lambda I)u_\lambda\| < \varepsilon(1 - |\lambda|)^2$.

Proof Properties (i) to (iv) were proved in Corollary 4.5.3. In order to check the last assertion, recall that $F \subset \Lambda_{\varepsilon,2}$ means that, for all $\lambda \in F$, there exists a vector $u_\lambda \in \mathcal{X}$ such that $\|u_\lambda\| \leq 1$ and $\|(T - \lambda I)u_\lambda\| < \varepsilon(1 - |\lambda|)^2$. Let

$\beta_\lambda, \gamma_\lambda \in \mathbb{C}$ such that $|\beta_\lambda| \le |\gamma_\lambda|$ and assume that $\| \sum_{\lambda \in F} \gamma_\lambda u_\lambda \| = 1$. By Theorem 4.4.8, we have $|\gamma_\lambda| \le 2Mb$.

In order to estimate $\| \sum_{\lambda \in F} \beta_\lambda u_\lambda \|$ let us introduce an analytic function as follows. By Corollary 4.4.7, there exists a function $h \in A(\mathbb{D})$ such that $\|h\|_\infty \le b$, $h(\lambda) = \beta_\lambda / \gamma_\lambda$ if $\gamma_\lambda \ne 0$ (and $\lambda \in F$) and $h(\lambda) = 0$ if $\gamma_\lambda = 0$ (and $\lambda \in F$). Then $h(T) \le Mb$. Moreover, $h(z) - h(\lambda) = (z - \lambda) h_\lambda(z)$, where $h_\lambda \in A(\mathbb{D})$ with $\|h_\lambda\|_\infty \le \frac{2b}{1-|\lambda|}$. Then we have

$$\begin{aligned}
\left\| \sum_{\lambda \in F} \beta_\lambda u_\lambda \right\| &= \left\| \sum_{\lambda \in F} h(\lambda) \gamma_\lambda u_\lambda \right\| \\
&\le \left\| \sum_{\lambda \in F} h(T) \gamma_\lambda u_\lambda \right\| + \left\| \sum_{\lambda \in F} (h(T) - h(\lambda)) \gamma_\lambda u_\lambda \right\| \\
&\le Mb \left\| \sum_{\lambda \in F} \gamma_\lambda u_\lambda \right\| + \sum_{\lambda \in F} \frac{2Mb}{1 - |\lambda|} \varepsilon (1 - |\lambda|)^2 |\gamma_\lambda| \\
&\le Mb + \frac{2Mb}{\pi} \sum_{\lambda \in F} (1 - |\lambda|) \\
&\le 2Mb,
\end{aligned}$$

using Remark 4.4.2. □

4.6 Invariant subspace results

We are now ready to combine the ideas of the previous sections to derive the main results on invariant subspaces due to Ambrozie and Müller. The main theme is a return to the S. Brown technique of factorization of Poisson kernels, as in Section 3.1.

Denote by $\mathcal{P}$ the normed space of all polynomials equipped with the sup-norm on $\mathbb{D}$. Let $\mathcal{P}^*$ be its dual with the usual dual norm. Let $\varphi \in \mathcal{P}^*$. By the Hahn–Banach theorem, φ can be extended without changing the norm to a functional on the space of all continuous functions on $\mathbb{T}$ with the sup norm. By the Riesz theorem, there exists a Borel measure μ on $\mathbb{T}$ whose total variation $\|\mu\|$ is equal to $\|\varphi\|$, and satisfying $\varphi(p) = \int p \, d\mu$ for all polynomials p. Of particular interest are the following functionals on $\mathcal{P}$:

(i) Let $\lambda \in \mathbb{D}$. Denote by $\mathcal{E}_\lambda$ the evaluation functional defined by $\mathcal{E}_\lambda(p) = p(\lambda)$ $(p \in \mathcal{P})$. Clearly $\|\mathcal{E}_\lambda\| = 1$.

(ii) For $f \in L^1(\mathbb{T})$, denote by $M_f \in \mathcal{P}^*$ the functional defined by

$$M_f(p) = \frac{1}{2\pi} \int_{-\pi}^{\pi} p(e^{it}) f(e^{it})\, dt, \quad p \in \mathcal{P}.$$

Then $\|M_f\| \leq \|f\|_1$. In particular, if $g = 1$, then $M_g(p) = p(0)$ for all p and M_g is the evaluation at the origin. More generally, if $\lambda \in \mathbb{D}$ and $g(e^{it}) = P_\lambda(t)$, then M_g is the evaluation at the point λ, which we shall denote by M_{P_λ}.

(iii) Let $T : \mathcal{X} \to \mathcal{X}$ be a polynomially bounded operator with polynomial bound M. For $x \in \mathcal{X}$ and $x^* \in \mathcal{X}^*$, let $x \otimes x^*$ be the functional defined by

$$(x \otimes x^*)(p) = \langle p(T)x, x^* \rangle, \quad p \in \mathcal{P}.$$

Since T is polynomially bounded, the norm of $x \otimes x^*$ is bounded by $M\|x\|\,\|x^*\|$. Of course the definition of $x \otimes x^*$ depends on T, but since we are going to consider a fixed operator T, this cannot lead to confusion.

We have the following Banach space generalization of Lemma 3.1.1.

Lemma 4.6.1 *Let T be a polynomially bounded operator on $\mathcal{X}$. Suppose that there exist $\lambda \in \mathbb{D}$, $x \in \mathcal{X}$ and $x^* \in \mathcal{X}^*$ such that $x \otimes x^* = M_{P_\lambda}$. Then either $T = \lambda I$ or T has a non-trivial hyperinvariant subspace.*

Proof First note that the equality $x \otimes x^* = M_{P_\lambda}$ implies that neither x nor x^* is zero. Since $\langle p(T)x, x^* \rangle = p(\lambda)$ for all polynomials p, in particular, $\langle (T - \lambda I)q(T)x, x^* \rangle = 0$ for all polynomials q. Let $\mathcal{M}$ be the closure of $\{(T - \lambda I)q(T)x : q \in \mathcal{P}\}$. Obviously $\mathcal{M} \neq \mathcal{X}$ since $x^* \neq 0$ and $\mathcal{M}$ contains $Tx - \lambda x$. If $Tx - \lambda x \neq 0$, then $\mathcal{M}$ is a non-trivial hyperinvariant subspace for T; otherwise, since $x \neq 0$, the point spectrum of T contains λ, and then $\mathbb{C}x$ is a non-trivial hyperinvariant subspace for T. □

Lemma 4.6.2 *Let T be a polynomially bounded operator on $\mathcal{X}$. Let $x \in \mathcal{X}$, $x^* \in \mathcal{X}^*$ such that $\|x\| = 1$, $\|x^*\| \leq 1$, $\langle x, x^* \rangle = 1$ and $\lambda \in \mathbb{D}$. Then*

$$\|x \otimes x^* - M_{P_\lambda}\| \leq 2M \frac{\|Tx - \lambda x\|}{1 - |\lambda|},$$

where M is the polynomial bound of T.

Proof For $p \in \mathcal{P}$ we have $p(z) - p(\lambda) = (z - \lambda)q(z)$, where $q \in \mathcal{P}$ satisfies $\|q\| \leq \frac{2\|p\|}{1-|\lambda|}$. It follows that

$$\begin{aligned}
|(x \otimes x^* - M_{P_\lambda})(p)| &= |\langle p(T)x, x^*\rangle - p(\lambda)| \\
&= |\langle p(T)x, x^*\rangle - p(\lambda)\langle x, x^*\rangle| \\
&= |\langle q(T)(T - \lambda I)x, x^*\rangle| \\
&\leq \|q(T)\| \|Tx - \lambda x\| \|x^*\| \\
&\leq M\|q\| \|Tx - \lambda x\| \\
&\leq 2M\|p\| \frac{\|Tx - \lambda x\|}{1 - |\lambda|},
\end{aligned}$$

and the result follows. □

Therefore, for $\lambda \in \Lambda_{\varepsilon,2}$, there exist x and x^* of norm 1 such that

$$\|x \otimes x^* - M_{P_\lambda}\| < 2M(1 - |\lambda|)\varepsilon.$$

Let b be the universal constant defined in Corollary 4.4.7. Let $c_3 \in (c_2, 1)$ where c_2 is the constant in Corollary 4.5.3, i.e., $c_2 \in (c_1, 1)$ where $c_1 = 1 - \frac{11}{8640}$.

Theorem 4.6.3 *Let $T \in \mathcal{L}(\mathcal{X})$ be a polynomially bounded operator whose polynomial bound is M. Suppose that $\mathbb{T} \subset \sigma(T)$ and that T has no non-trivial invariant subspace. Let $y^* \in \mathcal{X}^*$ and let $f \in L^1(\mathbb{T})$ be a non-negative function in $L^1(\mathbb{T})$ of norm 1. Then, for every positive integer n sufficiently large, there exist $x \in \mathcal{X}$ and $x^* \in \mathcal{X}^*$ such that $\|x\| \leq 2Mb\sqrt{2}$, $\|x^*\| \leq 1$ and $\|x \otimes (T^{*n}x^* + y^*) - M_f\| < c_3$.*

Proof Let $\varepsilon > 0$ be a positive real such that $\varepsilon < \frac{1}{2Mb\pi}$, $\varepsilon\|y^*\|^2 < 1$ and $8\sqrt{2}M^3b^2\pi\sqrt{\varepsilon} < c_3 - c_2$. Using Apostol's Theorem 4.1.3 and Proposition 4.5.4, for any n sufficiently large, there exist a finite set $F \in \mathbb{D}$ and positive numbers α_λ $(\lambda \in F)$ such that the intervals $(I_\lambda)_{\lambda \in F}$ are pairwise disjoint and

$$\frac{1}{2\pi} \int_{-\pi}^{\pi} \left| \sum_{\lambda \in F} \alpha_\lambda \lambda^n P_\lambda(t) - f(t) \right| dt \leq c_2.$$

There also exists a finite sequence of vectors $(u_\lambda)_{\lambda \in F}$ in $\mathcal{X}$ which are $2Mb$-unconditional, with $\|u_\lambda\| = 1$ and $\|(T - \lambda \mathrm{I})u_\lambda\| < \varepsilon(1 - |\lambda|)^2$.

We define the linear functional φ, on the linear span of $(u_\lambda)_{\lambda \in F}$, by $\varphi(u_\lambda) = \lambda^{-n}y^*(u_\lambda)$. By the Hahn–Banach theorem, we can extend it to a bounded functional on $\mathcal{X}$, which we denote by the same symbol φ. By Theorem 4.3.8, there exist complex numbers μ_λ and a functional $\psi \in \mathcal{X}^*$ such that

$$\left\| \sum_{\lambda \in F} \mu_\lambda u_\lambda \right\| \leq 2Mb\sqrt{2}, \qquad \|\psi - \varphi\| \leq 1 \qquad \text{and} \quad \psi(\mu_\lambda u_\lambda) = \alpha_\lambda,$$

for all $\lambda \in F$. By Theorem 4.4.8, we have $|\mu_\lambda| \leq 4\sqrt{2}M^2b^2$. Now we take

$$x = \sum_{\lambda \in F} \mu_\lambda u_\lambda \qquad \text{and} \quad x^* = \psi - \varphi.$$

Let g be the $L^1(\mathbb{T})$-function defined by $g(e^{it}) = \sum_{\lambda \in F} \alpha_\lambda \lambda^n P_\lambda(t)$. Thus we have

$$\|M_g - M_f\| \leq \|g - f\|_1 \leq c_2,$$

and for any polynomial $p \in \mathcal{P}$ we have

$$\begin{aligned} M_g p &= \frac{1}{2\pi} \int_{-\pi}^{\pi} g(e^{it}) p(e^{it})\, dt = \frac{1}{2\pi} \sum_{\lambda \in F} \alpha_\lambda \lambda^n \int_{-\pi}^{\pi} P_\lambda(t) p(e^{it})\, dt \\ &= \sum_{\lambda \in F} \alpha_\lambda \lambda^n p(\lambda). \end{aligned}$$

It follows that

$$\begin{aligned} \|x \otimes (T^{*n} x^* + y^*) - M_f\| &= \|T^n x \otimes x^* + x \otimes y^* - M_f\| \\ &\leq \|T^n x \otimes x^* + x \otimes y^* - M_g\| + \|M_g - M_f\| \\ &\leq \sup_{\|p\| \leq 1} |\langle p(T) T^n x, x^* \rangle + \langle p(T) x, y^* \rangle - M_g(p)| \\ &\quad + c_2. \end{aligned}$$

The definition of x implies that

$$\begin{aligned} \langle p(T) T^n x, x^* \rangle &= \sum_{\lambda \in F} \langle T^n p(T) u_\lambda, x^* \rangle \\ &= \sum_{\lambda \in F} \mu_\lambda \langle (T^n p(T) - \lambda^n p(\lambda)) u_\lambda, x^* \rangle \\ &\quad + \sum_{\lambda \in F} \mu_\lambda \langle \lambda^n p(\lambda) u_\lambda, x^* \rangle \end{aligned}$$

and

$$\begin{aligned} \langle p(T) x, y^* \rangle &= \sum_{\lambda \in F} \mu_\lambda \langle p(T) u_\lambda, y^* \rangle \\ &= \sum_{\lambda \in F} \mu_\lambda \langle (p(T) - p(\lambda) u_\lambda, y^* \rangle + \sum_{\lambda \in F} \mu_\lambda \langle p(\lambda) u_\lambda, y^* \rangle. \end{aligned}$$

Using the identities $\langle u_\lambda, y^* \rangle = \lambda^n \langle u_\lambda, \varphi \rangle$, $x^* + \varphi = \psi$ and $\langle \mu_\lambda u_\lambda, \psi \rangle = \alpha_\lambda$, we get

$$\begin{aligned}\sum_{\lambda \in F} \mu_\lambda \langle \lambda^n p(\lambda) u_\lambda, x^* \rangle + \sum_{\lambda \in F} \mu_\lambda \langle p(\lambda) u_\lambda, y^* \rangle &= \sum_{\lambda \in F} \mu_\lambda \langle \lambda^n p(\lambda) u_\lambda, x^* + \varphi \rangle \\ &= \sum_{\lambda \in F} \mu_\lambda \langle \lambda^n p(\lambda) u_\lambda, \psi \rangle \\ &= \sum_{\lambda \in F} \alpha_\lambda \lambda^n p(\lambda) = M_g(p).\end{aligned}$$

Therefore we have

$$\begin{aligned}&|\langle p(T)T^n x, x^* \rangle + \langle p(T)x, y^* \rangle - M_g(p)| \\ &\quad = \left| \sum_{\lambda \in F} \mu_\lambda \left(\langle (T^n p(T) - \lambda^n p(\lambda)) u_\lambda, x^* \rangle + \sum_{\lambda \in F} \langle (p(T) - p(\lambda)) u_\lambda, y^* \rangle \right) \right|.\end{aligned}$$

We estimate the right-hand side of the above equality in a standard way. Write $q(z) = \frac{z^n p(z) - \lambda^n p(\lambda)}{z - \lambda}$. Clearly $\|q\| \le 2\|p\|/(1 - |\lambda|)$, and then $\|q(T)\| \le 2M/(1 - |\lambda|)$. It follows that

$$\begin{aligned}\|(T^n p(T) - \lambda^n p(\lambda)) u_\lambda\| &= \|q(T)(T - \lambda) u_\lambda\| \\ &\le 2M(1 - |\lambda|)^{-1} \varepsilon (1 - |\lambda|)^2 \le 2M\varepsilon(1 - |\lambda|).\end{aligned}$$

Similarly, one obtains the estimate $\|(p(T) - p(\lambda)) u_\lambda\| \le 2M\varepsilon(1 - |\lambda|)$. Since $\|x^*\| \le 1$ and $\varepsilon \|y^*\|^2 < 1$, we obtain

$$\begin{aligned}&|\langle p(T)T^n x, x^* \rangle + \langle p(T)x, y^* \rangle - M_g(p)| \\ &\quad \le \sum_{\lambda \in F} |\mu_\lambda| (2M\varepsilon(1 - |\lambda|) + 2M\sqrt{\varepsilon}(1 - |\lambda|) \le 4\sqrt{2}M^2 b^2 4M\sqrt{\varepsilon}\frac{\pi}{2},\end{aligned}$$

since $|\mu_\lambda| \le 4\sqrt{2}M^2 b^2$ and using Remark 4.4.2. Therefore we get

$$\|x \otimes (T^{*n} x^* + y^*) - M_f\| \le 8\sqrt{2}M^3 b^2 \pi \sqrt{\varepsilon} + c_2 < c_3.$$

□

Corollary 4.6.4 *Let $T \in \mathcal{L}(\mathcal{X})$ be a polynomially bounded operator whose polynomial bound is M, such that $\mathbb{T} \subset \sigma(T)$. Suppose that T has no non-trivial invariant subspace and assume that $\lim_{n \to \infty} \|T^n x\| = 0$ for all $x \in \mathcal{X}$. Let $w \in \mathcal{X}$, $z^* \in \mathcal{X}^*$ and $\delta > 0$, and let f be a non-negative function in $L^1(\mathbb{T})$. Then there exist $u \in \mathcal{X}$ and $u^* \in \mathcal{X}^*$ such that*

(i) $\|u \otimes (u^* + z^*) - M_f\| \leq c_3 \|f\|_1$;
(ii) $\|w \otimes u^*\| < \delta$;
(iii) $\|u\| \leq 2Mb\sqrt{2}\|f\|_1^{1/2}$ *and* $\|u^*\| \leq M\|f\|_1^{1/2}$.

Proof We may assume that $\|f\|_1 \neq 0$, since otherwise the statement is trivial. Choose n large enough that $\|T^n w\| < \delta \|f\|_1^{-1/2} M^{-1}$. By Theorem 4.6.3 applied to the function $f/\|f\|_1$ and the functional $z^*/\|f\|_1^{1/2}$, there exist $v \in \mathcal{X}$ and $v^* \in \mathcal{X}^*$ with $\|v\| \leq 2Mb\sqrt{2}$, $\|v^*\| \leq 1$ and

$$\|v \otimes \left(T^{*n} v^* + z^*/\|f\|_1^{1/2}\right) - M_{f/\|f\|_1}\| \leq c_3.$$

Set $u = \|f\|_1^{1/2} v$ and $u^* \doteq \|f\|_1^{1/2} T^{*n} v^*$. Then (iii) is satisfied. Furthermore, we have

$$\begin{aligned}\|w \otimes u^*\| &= \|f\|_1^{1/2} \|w \otimes T^{*n} v^*\| = \|f\|_1^{1/2} \|T^n w \otimes v^*\| \\ &\leq \|f\|_1^{1/2} M \|T^n w\| < \delta,\end{aligned}$$

proving (ii). Finally, note that

$$\begin{aligned}&\|u \otimes (u^* + z^*) - M_f\| \\ &= \|f\|_1 \|v \otimes \left(T^{*n} v^* + z^*/\|f\|_1^{1/2}\right) - M_{f/\|f\|_1}\| \leq c_3 \|f\|_1,\end{aligned}$$

proving (i). □

The next lemma is the key to getting an approximation for an arbitrary function in $L^1(\mathbb{T})$, not necessarily non-negative.

Lemma 4.6.5 *Let f be a function in $L^1(\mathbb{T})$ and let n be a positive integer. Then there exist n non-negative functions in $L^1(\mathbb{T})$ such that*

$$\|f\|_1 = \sum_{j=0}^{n-1} \|g_j\|_1 \quad \textit{and} \quad \left\| f - \sum_{j=0}^{n-1} \nu^j g_j \right\|_1 \leq \frac{\pi}{n} \|f\|_1,$$

where $\mu = e^{2i\pi/n}$.

Proof We consider a partition of $\mathbb{C} \setminus \{0\}$ into sectors S_j $(j = 0, \ldots, n-1)$ defined by $S_j = \left\{re^{it} : r > 0, -\frac{\pi}{n} \leq t - \frac{2\pi j}{n} < \frac{\pi}{n}\right\}$. Setting $\Omega_j := f^{-1}(S_j)$, it follows that $\Omega_1, \ldots, \Omega_n, f^{-1}(0)$ is a partition of the unit circle. For all $j = 1, \ldots, n$, set $g_j = |f|\chi_{\Omega_j}$. Then, for $e^{it} \in \Omega_j$, we have

$$\begin{aligned}|\nu^j g_j(e^{it}) - f(e^{it})| &= |f(e^{it})|\, |\nu^j - e^{i \arg(f(e^{it}))}| \\ &\leq |f(e^{it})|\, |2\pi j/n - \arg f(e^{it})| \\ &\leq |f(e^{it})| \frac{\pi}{n}.\end{aligned}$$

Therefore, we get

$$\int_{-\pi}^{\pi} \left| f(e^{it}) - \sum_{j=0}^{n-1} \nu^j g_j(e^{it}) \right| dt = \sum_{j=0}^{n-1} \int_{S_j} |f(e^{it}) - \nu^j g_j(e^{it})| \, dt$$
$$\leq \frac{\pi}{n} \int_{-\pi}^{\pi} |f(e^{it})| \, dt.$$

Obviously we have also $\|f\|_1 = \sum_{j=0}^{n-1} \|g_j\|_1$. □

Theorem 4.6.6 *Let $T \in \mathcal{L}(\mathcal{X})$ be a polynomially bounded operator whose polynomial bound is M, satisfying $\sigma(T) \supset \mathbb{T}$. Suppose that T has no non-trivial invariant subspace and assume that $\lim_{n\to\infty} \|T^n x\| = 0$ for all $x \in \mathcal{X}$. There exist $c \in (0, 1)$, $C_1, C_2 > 0$ such that, given $w \in \mathcal{X}$, $z^* \in \mathcal{X}^*$, $\delta > 0$ and f a function in $L^1(\mathbb{T})$, then there exist $u \in \mathcal{X}$ and $u^* \in \mathcal{X}^*$ such that*

(i) $\|u \otimes (u^* + z^*) - M_f\| \leq c\|f\|_1$;
(ii) $\|w \otimes u^*\| < \delta$;
(iii) $\|u^*\| \leq C_1\|f\|_1^{1/2}$ *and* $\|u\| \leq C_2\|f\|_1^{1/2}$.

Proof Let c_3 be the constant in Corollary 4.6.4. Since $c_3 < 1$, there exists a positive integer m_0 such that $c := c_3 + \frac{\pi}{m_0} + \frac{1}{m_0} < 1$.

By Lemma 4.6.5, there exist m_0 non-negative functions in $L^1(\mathbb{T})$ such that

$$\|f\|_1 = \sum_{j=0}^{m_0-1} \|g_j\|_1 \quad \text{and} \quad \left\| f - \sum_{j=0}^{m_0-1} \nu^j g_j \right\|_1 \leq \frac{\pi}{m_0}\|f\|_1,$$

where $\mu = e^{2i\pi/m_0}$. By Theorem 4.6.3, for all $j = 0, \ldots, m_0 - 1$ and for all n sufficiently large, there exist $x_j \in \mathcal{X}$ and $x_j^* \in \mathcal{X}^*$ such that

$$\|x_j\| \leq 2Mb\sqrt{2}\|g_j\|_1^{1/2}, \quad \|x_j^*\| \leq \|g_j\|_1^{1/2}$$

and

$$\|x_j \otimes \left(T^{*n} x_j^* + z^*\right) - M_{g_j}\| < c_3.$$

Let $u = \sum_{j=0}^{m_0-1} \nu^j x_j$ and $u^* = \sum_{j=0}^{m_0-1} T^{*n} x_j^*$. Then we have

$$\|u\| \leq 2Mb\sqrt{2} \sum_{j=0}^{m_0-1} \|g_j\|_1^{1/2} \leq 2Mb\sqrt{2}\sqrt{m_0}\|f\|_1^{1/2}$$

and

$$\|u^*\| \leq M \sum_{j=0}^{m_0-1} \|g_j\|_1^{1/2} \leq M\sqrt{m_0}\|f\|_1^{1/2}.$$

Note that

$$\begin{aligned}\|u \otimes (u^* + z^*) - M_f\|_1 &\le \sum_{j=0}^{m_0-1} \|\nu^j x_j \otimes \left(T^{*n} x_j^* + z^*\right) - \nu^j M_{g_j}\|_1 \\ &\quad + \sum_{k=0}^{m_0-1} \left\| \left(\sum_{j=0, j\neq k}^{m_0-1} \nu^j x_j \right) \otimes T^{*n} x_k^* \right\|_1 + \frac{\pi}{m_0} \|f\|_1 \\ &\le c_3 \sum_{j=0}^{m_0-1} \|g_j\|_1 + \sum_{k=0}^{m_0-1} \sum_{j=0, j\neq k}^{m_0-1} \|T^n x_j\| \, \|x_k^*\| \\ &\quad + \frac{\pi}{m_0} \|f\|_1.\end{aligned}$$

Choose n large enough so that $\|T^n x_j\| \le \frac{1}{m_0^3} \|g_j\|_1^{1/2}$. It follows that

$$\|u \otimes (u^* + z^*) - M_f\|_1 \le \left(c_3 + m_0^2/m_0^3 + \pi/m_0\right) \|f\|_1 \le c\|f\|_1.$$

Finally we have also

$$\begin{aligned}\|w \otimes u^*\|_1 &= \left\| \sum_{j=0}^{m_0-1} T^n w \otimes x_j^* \right\| \\ &\le \|T^n w\| \sum_{j=0}^{m_0-1} \left\|x_j^*\right\| \\ &\le \|T^n w\| \sum_{j=0}^{m_0-1} \|g_j\|_1^{1/2} \\ &\le \|T^n w\| \sqrt{m_0} \|f\|_1^{1/2} < \delta,\end{aligned}$$

provided n is large enough, since $\lim_{n\to\infty} \|T^n x\| = 0$ for all $x \in \mathcal{X}$. This ends the proof of the theorem, taking $C_1 = M\sqrt{m_0}$ and $C_2 = 2Mb\sqrt{2}\sqrt{m_0}$. □

Theorem 4.6.7 *Let T be a polynomially bounded operator on a complex space $\mathcal{X}$. Assume that the spectrum of T contains $\mathbb{T}$ and $\lim_{n\to\infty} \|T^n x\| = 0$ for all $x \in \mathcal{X}$. Then T has non-trivial invariant subspaces.*

Proof Let B be the bilinear map defined on $\mathcal{X}^* \times \mathcal{X}$ by $B(x^*, x) = x \otimes x^*$. By Theorem 2.4.3, the map B takes values in the normed space $\mathcal{Z} = \{M_f : f \in L^1(\mathbb{T})\}$. If T has no non-trivial invariant subspace, then, by Theorem 4.6.6, B has the N-RVAP defined in Section 3.5.2, taking $N = 1$, $c \in (c_3, 1)$, $C_1 = M$ and $C_2 = 2Mb\sqrt{2}$, where M is the polynomial bound of T and b is the constant in Corollary 4.4.7. Therefore, by Theorem 3.5.6, for all $\lambda \in \mathbb{D}$, there exists $x_\lambda \in \mathcal{X}$ and $x_\lambda^* \in \mathcal{X}^*$ such that $M_{P_\lambda} = x_\lambda \otimes x_\lambda^*$. By Lemma 4.6.1, T has non-trivial invariant subspaces. □

In the case where T does not satisfy any 'vanishing condition', we have nevertheless the following result.

Theorem 4.6.8 *Let T be a polynomially bounded operator on a complex Banach space $\mathcal{X}$. Assume that the spectrum of T contains $\mathbb{T}$. Then T^* has non-trivial invariant subspaces. In particular, if $\mathcal{X}$ is reflexive, then T has a non-trivial invariant subspace.*

Proof By Lemma 4.2.7, setting $\mathcal{X}_1 := \{x \in \mathcal{X} : \|T^n x\| \to 0\}$ and $\mathcal{Y}_1 := \{x^* \in \mathcal{X}^* : \|T^{*n} x^*\| \to 0\}$, we may assume that $\mathcal{X}_1$ and $\mathcal{Y}_1$ are trivial. If $\mathcal{X}_1 = \mathcal{X}$, by Theorem 4.6.7, T has a non-trivial invariant subspace, and so has T^*. If $\mathcal{Y}_1 = \mathcal{X}^*$, again by Theorem 4.6.7, T^* has a non-trivial invariant subspace. In the remaining case, where $\mathcal{X}_1$ and $\mathcal{Y}_1$ are reduced to 0, Theorem 4.2.9 provides the desired result. □

Exercises

1. Show that if $T \in \mathcal{L}(\mathcal{X})$ is power-bounded, then T satisfies the Kreiss condition. (Hint: develop $(\lambda I - T)^{-1}$ as a power series.)
2. Let f be a continuous function on $\mathbb{T}$. Prove that for all $\varepsilon > 0$ there exist $c_1, \ldots, c_n > 0$ and $\lambda_1, \ldots, \lambda_n \in \mathbb{D}$ such that
$$\left\| f - \sum_{k=1}^{n} c_\lambda P_\lambda \right\|_\infty < \varepsilon.$$
(Hint: take $r \in (0, 1)$ such that $\|f - f_r\|_\infty < \frac{\varepsilon}{2}$, where $f_r(e^{i\theta}) := \frac{1}{2\pi}\int_0^{2\pi} f(e^{it}) P_{re^{i\theta}}(t)\, dt$.)
3. Let f be a non-zero function in $L^1(\mathbb{T})$. Using the previous exercise, check that the following assertions are equivalent:
   (a) The function f is lower semi-continuous and positive on $\mathbb{T}$.
   (b) For every dominating sequence $\lambda = (\lambda_n)_n$ in $\mathbb{D}$ (see Definition 3.1.9), there exists a summable sequence $(c_n)_n$ of positive reals such that $f = \sum_n c_n P_{\lambda_n}$.
4. Using Lemma 4.6.5, give the analogue of Corollary 4.5.3 for a complex and integrable function f on $(-\pi, \pi]$.
5. Verify that Zenger's theorem (Theorem 4.3.1) in the context of a complex Hilbert space asserts that if $x_1, \ldots, x_n$ are linearly independent and

of norm 1, then, for all $c_1, \dots, c_n > 0$ such that $\sum_{i=1}^n c_i = 1$, there exist $\mu_1, \dots, \mu_n \in \mathbb{C}$ such that

$$\left\| \sum_{i=1}^n \mu_i x_i \right\| = 1 \quad \text{and} \quad \left\langle \mu_i x_i, \sum_{k=1}^n \mu_k x_k \right\rangle = c_i.$$

6. Let $\mathcal{H}$ be a complex Hilbert space and $T \in \mathcal{L}(\mathcal{H})$. Suppose that $\lambda_1, \dots, \lambda_n$ are distinct eigenvalues of T and let $x_1, \dots, x_n$ be some normalized eigenvectors associated with $\lambda_1, \dots, \lambda_n$, respectively. Let $c_1, \dots, c_n > 0$ be such that $\sum_{i=1}^n c_i = 1$. Using Zenger's theorem in the context of a complex Hilbert space, prove that there exists $x \in \mathcal{H}$ of norm 1 such that

$$x \otimes x = \sum_{i=1}^n c_i x_i \otimes x_i \quad \text{and} \quad x \overset{T}{\cdot} x = \sum_{i=1}^n c_i P_{\lambda_i}.$$

This result is *Westwood's theorem.*

Comments

The Kreiss condition [138] originates in in the context of differential equations. This condition is weaker than power boundedness [188].

Theorem 4.1.3 was proved by Apostol [12] for $k = 1$ and T a contraction on a Hilbert space. It was observed by S. Brown [51] that the same proof also works for general exponents $k \geq 1$. The existing proofs of Theorem 4.1.3 can be found in Apostol [12], Chevreau and Pearcy [79], Brown [51] and Bercovici [33] in the context of contractions on Hilbert spaces. Liu [144] gives a version for contractions on a general Banach space. The proof remains essentially unchanged for operators satisfying the Kreiss condition.

The classical version of Zenger's theorem [206] can also be found in Bonsall and Duncan [46, pp. 18-20]. The idea of using Zenger's theorem in the S. Brown technique appears in Eschmeier [100] and is implicitly already present in Apostol [13]. The variants of Zenger's theorem presented in Section 4.3.2 come from Ambrozie and Müller [8].

Section 4.2 comes from Colojoară and Foiaş [82]. It contains the tools for the proof of Theorem 4.2.9, proved [82, p. 136] in the case where $\mathcal{X}$ is a reflexive Banach space, and proved in the general situation of a Banach space [8, theorem 7.5].

Sections 4.4, 4.5 and 4.6 are based on the deep work of Ambrozie and Müller [8].

Theorem 4.6.8 is a non-trivial generalization of the following previous results: Brown, Chevreau and Pearcy [53] proved the existence of invariant subspaces for contractions on a Hilbert space with dominating spectrum, and finally obtained the same result under the relaxed hypothesis that the spectrum contains the unit circle [54].

Independently, Rejasse [173] and Ambrozie and Müller (unpublished) proved the following result: every polynomially bounded operator on a reflexive Banach space such that its spectrum contains the unit circle has a non-trivial hyperinvariant subspace or is reflexive.

Westwood's theorem (Exercise 6) is theorem 2.2 in Westwood [202].

5
Beurling algebras

As far back as 1952, Wermer [201] proved the following striking result on the existence of invariant subspaces, by means of a functional calculus construction.

Theorem 5.0.1 (Wermer) *Let $\mathcal{X}$ be a Banach space and $T \in \mathcal{L}(\mathcal{X})$ be invertible. Suppose that T satisfies one of the following two conditions:*

1. *$\sigma(T)$ contains more than one point, and*

$$\sum_{n=-\infty}^{\infty} \frac{\log \|T^n\|}{1+n^2} < \infty. \tag{5.1}$$

2. *There is an integer $k \geq 0$ such that $\|T^n\| = O(|n|^k)$ as $n \to \pm\infty$.*

Then either T is a multiple of the identity or it has a non-trivial hyperinvariant subspace.

Our approach to this result will be by means of a stronger variant subsequently introduced by Atzmon [20]. The common feature is the definition of a functional calculus, mapping an algebra A_ρ of functions defined on $\mathbb{T}$ into $\mathcal{L}(\mathcal{X})$.

Another application of such a functional calculus is to the theory of *Bishop operators*. For $\alpha \in (0, 1)$, Bishop defined the operator T_α on $L^2([0, 1])$ by

$$T_\alpha f(x) = xf(\{x+\alpha\}), \tag{5.2}$$

where the curly brackets denote fractional part; that is, $\{x\} = x - \lfloor x \rfloor$ and $\lfloor x \rfloor$ is the integer part of x.

As explained by Davie [89], in the case of irrational α, these operators were proposed as examples of Hilbert space operators without non-trivial invariant subspaces. However, the work of Davie [89], and subsequent extensions due to Blecher and Davie [42], MacDonald [147, 148] and Flattot [107], have shown that for a large class of α, including all numbers that are not Liouville numbers, these operators have non-trivial hyperinvariant subspaces. We shall present Davie's work, together with some modern refinements.

We conclude the chapter with a discussion of the case of rational α, where a complete description of the invariant subspace lattice can be given.

5.1 Properties of Beurling algebras

In order to proceed, we make the following definition.

Definition 5.1.1 Let $(\rho_n)_{n\in\mathbb{Z}}$ be a sequence in $[1,\infty)$. We define

$$A_\rho = \{f \in \mathcal{C}(\mathbb{T}) : \|f\|_\rho < \infty\},$$

where

$$\|f\|_\rho = \sum_{n\in\mathbb{Z}} |\hat{f}(n)|\rho_n,$$

and $(\hat{f}(n))_{n\in\mathbb{Z}}$ is the sequence of the Fourier coefficients of f.

Proposition 5.1.2 *The space A_ρ endowed with the norm $\|.\|_\rho$ is a Banach space. Moreover, if* $(\log(\rho_n))_{n\in\mathbb{Z}}$ *is* sub-additive, *i.e., if $\rho_{n+m} \le \rho_n\rho_m$ for all $n, m \in \mathbb{Z}$, then A_ρ is a unital Banach algebra under pointwise multiplication.*

Proof This is routine checking, and we leave it as an exercise. □

The weighted convolution algebra $\ell^1(\mathbb{Z}, (\rho_n)_n)$ is often called a *Beurling algebra*. We shall also use this terminology for the function algebra A_ρ, which is isometrically isomorphic to the weighted convolution algebra.

Remark 5.1.3 Note that, when the Banach space A_ρ is regarded as a weighted ℓ^1 space, its dual space is easy to identify, as a weighted ℓ^∞ space, namely

$$A_\rho^* = \{g = (g_n)_n : \|g\| = \sup_n |g_n|/\rho_n < \infty\}.$$

Thus, at the level of function algebras, an element $\Phi \in A_\rho^*$ can be identified with a sequence $(\widehat{\Phi}(n))_n$ satisfying

$$\sup_n |\widehat{\Phi}(n)|/\rho_n < \infty,$$

obtained by evaluation as $\widehat{\Phi}(n) = \langle g, e_n\rangle$, where e_n is the mapping $z \mapsto z^n$.

Definition 5.1.4 We say that a sequence $(\rho_n)_{n\in\mathbb{Z}}$ of real numbers is a *Beurling sequence* if $\rho_0 = 1$, $\rho_n \geq 1$ for all $n \in \mathbb{Z}$, $\rho_{m+n} \leq \rho_m \rho_n$ for all $m, n \in \mathbb{Z}$, and ρ_n satisfies the *Beurling condition*

$$\sum_{n\in\mathbb{Z}} \frac{\log \rho_n}{1+n^2} < \infty. \tag{5.3}$$

The following result is the analogue of Wiener's Theorem 1.2.12; we include a short proof. It uses the Gelfand theory sketched in Theorem 1.2.10.

Proposition 5.1.5 *Let $(\rho_n)_{n\in\mathbb{Z}}$ have the properties that $\rho_n \geq 1$ for each n and that $(\log \rho_n)_{n\in\mathbb{Z}}$ is sub-additive. If $\log \rho_n/n \to 0$ as $n \to \pm\infty$, then the maximal ideal space of A_ρ can be identified with $\mathbb{T}$. In particular, if $(\rho_n)_{n\in\mathbb{Z}}$ satisfies condition (5.3), then a function $f \in A_\rho$ is invertible if and only if $f(e^{i\theta}) \neq 0$ for all $\theta \in [0, 2\pi]$.*

Proof Let $\phi : A_\rho \to \mathbb{C}$ be a homomorphism, so that, writing z for the identity function on $\mathbb{T}$, there is a $\beta \in \mathbb{C}$ such that $\phi(z^n) = \beta^n$ for each $n \in \mathbb{Z}$. Since $\|\phi\| = 1$, we conclude that $|\beta^n| \leq \|z^n\|_\rho = \rho_n$. Now, $\log \rho_n/n \to 0$ as $n \to \pm\infty$, and this implies that $|\beta| = 1$. By the density of the trigonometric polynomials in A_ρ we conclude that $\phi(f) = f(\beta)$ for all $f \in A_\rho$. □

We now recall the definition of regularity in the context of function algebras.

Definition 5.1.6 A function algebra B on a compact space X is said to be *regular* if, for all $p \in X$ and all compact subsets K of X with $p \notin K$, there exists $f \in B$ such that $f(p) = 1$ and $f = 0$ on K.

One advantage of regularity in a function algebra on $\mathbb{T}$ is that it enables one to construct two non-zero functions whose product is identically zero; for example, we may find functions supported on disjoint neighbourhoods of 1 and -1. Combined with functional calculus arguments, this gives a strategy for obtaining invariant subspaces.

Theorem 5.1.7 *Let $(\rho_n)_{n\in\mathbb{Z}}$ be a sequence in $[1, \infty)$ such that $(\log(\rho_n))_{n\in\mathbb{Z}}$ is sub-additive. Suppose also that $(\rho_n)_{n\in\mathbb{Z}}$ satisfies the Beurling condition. Then the Banach algebra A_ρ is regular.*

Proof We shall show that for each $p = e^{it_0} \in \mathbb{T}$ and $\varepsilon \in (0, \pi)$ there is a function $f \in B$ such that $f(p) \neq 0$ and $f(q) = 0$ for all $q \in \mathbb{T}$ with $|p - q| \geq \varepsilon$. This clearly implies regularity.

It will be convenient to write $\sigma_n = (1 + n^2)\rho_n$. Clearly

$$\sum_{n\in\mathbb{Z}} \frac{\log \sigma_n}{1+n^2} < \infty. \tag{5.4}$$

First we extend σ to a function defined a.e. on $\mathbb{R}$ by setting

$$\sigma(x) = \sigma(n) \qquad \text{for} \quad x \in \left(n - \frac{1}{2}, n + \frac{1}{2}\right),$$

so that

$$\int_{-\infty}^{\infty} \frac{\log \sigma(x)}{1 + x^2}\, dx < \infty.$$

This implies that there is a function $F \in H^1(\mathbb{C}_+) \cap H^2(\mathbb{C}_+)$ such that $|F(ix)| = 1/\sigma(x)$ a.e., for example the outer function

$$F(s) = \exp\left(\frac{1}{\pi} \int_{-\infty}^{\infty} \frac{ys + i}{y + is} \log \frac{1}{\sigma(y)} \frac{dy}{1 + y^2}\right).$$

Note that F is the bilateral Laplace transform of a continuous function g in $L^2(\mathbb{R})$, supported on $(0, \infty)$: see Theorem 1.4.1.

Consider now the product $h : t \mapsto g(t - a)g(b - t)$, where $a, b \in \mathbb{R}$ satisfy $a < b$. This clearly lies in $C(\mathbb{R})$ and is supported on $[a, b]$. By a suitable choice of a and b we can ensure that $h(t_0) \neq 0$ but $h(t) = 0$ for all $t \in \mathbb{R}$ with $|t - t_0| \geq \varepsilon$.

We now transfer this to the unit circle, to obtain a continuous function f, defining it by $f(e^{it}) = 2\pi h(t)$ for $|t - t_0| < \pi$. It remains to show that $f \in A_\rho$. Note that

$$\hat{f}(n) = \frac{1}{2\pi} \int_{t_0-\pi}^{t_0+\pi} e^{-int} f(e^{it})\, dt = \int_{-\infty}^{\infty} e^{-int} h(t)\, dt.$$

Now, by the fact that the (bilateral) Laplace transform of a product is the product of the Laplace transforms, we have, writing $g_1(t) = g(t - a)$ and $g_2(t) = g(b - t)$,

$$\hat{f}(n) = \int_{-\infty}^{\infty} \mathcal{L}g_1(iy)\mathcal{L}g_2(in - iy)\, dy.$$

Since

$$\mathcal{L}g_1(iw) = \int_{-\infty}^{\infty} g(t - a)e^{-iwt}\, dt = F(iw)e^{-iwa}$$

and

$$\mathcal{L}g_2(iw) = \int_{-\infty}^{\infty} g(b - t)e^{-iwt}\, dt = F(-iw)e^{iwb},$$

we find that

$$|\hat{f}(n)| \leq \int_{-\infty}^{\infty} |F(iy)|\,|F(-i(n-y))|\,dy = \sum_{k=-\infty}^{\infty} \frac{1}{\sigma(k)} \frac{1}{\sigma(n-k)}$$
$$= \sum_{k=-\infty}^{\infty} \frac{1}{(1+k^2)(1+(n-k)^2)} \frac{1}{\rho_k \rho_{n-k}},$$

and so

$$\sum_{n\in\mathbb{Z}} |\hat{f}(n)|\rho_n \leq \sum_{n=-\infty}^{\infty} \sum_{k=-\infty}^{\infty} \frac{1}{(1+k^2)(1+(n-k)^2)} \frac{\rho_n}{\rho_k \rho_{n-k}}$$
$$\leq \sum_{n=-\infty}^{\infty} \sum_{k=-\infty}^{\infty} \frac{1}{(1+k^2)(1+(n-k)^2)}$$
$$= \sum_{m=-\infty}^{\infty} \sum_{k=-\infty}^{\infty} \frac{1}{(1+k^2)(1+m^2)} < \infty,$$

and thus $f \in A_\rho$, as required. □

We shall also have recourse to one other technical device, which we now define.

Definition 5.1.8 Let A_ρ be a Beurling algebra on $\mathbb{T}$, and let $\Phi \in A_\rho^*$. Let the coefficients $\widehat{\Phi}(n)$ be defined as in Remark 5.1.3. Then the *Carleman transform* of Φ is defined by

$$\widetilde{\Phi}(z) = \begin{cases} \sum_{n=1}^{\infty} \widehat{\Phi}(n) z^{n-1} & \text{if } |z| < 1, \\ -\sum_{n=-\infty}^{0} \widehat{\Phi}(n) z^{n-1} & \text{if } |z| > 1. \end{cases}$$

Note that $|\widehat{\Phi}(n)| \leq \rho_n \|\Phi\|$ for each n, and so, if $(\rho_n)_n$ is a Beurling sequence, the function $\widetilde{\Phi}$ is defined and analytic in both $\mathbb{D}$ and $\mathbb{C} \setminus \overline{\mathbb{D}}$.

To motivate the expressions considered here and in Theorem 5.2.1 below, we note that if $\sigma(T) \subset \mathbb{T}$, then the resolvent of T, given by

$$R(z) = (T - zI)^{-1} \qquad (z \in \mathbb{C} \setminus \sigma(T)),$$

has the following Laurent series development:

$$R(z) = \begin{cases} \sum_{n=1}^{\infty} T^{-n} z^{n-1} & \text{if } |z| < 1, \\ -\sum_{n=-\infty}^{0} T^{-n} z^{n-1} & \text{if } |z| > 1. \end{cases} \tag{5.5}$$

5.2 Theorems of Wermer and Atzmon

Theorem 5.2.1 (Atzmon) *Let $T \in \mathcal{L}(\mathcal{X})$ and suppose that there exist sequences $(x_n)_{n\in\mathbb{Z}}$ in $\mathcal{X}$ and $(f_n)_{n\in\mathbb{Z}}$ in $\mathcal{X}^*$ with $x_0 \neq 0$ and $f_0 \neq 0$ such that*

$$Tx_n = x_{n+1} \qquad \text{and} \qquad T^* f_n = f_{n+1} \tag{5.6}$$

for all $n \in \mathbb{Z}$. Suppose further that one of the following conditions holds:

1. *The sequence $(\|f_n\|)_{n\in\mathbb{Z}}$ is dominated by a Beurling sequence and, for some integer $k \geq 0$,*
$$\|x_n\| = O(|n|^k) \qquad \text{as} \quad n \to \pm\infty.$$
2. *The sequence $(\|x_n\|)_{n\in\mathbb{Z}}$ is dominated by a Beurling sequence and, for some integer $k \geq 0$,*
$$\|f_n\| = O(|n|^k) \qquad \text{as} \quad n \to \pm\infty.$$
3. *Both the sequences $(\|x_n\|)_{n\in\mathbb{Z}}$ and $(\|f_n\|)_{n\in\mathbb{Z}}$ are dominated by Beurling sequences, and there are at least two points λ on $\mathbb{T}$ at which the following vector-valued functions G_1 and G_2 defined on $\mathbb{C} \setminus \mathbb{T}$ do not both possess analytic continuations into a neighbourhood of λ:*

$$G_1(z) = \begin{cases} \sum_{n=1}^{\infty} x_{-n} z^{n-1} & \text{if } |z| < 1, \\ -\sum_{n=-\infty}^{0} x_{-n} z^{n-1} & \text{if } |z| > 1; \end{cases} \tag{5.7}$$

$$G_2(z) = \begin{cases} \sum_{n=1}^{\infty} f_{-n} z^{n-1} & \text{if } |z| < 1, \\ -\sum_{n=-\infty}^{0} f_{-n} z^{n-1} & \text{if } |z| > 1. \end{cases} \tag{5.8}$$

Then either T is a multiple of the identity or it has a non-trivial hyperinvariant subspace.

It is easily checked that the Laurent series defining G_1 and G_2 converge absolutely in $\mathbb{C} \setminus \mathbb{T}$ since $(\|x_n\|)_{n\in\mathbb{Z}}$ and $(\|f_n\|)_{n\in\mathbb{Z}}$ are dominated by Beurling sequences.

The theorem may be rewritten as follows for invertible operators.

Corollary 5.2.2 *Let $T \in \mathcal{L}(\mathcal{X})$ be invertible, and $x_0 \in \mathcal{X}$ and $f_0 \in \mathcal{X}^*$ be non-zero vectors. If the sequences $(x_n = T^n x_0)_{n\in\mathbb{Z}}$ and $(f_n = T^{*n} f_0)_{n\in\mathbb{Z}}$ satisfy one of the hypotheses of Theorem 5.2.1, then either T is a multiple of the identity or it has a non-trivial hyperinvariant subspace.*

The proof of Theorem 5.2.1 relies on the construction of a weak form of a functional calculus. For a function $g \in C(\mathbb{T})$ such that

$$\sum_{n\in\mathbb{Z}} |\hat{g}(n)| \|x_n\| < \infty, \tag{5.9}$$

we define

$$\Phi_x(g) = \sum_{n\in\mathbb{Z}} \hat{g}(n) x_n.$$

The series for $\Phi_x(g)$ converges in norm in $\mathcal{X}$. Likewise we write $\Phi_f(h) = \sum_{n\in\mathbb{Z}} \hat{h}(n) f_n$ for functions $h \in C(\mathbb{T})$ such that

$$\sum_{n\in\mathbb{Z}} |\hat{h}(n)| \|f_n\| < \infty; \tag{5.10}$$

in this case the series for $\Phi_f(h)$ converges in $\mathcal{X}^*$. The first step towards obtaining hyperinvariant subspaces is the following result, which stresses what will be a common theme: if two non-zero functions have a pointwise zero product, then a functional calculus argument can sometimes be used to produce hyperinvariant subspaces.

Lemma 5.2.3 *Let $T \in \mathcal{L}(\mathcal{X})$ and let $(x_n)_n$ and $(f_n)_n$ be sequences such that (5.6) holds. If $g, h \in C(\mathbb{T})$ are functions satisfying (5.9) and (5.10) such that $g.h = 0$, then*

$$\langle A\Phi_x(g), \Phi_f(h)\rangle = 0$$

for all operators $A \in \mathcal{L}(\mathcal{X})$ that commute with T.

Proof Suppose that $AT = TA$ and $j, k \in \mathbb{Z}$; for $k \geq 0$ we have

$$\langle Ax_j, f_k\rangle = \langle Ax_j, T^{*k} f_0\rangle = \langle AT^k x_j, f_0\rangle = \langle Ax_{j+k}, f_0\rangle,$$

and for $k < 0$ we have

$$\langle Ax_j, f_k\rangle = \langle AT^{-k} x_{j+k}, f_k\rangle = \langle Ax_{j+k}, T^{*-k} f_k\rangle = \langle Ax_{j+k}, f_0\rangle,$$

once more. We conclude that

$$\begin{aligned}\sum_{j,k\in\mathbb{Z}} |\hat{g}(j)\hat{h}(k)\langle Ax_{j+k}, f_0\rangle| &= \sum_{j,k\in\mathbb{Z}} |\hat{g}(j)\hat{h}(k)\langle Ax_j, f_k\rangle| \\ &\leq \|A\| \Big(\sum_{j\in\mathbb{Z}} |\hat{g}(j)| \|x_j\|\Big)\Big(\sum_{k\in\mathbb{Z}} |\hat{h}(k)| \|f_k\|\Big) \\ &< \infty.\end{aligned}$$

This gives us that

$$\sum_{n\in\mathbb{Z}} |\widehat{g.h}(n)\langle Ax_n, f_0\rangle| < \infty$$

and

$$\langle A\Phi_x(g), \Phi_f(h)\rangle = \sum_{j,k\in\mathbb{Z}} \hat{g}(j)\hat{h}(k)\langle Ax_j, f_k\rangle = \sum_{n\in\mathbb{Z}} \widehat{g.h}(n)\langle Ax_n, f_0\rangle,$$

using the absolute convergence of the series. Thus, if $g.h = 0$, we have $\langle A\Phi_x(g), \Phi_f(h)\rangle = 0$. □

Note that the conclusions of Lemma 5.2.3 are what is needed for the construction of hyperinvariant subspaces, in the following sense.

Proposition 5.2.4 *Let $T \in \mathcal{L}(\mathcal{X})$. Then T has a non-trivial hyperinvariant subspace if and only if there exist non-zero vectors $x \in \mathcal{X}$ and $f \in \mathcal{X}^*$ such that $\langle Ax, f\rangle = 0$ for all operators A commuting with T.*

Proof If $\mathcal{Y}$ is such a hyperinvariant subspace, then we may take any non-zero $x \in \mathcal{Y}$ and (using the Hahn–Banach theorem) a non-zero $f \in \mathcal{X}^*$ such that $f(y) = 0$ for all $y \in \mathcal{Y}$.

Conversely, if we are given non-zero x and f as in the statement of the proposition, then we may take $\mathcal{Y}$ to be the closed linear span of all vectors Ax where A commutes with T. Since $f(y) = 0$ for all $y \in Y$, it is easy to see that $\mathcal{Y}$ is a non-trivial hyperinvariant subspace for T. □

Proof of Theorem 5.2.1, case 3 The maximum of two Beurling sequences is still a Beurling sequence, and so we may suppose without loss of generality that the Beurling sequence $(\rho_n)_{n\in\mathbb{Z}}$ dominates both $(\|x_n\|)_{n\in\mathbb{Z}}$ and $(\|f_n\|)_{n\in\mathbb{Z}}$. By Lemma 5.2.3 and Proposition 5.2.4, it is sufficient to find $g_0, h_0 \in A_\rho$ such that $g_0.h_0 = 0$ and $\Phi_x(g_0)$ and $\Phi_f(h_0)$ are both non-zero.

Consider the vector-valued functions G_1 and G_2 defined in (5.7) and (5.8). As already noted, these are both analytic in $\mathbb{C} \setminus \mathbb{T}$, because their power series converge locally uniformly there. If they possessed no singularities on $\mathbb{T}$, then they would be analytic and bounded on $\mathbb{C}$, hence constant by Liouville's theorem. Since they tend to zero at infinity, but x_0 and f_0 are non-zero, this is impossible.

We may thus find points $\lambda_1 \neq \lambda_2$ on $\mathbb{T}$ such that G_1 has a singularity at λ_1 and G_2 has a singularity at λ_2. In fact, more is true: there exist (non-zero) $x \in \mathcal{X}$ and $f \in \mathcal{X}^*$ such that the scalar functions $z \mapsto \langle G_1(z), f\rangle$ and $z \mapsto \langle x, G_2(z)\rangle$ have singularities at λ_1 and λ_2, respectively. For, on any disc D on which the scalar-valued functions $\langle G_1, f\rangle$ are analytic for all f, we also have that the vector-valued function G_1 is analytic ('weak analytic implies analytic'); there are many ways to see this, for example by using the Cauchy integral formula,

$$G_1(z_0) = \frac{1}{2\pi i} \int_\Gamma \frac{G_1(z)\, dz}{z - z_0},$$

for $z_0 \in \mathbb{D}$ and Γ a small circle centred at z_0, lying in D. This formula holds in the weak and hence the strong sense, implying analyticity of G_1 on D. A similar argument holds for G_2.

Now every $\Psi \in A_\rho^*$ has a well-defined support $\operatorname{supp} \Psi$, namely the complement of the largest open subset U of $\mathbb{T}$ such that $\langle f, \Psi \rangle = 0$ for all $f \in A_\rho$ supported on U. To see this, suppose that $\langle f, \Psi \rangle = 0$ for all f supported on either of two open sets U_1 and U_2. For a function f supported on $U_1 \cup U_2$ we may use the regularity of A_ρ to find functions $f_1, f_2 \in A_\rho$ such that $f = f_1 + f_2$, with f_1 supported on U_1 and f_2 on U_2 (we leave this for the Exercises, with a hint). Then

$$\langle f, \Psi \rangle = \langle f_1, \Psi \rangle + \langle f_2, \Psi \rangle = 0.$$

Thus the set of open subsets with this vanishing property is closed under finite unions, and, by the compactness of $\mathbb{T}$, under arbitrary unions. Thus $\operatorname{supp} \Psi$ is well defined.

We now define $\Psi_1, \Psi_2 \in A_\rho^*$ by

$$\widehat{\Psi}_1(n) = \langle x_{-n}, f \rangle \qquad \text{and} \qquad \widehat{\Psi}_2(n) = \langle x, f_{-n} \rangle \qquad (n \in \mathbb{Z}).$$

Now λ_1 is a singularity of the Carleman transform of the corresponding Ψ_1, and hence, if we let $\mathcal{J}$ be the ideal of functions g such that $g.\Psi_1 = 0$, we have that

$$\lambda_1 \in \mathcal{K} := \{\lambda \in \mathbb{T} : h(\lambda) = 0 \text{ for all } h \in \mathcal{J}\}.$$

This is because we may express the analytic function $\widetilde{\Psi}_1(z)$ by the formula

$$\widetilde{\Psi}_1(z) = \langle (e_1 - z e_0 + \mathcal{J})^{-1}, \Psi_{1\mathcal{J}} \rangle \tag{5.11}$$

for $z \in \mathbb{C} \setminus \mathcal{K}$, where e_0 and e_1 are as in Definition 5.1.8. Here $\Psi_{1\mathcal{J}}$ denotes the induced mapping on $A_\rho/\mathcal{J}$ defined by $\langle b + \mathcal{J}, \Psi_{1\mathcal{J}} \rangle = \langle b, \Psi_1 \rangle$. (This is easily verified using the formulae in (5.5).) Thus no point in the complement of $\mathcal{K}$ can be a singularity of the Carleman transform.

Next, $\mathcal{K} \subset \operatorname{supp} \Psi_1$, for if $\lambda \notin \operatorname{supp} \Psi_1$ we may find a function g with $g(\lambda) \neq 0$ but where g is supported on an arc containing λ that is disjoint from $\operatorname{supp} \Phi_1$. Thus $g \in \mathcal{J}$ and so $\lambda \notin \mathcal{K}$. Similarly $\lambda_2 \in \operatorname{supp} \Psi_2$.

Let Γ_1 and Γ_2 be disjoint open arcs in $\mathbb{T}$ with $\lambda_k \in \Gamma_k$ for $k = 1, 2$. Since $\lambda_k \in \operatorname{supp} \Psi_k$ there exist functions $g_0, h_0 \in A_\rho$ supported on Γ_1 and Γ_2, respectively, such that $\langle g_0, \Psi_1\rangle \neq 0$ and $\langle h_0, \Psi_2\rangle \neq 0$. Now $\langle g_0, \Psi_1\rangle = \langle \Phi_x(g_0), f\rangle$, and $\langle h_0, \Psi_2\rangle = \langle x, \Phi_f(h_0)\rangle$, so that $\Phi_x(g_0)$ and $\Phi_f(h_0)$ are non-zero, as required.

Proof of Theorem 5.2.1, cases 1 and 2 If G_1 or G_2 has more than one singularity, then the result follows from (3) since $((1+|n|)^k)_{n\in\mathbb{Z}}$ is a Beurling sequence. Now each function has at least one singularity on $\mathbb{T}$, so it remains to analyse the case when each has exactly one singularity of $\mathbb{T}$.

Suppose that G_1 has a unique singularity on $\mathbb{T}$. Without loss of generality (replacing T by a unimodular constant multiple of itself and making the obvious changes to the sequences $(x_n)_n$ and $(f_n)_n$), we may suppose that this is at $\lambda = 1$.

We shall require the following technical lemma from Hille and Phillips [123, theorem 3.13.5, p. 102], of which we give merely a sketch proof.

Lemma 5.2.5 *Let $F : \mathbb{C}\setminus\{1\} \to \mathcal{X}$ be analytic, such that*

$$F(z) = \begin{cases} \sum_{n=0}^{\infty} a_n z^n & \text{if } |z| < 1, \\ -\sum_{n=0}^{\infty} b_n z^{-n} & \text{if } |z| > 1, \end{cases}$$

where $\|a_n\|$ and $\|b_n\|$ are both $o(n^\alpha)$ as $n \to \infty$. Then $z = 1$ is a pole of F of order less than $\alpha + 1$ (i.e., F is a polynomial in $1/(z-1)$ of degree less than $\alpha + 1$).

Proof The hypotheses imply estimates

$$\|F(w)\| \le 2\varepsilon(1-|z|)^{-1-\alpha}$$

for $r_\varepsilon < |z| < 1$, say, and

$$\|F(w)\| \le 2\varepsilon(1-|z|^{-1})^{-1-\alpha}$$

for $1 < |z| < r'_\varepsilon$, say. Write $1 - z = \left(w + \frac{1}{2}\right)^{-1}$, and let $G(w) = F(z)$. Now G is an entire function of w, and, writing $w = Re^{i\theta}$, a simple calculation shows that

$$(1-|z|)^{-1} \le \frac{9}{4}\frac{R}{\cos\theta} \quad \text{for} \quad -\pi/2 < \theta < \pi/2, \quad (0 \le |z| < 1) \quad \text{and} \quad R \ge 1,$$

while

$$(1-|z|^{-1})^{-1} \le \frac{9}{4}\frac{R}{\cos\theta} \quad \text{for} \quad \pi/2 < \theta < 3\pi/2, \quad (1 < |z| \le \infty) \quad \text{and} \quad R \ge 1.$$

Since $z \to 1$ as $R \to \infty$, we conclude that $R^{-\alpha-1}|\cos\theta|^{\alpha+1}\|G(Re^{i\theta})\| \to 0$ as $R \to \infty$, uniformly in θ. By Cauchy's integral formula, G is a polynomial in w, and hence $(1-z)^{-1}$, of degree less than $\alpha + 1$, as required. □

Given that G_1 is a polynomial in $(1-z)^{-1}$ of degree at most k, with coefficients in $\mathcal{X}$, it follows that the function $p(z) = (1-z)^{k+1}G_1(1/z)/z$ is a polynomial of degree at most k. We have

$$p(z) = (1-z)^{k+1}\sum_{n=0}^{\infty}(T^n x_0)z^n,$$

at least for $|z| < 1$. Now the coefficient of z^{k+1} in this series is

$$\begin{aligned}\sum_{r=0}^{k+1}\binom{k+1}{r}(-1)^r T^{k+1-r}x_0 &= (-1)^{k+1}\sum_{s=0}^{k+1}\binom{k+1}{s}(-T)^s x_0\\ &= (-1)^{k+1}(\mathrm{I}-T)^{k+1}x_0,\end{aligned}$$

using the substitution $s = k+1-r$; since this coefficient vanishes we conclude that $(\mathrm{I}-T)^{k+1}x_0 = 0$. This establishes case 1.

Similarly, if G_2 has a single singularity at (without loss of generality) $z = 1$, we deduce that $(\mathrm{I}-T^*)^{k+1}f_0 = 0$. Now case 2 follows, since either $T = \mathrm{I}$ or the range of $T - \mathrm{I}$ fails to be dense, and so its closure is a non-trivial hyperinvariant subspace for T. □

Proof of Theorem 5.0.1 The easiest case to dispose of is when T satisfies condition 2 of Theorem 5.0.1. In this case the result follows immediately from part 1 of Theorem 5.2.1.

Assume now that condition 1 of Theorem 5.0.1 holds, and observe that this implies immediately that the sequence $(\rho_n)_n = (\|T^n\|)_n$ is a Beurling sequence. We intend to use part 3 of Theorem 5.2.1, and to do this we consider the resolvent function $R(z) = (T - z\mathrm{I})^{-1}$ defined on $\mathbb{C} \setminus \sigma(T)$.

Now we claim that $\sigma(T)$ is a subset of the circle. Note that $\frac{1}{n}\log\|T^n\|$ is a decreasing sequence, and it must tend to zero, since if it were larger than some $\varepsilon > 0$, the series $\sum_{n=1}^{\infty}\frac{\log\|T^n\|}{1+n^2}$ would diverge by comparison with $\sum_{n=1}^{\infty}\frac{\varepsilon}{n}$. We see that the spectral radius of T (and similarly of T^{-1}) is equal to 1.

It therefore follows that the Laurent series (5.5) hold for $R(z)$. Now we claim that for some choice of $x_0 \in \mathcal{X}$ every $\lambda \in \sigma(T)$ is a singular point of the function $G_1(z) = R(z)x_0$ defined by (5.7); here, as before, $x_n = T^n x_0$. Our claim holds because if $(\lambda_n)_n$ is a sequence in $\mathbb{D}$ such that $\lambda_n \to \lambda$ as $n \to \infty$, then we have $\|(T - \lambda_n \mathrm{I})^{-1}\| \to \infty$: otherwise, the identity

$$T - \lambda \mathrm{I} = (T - \lambda_n \mathrm{I})(\mathrm{I} + (\lambda_n - \lambda)(T - \lambda_n \mathrm{I})^{-1})$$

would imply the invertibility of $T - \lambda \mathrm{I}$ as soon as

$$|\lambda_n - \lambda| \, \|(T - \lambda_n \mathrm{I})^{-1}\| < 1.$$

By the Banach–Steinhaus uniform boundedness theorem (Theorem 1.1.7), there is an $x_0 \in \mathcal{X}$ such that the sequence $((T - \lambda_n \mathrm{I})^{-1} x_0)_n$ is unbounded, which implies that $R(z)x_0$ fails to be analytic at λ.

Now $\sigma(T)$ consists of more than one point, and so (considering also T^*) we may find distinct λ_1 and λ_2 in $\sigma(T)$, and non-zero $x_0 \in \mathcal{X}$ and $f_0 \in \mathcal{X}^*$, such that λ_1 is a singular point of G_1 and λ_2 a singular point of G_2 (defined analogously, starting with f_0 and using T^*). The result now follows using part 3 of Theorem 5.2.1. □

5.3 Bishop operators

5.3.1 Davie's functional calculus

Recall that the Bishop operator T_α on $L^2(0, 1)$ is defined by (5.2). More generally, we may wish to consider weighted composition operators $T_{\alpha,w}$ on $L^2([0, 1])$ defined by

$$T_{\alpha,w} f(t) = w(t) f(\{t + \alpha\}),$$

where $w \in L^\infty([0, 1])$. The techniques we shall present show the existence of hyperinvariant subspaces for any Bishop-type operator $T_{\alpha,w}$ with weight w of the form $w(t) = \prod_{k=1}^{K} \{t - \beta_k\}^{\gamma_k}$ with $\gamma_k > 0$ for each k, provided, that is, that α is either rational or from a very large class of irrationals that we shall call *ρ-regular*. In particular, finite products $T_{\alpha_1} \cdots T_{\alpha_K}$ of Bishop operators satisfy this condition as long as $\alpha_1 + \cdots + \alpha_K$ is either rational or ρ-regular.

The presentation of Davie's results and subsequent generalizations is fairly technical, and we shall therefore concentrate on explaining the method in the context of a single operator T_α.

There is one easy case which we should mention before proving a deeper result.

Proposition 5.3.1 *Let $\alpha \in (0, 1)$ be rational. Then T_α has a non-trivial hyperinvariant subspace.*

Proof Since α is rational, there exists a smallest positive integer, say m, such that $m\alpha$ is an integer. Hence T_α^m is an operator of multiplication on $L^2[0, 1]$, and thus a normal operator. Hence, unless T_α^m is a multiple of the identity (this case is left as an exercise), it has hyperinvariant subspaces by Theorem 3.4.1. Clearly, these are hyperinvariant subspaces for T_α as well. □

We shall discuss 'rational' Bishop operators in more detail in Section 5.4. Not only do they have many invariant subspaces, but it is even possible to give a fairly clear description of them.

For 'irrational' Bishop operators, it is still not known whether invariant subspaces exists for all α, so we begin with a definition. At this point we refer the reader to Section 1.5 for the basic ideas in the theory of approximation of numbers by rationals.

Definition 5.3.2 Let $(\rho_n)_n$ be a sequence greater than 1 such that $(\log \rho_n)_n$ is sub-additive and satisfies the Beurling condition, $\sum_n \frac{\log \rho_n}{1+n^2} < \infty$. An irrational real α is said to be *ρ-regular* if there exist $m_0 \in \mathbb{N}$ and two functions h_1, h_2 satisfying

$$\frac{h_1(m)\log(\rho_m)}{m\log m} \to \infty \qquad \text{and}$$
$$\frac{h_2(m)\log m}{\log(\rho_m)} \to 0, \qquad \text{as} \quad m \to \infty,$$

such that, for all $m > m_0$, there exist $p, q \in \mathbb{N}$, $(p, q) = 1$, satisfying

$$\left|\alpha - \frac{p}{q}\right| \le \frac{1}{q^2} \quad \text{and}$$
$$h_1(m) \le q \le h_2(m).$$

Davie [89] made the choices $\log \rho_n = |n|^\rho$ with $\frac{1}{2} < \rho < 1$, $h_1(m) = m^{\rho'}$ with $0 < \rho' < \frac{1}{2}$ and $\rho + \rho' > 1$, and $h_2(m) = m^{1/2}$, which characterise the non-Liouville numbers. Typically, in order to obtain a larger class of irrationals, including some Liouville numbers, we will make choices such that

$$\log \rho_n \asymp |n|/(\log|n|)^{3/2}, \quad h_1(m) = (\log m)^3 \quad \text{and} \quad h_2(m) = m/(\log m)^3. \tag{5.12}$$

Theorem 5.3.3 *Let $\alpha \in [0, 1]$ be a ρ-regular irrational number, with condition (5.12) being satisfied. Then T_α has a non-trivial hyperinvariant subspace.*

Proof The idea is to construct a functional calculus for $T = eT_\alpha$, by defining $\phi(T)$ for $\phi \in A_\rho$ as a linear mapping from L_ρ to itself, where

$$L_\rho = \left\{ f : f \text{ measurable on } [0,1] \text{ and } \sup_{m\in\mathbb{Z}} \frac{T^m|f|(x)}{\rho(|m|)} < \infty \text{ a.e.} \right\}$$

by

$$\phi(T)f = \sum_{n\in\mathbb{Z}} \hat{f}(n) T^n f.$$

Although $\phi(T)$ is not continuous, we have the following immediate properties:

- If $\|\phi_n - \phi\|_\rho \to 0$, with ϕ_n and ϕ in A_ρ, then $\phi_n(T)f \to \phi(T)f$ almost everywhere for each $f \in L_\rho$.
- $(\phi\psi)(T) = \phi(T)\psi(T)$ for each ϕ and ψ in A_ρ.

The choices of ρ and α that we have made will enable us to establish the following less elementary properties:

I. $L^2[0,1] \subset L_\rho$, and if $f_n \to f$ in $L^2[0,1]$ then for each $\phi \in A_\rho$ it follows that $\phi(T)f_n \to \phi(T)f$ in measure.

II. There exists a dense subspace $S \subset L^2[0,1]$ such that

   1. If $f \in S$ and $\phi \in A_\rho$, then $\phi(T)f \in L^2[0,1]$.
   2. If $f \in S$ and $\phi_n \to \phi$ in norm in A_ρ then $\phi_n(T)f \to \phi(T)f$ in norm in $L^2[0,1]$.

It will be useful to note the following formulae for $m \geq 1$:

$$(T^m f)(x) = e^m \left(\prod_{j=0}^{m-1} \{x + j\alpha\} \right) f(\{x + m\alpha\})$$

and

$$(T^{-m} f)(x) = \frac{e^{-m}}{\left(\prod_{j=1}^{m} \{x - j\alpha\}\right)} f(\{x - m\alpha\}) = F_m(x) f(\{x - m\alpha\}), \quad \text{say.}$$

For $t > 0$ and $n \geq 1$ we set

$$E_{t,n} = \left\{ x \in [0,1] : \{x - n\alpha\} \geq t/n^2 \right\}$$

and $E_t = \bigcap_{n\geq 1} E_{t,n}$. As t decreases to 0, the set E_t increases and $\bigcup_{t>0} E_t$ has measure 1.

Recall that m_0 was given in Definition 5.3.2. We have the following estimates:

for $m > m_0$,

$$\|T^{-m} f\|_{L^2(E_t)} \leq \exp\left\{\frac{4m}{(\log m)^3}\log(m^2/t)\right\}\|f\|_{L^2[0,1]} \tag{5.13}$$

and

$$\|T^m f\|_{L^2(E_t)} \leq \exp\left\{\frac{4m}{(\log m)^2}\right\}\|f\|_{L^2[0,1]}; \tag{5.14}$$

for $1 \leq m \leq m_0$,

$$\|T^{-m} f\|_{L^2(E_t)} \leq \frac{m^{2m}}{t^m}\|f\|_{L^2[0,1]} \tag{5.15}$$

and

$$\|T^m f\|_{L^2(E_t)} \leq e^m\|f\|_{L^2[0,1]}. \tag{5.16}$$

To obtain (5.13) and (5.14), we write $m = qr + s$, where $0 \leq s < q$ and q is chosen to satisfy $|\alpha - p/q| < q^{-2}$ as in Definition 5.3.2, with

$$(\log m)^3 \leq q \leq \frac{m}{(\log m)^3},$$

and hence $r \leq m/(\log m)^3$ as well.

Suppose that $\{y - k\alpha\} \geq \delta > 0$ for $k = 1, \ldots, q$. We have that $\{kp/q\}$ takes all the values $0, 1/q, \ldots, (q-1)/q$ in some order as k ranges from 0 to $q-1$, and $|k\alpha - kp/q| \leq 1/q$, so $\{y - k\alpha\} \geq |y - kp/q| - 2/q$. We deduce that

$$F_q(y) \leq \frac{e^{-q}}{\delta^3(1/q)\ldots(q-3)/q} \leq \frac{e^{-q}q^q}{\delta^3 q!} \leq \delta^{-3}.$$

Then $T^{-m}f(x) = F_m(x)f(\{x - m\alpha\})$ with

$$F_m(x) = F_q(x)F_q(\{x - q\alpha\})\ldots F_q(\{x - (r-1)q\alpha\})F_s(\{x - rq\alpha\}),$$

and so, with $\delta = t/m^2$, we have $F_m(x) \leq (m^2/t)^{3r+s}$, which easily gives (5.13).

Likewise, $T^q f(x) = G_q(x)f(\{x + q\alpha\})$, where

$$G_q(x) = e^q x\{x + \alpha\}\ldots\{x + (q-1)\alpha\} \leq e^q(3/q)(4/q)\ldots(q/q) \leq q^3,$$

since $\{x + k\alpha\} \leq |x - kp/q| + 2/q$. Thus $\|T^m\| \leq q^{3r}e^s$, which implies (5.14).

Estimate (5.15) is easier, and uses the fact that $\|T^{-m}f\|_{L^2(E_t)} \leq \delta^{-m}\|f\|_2$. Finally (5.16) is immediate.

By means of (5.13)–(5.16) we can estimate

$$A = \int_{E_t} \sum_{m \in \mathbb{Z}} \frac{|T^m f(x)|^2}{\rho(m)^2} \, dx,$$

for it now follows that there is a constant C_t such that $A \le C_t \|f\|^2_{L^2[0,1]}$. This implies assertion I above.

Assertion II is proved using the following definition of S.

For $t > 0$ let G_t be the set of x in $[0, 1]$ such that $\{x + n\alpha\} \le t/(1 + n^2)$ for some $n \ge 0$. Let S_t be the subspace of functions in $L^2[0, 1]$ that vanish on G_t, and let $S = \bigcup_{t>0} S_t$. Note that $m(G_t) \to 0$ as $t \to 0$, and so S is dense.

Now, if x satisfies $\{x - k\alpha\} \le t/(m + 1)^2$ for some $0 \le k \le m$, then $x - m\alpha \in G_t$ and so $f(x - m\alpha) = 0$ for all $f \in S_t$.

Thus, similar calculations to those in step I show that there is a constant C'_t such that

$$\int_0^1 \sum_{m \in \mathbb{Z}} \frac{|T^m f(x)|^2}{\rho(m)^2} \, dx \le C'_t \|f\|_2^2$$

for $f \in S_t$.

The remainder of the argument follows along standard lines. Since A_ρ is regular, the following assertion holds easily:

III. If $\phi \in A_\rho$ with $\phi \ge 0$, and if $\phi(T)f = 0$ for all $f \in L^2[0, 1]$, then $\phi = 0$.

Choose non-negative functions $\phi, \psi \in A_\rho$ such that $\phi.\psi = 0$ but neither ϕ nor ψ is identically zero. Define $\mathcal{M}$ to be the set of all $f \in L^2[0, 1]$ such that $\psi(T)Vf = 0$ for every V that commutes with T (such V also commute with $\psi(T)$, by assertion II). Now, by assertion I, $\mathcal{M}$ is closed, and it is invariant under the commutant of T; it is non-trivial by assertion III. □

5.3.2 The point spectrum

The results of the previous section would be elementary if it were the case that the operators considered had some point spectrum (eigenvalues). We shall see that for a large class of weighted composition operators this is not the case. This section illustrates the use of certain number-theoretic methods, in particular, Dirichlet's theorem.

Proposition 5.3.4 *Let ω be a positive function on a sub-interval I of the real line. Suppose that there exists $\delta > 0$ such that, for all $c, d \in I$,*

$$\left| \frac{\log(\omega(d)) - \log(\omega(c))}{d - c} \right| \geq \delta. \tag{5.17}$$

Then, for all $\beta \in (0, 1/2)$, we have

$$\mu\{t \in I : |1 - \omega(t)| \leq \beta\} \leq \frac{4\beta}{\delta}.$$

Proof First take $s \neq t$ such that $\omega(s), \omega(t) \in [1 - \beta, 1 + \beta]$. By hypothesis, we have

$$\delta|s - t| \leq |\log \omega(s) - \log \omega(t)| \leq \log \frac{1+\beta}{1-\beta}.$$

Now, using the mean value theorem, for all $u > v$, we have

$$\log(1 + u) - \log(1 + v) \leq \frac{1}{1+v}(u - v).$$

Therefore we get

$$|s - t| \leq \frac{1}{1-\beta} \frac{2\beta}{\delta} \leq \frac{4\beta}{\delta}.$$

□

Corollary 5.3.5 *Let $\delta > 0$ and suppose that ω is differentiable and satisfies $|\omega'(t)/\omega(t)| \geq \delta$ on a real sub-interval I. Then, for all $\beta \in (0, 1/2)$, we have*

$$\mu\{t \in I : |1 - \omega(t)| \leq \beta\} \leq \frac{4\beta}{\delta}.$$

Proof The result follows from Proposition 5.3.4, since we clearly have (5.17) by the mean value theorem. □

Example 5.3.6 Let $\mathcal{P} \subset L^\infty([0, 1])$ be the class of functions ω on $[0, 1]$ such that

$$\omega(x) = C \prod_{i=0}^{n} (x - x_i)^{s_i}, \qquad x \in [0, 1],$$

for some constant C, fixed indices $s_0, \ldots, s_n \in \mathbb{R}_+$, and points $x_0, \ldots, x_n$ in $[0, 1]$. This is called the class of *generalized polynomials*, and operators of Bishop type associated with a weight in $\mathcal{P}$ are studied in Androulakis and Flattot [10]. If $\omega \in \mathcal{P}$, then ω satisfies the conditions of Corollary 5.3.5. In particular, $\omega \in \mathcal{P}$ defined by $\omega(t) = \sqrt{t}$ is not an admissible function in the sense of Chalendar, Flattot and Guillotin-Plantard [63] or Chalendar and Partington [74].

Proposition 5.3.7 *Let* $0 = a_0 < a_1 < \cdots < a_N = 1$ *and suppose that* ω *is a positive and increasing function on each interval* (a_k, a_{k+1}) *for* $k = 0, \ldots, N-1$, *satisfying (5.17) on each interval. Let* $\beta \in (0, 1/2)$, $\alpha \in \mathbb{R} \setminus \mathbb{Q}$ *and* $\lambda \in \mathbb{C} \setminus \{0\}$. *We write, for all* $n \geq 1$,

$$F_n(t) = \frac{1}{\lambda^n} \prod_{j=1}^{n} \omega(\{t - j\alpha\}).$$

Then

$$\mu\{t : |1 - |F_n(t)|| \leq \beta\} \leq \frac{4\beta N}{\delta}.$$

Proof First, note that we can partition $(0, 1)$ into at most nN sub-intervals such that if c and d belong to the same sub-interval, then $\{c - j\alpha\}$ and $\{d - j\alpha\}$ are in the same sub-interval of the original partition for each $j = 1, \ldots, N$. It follows that

$$\begin{aligned}
&\frac{\log(|F_n|(d)) - \log(|F_n|(c))}{d - c} \\
&= \frac{\sum_{j=1}^{n} \log(\omega(\{d - j\alpha\})) - \sum_{j=1}^{n} \log(\omega(\{c - j\alpha\}))}{d - c} \\
&= \sum_{j=1}^{n} \frac{\log(\omega(\{d - j\alpha\})) - \log(\omega(\{c - j\alpha\}))}{d - c} \\
&\geq n\delta.
\end{aligned}$$

Thus, on each sub-interval, $|F_n|$ satisfies the hypothesis of Proposition 5.3.4, with δ replaced by $n\delta$. Using Proposition 5.3.4, we get

$$\mu\{t \in (0, 1) : |1 - |F_n(t)|| \leq \beta\} \leq \frac{4\beta}{n\delta} Nn = \frac{4\beta N}{\delta}.$$

□

We are now ready to show that a large class of weighted composition operators lack any point spectrum.

Theorem 5.3.8 *Let* $\alpha \in \mathbb{R} \backslash \mathbb{Q}$ *and suppose that* ω *satisfies the hypothesis of Proposition 5.3.7. Then*

$$\sigma_p(T_{\alpha,\omega}) = \emptyset.$$

Proof Suppose that the point spectrum of $T_{\alpha,\omega}$ is not empty. Then there exist $\lambda \in \mathbb{C}$ and $f \in L^2([0, 1])$ such that $T_{\alpha,\omega} f = \lambda f$. If $\lambda = 0$, then, since ω is positive, it follows that $f = 0$ on $[0, 1]$, which is impossible.

Now, suppose that $\lambda \neq 0$. By Dirichlet's Theorem 1.5.2, there are two sequences $(p_k)_{k\geq 1}$, $(q_k)_{k\geq 1}$, such that

$$\left|\alpha - \frac{p_k}{q_k}\right| \leq \frac{1}{q_k^2}$$

and $\lim_{k\to\infty} q_k = \infty$.

Using Lusin's theorem (Theorem 1.1.10), it follows that

$$f(x) - f(\{x - q_k\alpha\}) \to 0$$

in measure as $k \to \infty$. So, for all $\beta > 0$ and $\eta > 0$, we have

$$\mu(\{x \in [0,1] : |f(\{x - q_k\alpha\}) - f(x)| > \beta\}) \leq \eta$$

for k sufficiently large.

Suppose that $\beta \in \left(0, \frac{1}{2}\right)$ such that $\beta < \frac{\delta}{5N}$ and $\eta = 1 - \frac{5N\beta}{\delta}$, where δ is as given in condition (5.17). We can construct a subsequence of $(q_k)_k$, which we continue to call $(q_k)_k$, such that, for all $k \geq 1$, we have

$$\mu\left(\left\{x \in [0,1] : |f(\{x - q_k\alpha\}) - f(x)| \leq \frac{\beta}{k}\right\}\right) > 1 - \eta.$$

By hypothesis, we have

$$f(\{x - q_k\alpha\}) = \lambda^{-q_k} T_{\alpha,\omega}^{q_k} f(\{x - q_k\alpha\}) = F_k(x) f(x),$$

where $F_k(x) = \lambda^{-q_k} \prod_{j=1}^{q_k} \omega(\{x - j\alpha\})$.

Using Proposition 5.3.7, we know that $\mu\{t : |1 - |F_n(t)|| \leq \beta\} \leq \dfrac{4\beta N}{\delta}$.

Since

$$|f(x) - f(\{x - q_k\alpha\})| = |1 - F_k(x)|\,|f(x)|$$

and $|f(\{x - q_k\alpha\}) - f(x)| \leq \frac{\beta}{k}$ on a set of measure greater than $1 - \eta$, it follows that for all $k \geq 1$ we have $|f(x)| \leq \frac{1}{k}$ on a set of measure greater than $\frac{\beta N}{\delta}$.

This implies immediately that $f = 0$ on a set of positive measure, which is invariant under translation by α (modulo 1). But the transformation

$x \mapsto \{x+\alpha\}$ is *ergodic*, in the sense that there are no (a.e.) invariant subsets except those of measure 0 or 1 (we leave this as an exercise). Hence $f = 0$ a.e. on $[0, 1]$. □

Remark 5.3.9 Blecher and Davie [42] study Bishop-type operators whose continuous and positive weight ω satisfies the following condition:

$$\int_0^1 \frac{\gamma(t)}{t}\, dt < \infty, \qquad \text{where} \qquad \gamma(t) = \sup_{|x-y|\le t} |\log \omega(x) - \log \omega(y)|. \tag{5.18}$$

Unfortunately, there is no link between condition (5.18) and condition (5.17). Indeed, for $\omega(t) = e^{t^k}$ where $k > 1$, (5.17) is not satisfied but (5.18) is satisfied since $\gamma(t) \le kt$. On the other hand, for $\omega(t) = e^{t^k}$ where $0 < k < 1$, (5.17) is satisfied with $\delta = k$ but (5.18) is not satisfied since $\gamma(t) \ge kt^{k-1}$.
Nevertheless, since (5.18) is satisfied when

$$\sup_{a,b\in[0,1]a\neq 0} \frac{|\log \omega(a) - \log \omega(b)|}{|a-b|} < \infty,$$

it follows that, whenever there exist positive numerical constants C_1, C_2 such that

$$C_1 \le \frac{\log \omega(a) - \log \omega(b)}{a-b} \le C_2,$$

then $\sigma_p(T_{\alpha,\omega}) = \emptyset$ and $T_{\alpha,\omega}$ has non-trivial hyperinvariant subspaces for non-Liouville irrational α.

5.4 Rational Bishop operators

For an operator T on a Hilbert space $\mathcal{H}$, and for $f \in \mathcal{H}$, we write orb(T, f) for the closure in $\mathcal{H}$ of the linear hull of $\{T^n f : n = 0, 1, 2, \ldots\}$; clearly, this is the smallest closed T-invariant subspace of $\mathcal{H}$ containing f. Recall that a function f is said to be *cyclic* if orb$(T, f) = \mathcal{H}$.

For $\alpha \in (0, 1)$ rational, say $\alpha = p/q$ in its lowest terms, it is easy to see that the Bishop operator $T = T_\alpha$ has many invariant subspaces; for example, the set $\chi_K L^2[0, 1]$ is one such, whenever K is invariant under rotation by $1/q$ (here χ_K denotes the characteristic function of a set K). However, it is distinctly less straightforward to give a full classification of the invariant subspaces, or even of the cyclic vectors, of T_α: a similar phenomenon occurs in the study of invariant subspaces for the unilateral and bilateral shifts, where it is easy to exhibit invariant subspaces but their full classification requires the theorems of Beurling and Wiener (Theorems 1.4.20 and 1.4.22).

We begin by giving a characterization of the cyclic vectors of T_α when α is rational; we then go further and obtain a description of the lattice of invariant subspaces. Finally, we illustrate the ideas by considering the case $\alpha = 1/2$, where simpler expressions can be obtained.

5.4.1 Cyclic vectors

Let $\alpha = p/q$ in $(0, 1)$, where p and q are positive integers with no common divisor. Set $T = T_\alpha$ and

$$w(t) = t\{t+\alpha\}\cdots\{t+(q-1)\alpha\} = t\{t+1/q\}\cdots\{t+(q-1)/q\}.$$

The function w is obviously a $1/q$-periodic function in $L^\infty([0, 1])$.

Note that

$$T^{nq+r} f = w^n f_{r+1}, \text{ for all } n \geq 0,\ r \in \{0, \ldots, q-1\},$$

where $f_1 = f$, $f_2 = Tf, \ldots, f_q = T^{q-1} f$.

For the following discussion it will also be useful to consider the *determinant* of f associated with $\alpha = p/q$.

Definition 5.4.1 Let $f \in L^2([0, 1])$ and $f_k = T^{k-1} f$, $k \in \{1, \ldots, q\}$. The *determinant of f associated with $\alpha = p/q$* is the $1/q$-periodic function in $L^{2/q}([0, 1])$ defined by

$$\Delta(f, p/q) = \begin{vmatrix} f_1(t) & \cdots & f_q(t) \\ f_1(\{t+p/q\}) & \cdots & f_q(\{t+p/q\}) \\ \vdots & \vdots & \vdots \\ f_1(\{t+(q-1)p/q\} & \cdots & f_q(\{t+(q-1)p/q\}) \end{vmatrix}$$

Note that, since p and q have no common divisor, $|\Delta(f, p/q)| = |\Delta(f, 1/q)|$.

The next lemma shows explicitly that a dense set of functions can be decomposed using the functions f_k $(1 \leq k \leq q)$ and $1/q$-periodic functions.

Lemma 5.4.2 *Let n be a positive integer and $f \in L^2([0, 1])$. Let $h \in L^\infty([0, 1])$ be such that $h(t) = 0$ whenever $t \in \Omega_{n,f}$, where*

$$\Omega_{n,f} = \{s \in [0,1] : |\Delta(f,\alpha)| < 1/n\} \cup \bigcup_{1\le r\le q, 0\le k\le q-1} \{s \in [0,1] : |f_r(\{s+k/q\})| > n\}.$$

Then there exist $1/q$*-periodic functions* $h_1, \ldots, h_q$ *in* $L^\infty([0,1])$ *such that*

$$h = \sum_{k=1}^{q} h_k f_k.$$

Proof If $h = \sum_{k=1}^{q} h_k f_k$, where $h_1, \ldots, h_q$ are $1/q$-periodic functions in $L^\infty([0,1])$, it follows that, for all $t \in [0, 1/q)$,

$$\begin{aligned} h(t) &= h_1(t) f_1(t) + \cdots + h_q(t) f_q(t), \\ h(t+1/q) &= h_1(t) f_1(t+1/q) + \cdots + h_q(t) f_q(t+1/q), \\ &\vdots \\ h(t+(q-1)/q) &= h_1(t) f_1(t+(q-1)/q) + \cdots + h_q(t) f_q(t+(q-1)/q). \end{aligned} \tag{5.19}$$

For t in the complement of $\Omega_{n,f}$, one can easily check that a solution of the above system exists consisting of bounded functions. On $\Omega_{n,f}$, one can take h_k, $1 \le k \le q$, equal to zero, for example. □

Lemma 5.4.3 *If* $F \in L^2([0,1/q])$ *is non-zero almost everywhere, then the set* $\{g(w)F : g \in \mathbb{C}[X]\}$ *is dense in* $L^2([0,1/q])$.

Proof Suppose that a function $G \in L^2([0,1/q])$ is orthogonal to $g(w)F$, for all polynomials g. So

$$\int_0^{1/q} \overline{F(t)} G(t) g(w(t))\, dt = 0, \quad g \in \mathbb{C}[X].$$

Now, denote by $\tilde{w}$ the continuous function on $[0,1/q]$ satisfying $\tilde{w}(t) = w(t)$ for $t \in [0,1/q)$. Then we have

$$\int_0^{1/q} \overline{F(t)} G(t) g(\tilde{w}(t))\, dt = 0, \quad g \in \mathbb{C}[X].$$

Applying the Stone–Weierstrass theorem to $C([0,1/q])$, the set of continuous functions on the compact set $[0,1/q]$ and the algebra $\mathcal{B} = \{g(\tilde{w}) : g \in \mathbb{C}[X]\}$, we deduce that $\mathcal{B}$ is dense in $C([0,1/q])$ for the supremum norm, and therefore

$$\int_0^{1/q} \overline{F(t)} G(t) k(t)\, dt = 0, \quad k \in C([0,1/q]).$$

It follows that the $L^1([0, 1/q])$-function $\overline{F}G$ is zero almost everywhere on $[0, 1/q]$, and thus G is zero almost everywhere on $[0, 1/q]$. The result follows. □

Theorem 5.4.4 *Let $T = T_{p/q}$ where $p < q$ and p and q are coprime. A function $f \in L^2([0, 1])$ is cyclic for T if and only if $\Delta(f, 1/q)$ is non-zero almost everywhere on* $[0, 1]$.

Proof Note first that the set of functions h that satisfy the conditions of Lemma 5.4.2 for some value of n is dense in $L^2([0, 1])$. Using the previous notation, recall that for all $r \in \{0, \ldots, q-1\}$ and $n \geq 0$, we have $T^{nq+r} f = w^n f_{r+1}$. Therefore, the cyclicity of f is satisfied if, for each $1/q$-periodic function $h \in L^\infty([0, 1])$ and $\varepsilon > 0$, there exists a polynomial g such that

$$\|g(w) f_k - h f_k\|_2 < \varepsilon. \tag{5.20}$$

Fix $k \in \{1, \ldots, q\}$ and set

$$F(t) = |f_k(t)| + |f_k(\{t + 1/q\})| + \cdots + |f_k(\{t + (q-1)/q\})|.$$

Using Lemma 5.4.3, there exists a polynomial g such that

$$\|(g(w) - h)F\|_{L^2([0,1/q))} < \frac{\varepsilon}{q}.$$

It follows straightaway that (5.20) is satisfied.

Now, suppose that $\Delta(f, 1/q)$ is zero on a set S of positive measure. Then there exist $a_1, \ldots, a_{q-1}$ not all zero, such that

$$\sum_{k=0}^{q-1} a_k \phi(\{t + kp/q\}) = 0 \qquad \text{a.e. on} \quad S$$

for all ϕ in the cyclic subspace generated by $\{T^n f : n \geq 0\}$. It follows that f cannot be cyclic. □

An example of the use of this theorem can be found in the Exercises.

5.4.2 The lattice of invariant subspaces

Let q be a positive integer and denote by $\mathrm{Proj}(\mathbb{C}^q)$ the set of orthogonal projections on $\mathbb{C}^q$. The next theorem gives the description of the lattice of invariant subspaces of each rational Bishop operator $T = T_{p/q}$.

Theorem 5.4.5 *Let $P : [0, 1/q] \to \mathrm{Proj}(\mathbb{C}^q)$ be measurable and such that*

$$V(t)P(t)\mathbb{C}^q \subset P(t)\mathbb{C}^q \qquad \text{a.e. on} \qquad [0, 1/q], \tag{5.21}$$

where

$$\begin{aligned} &V(t)(a_1, \ldots, a_q) \\ &= \left(ta_2, \{t + \frac{p}{q}\}a_3, \{t + 2\frac{p}{q}\}a_4, \ldots, \{t + (q-2)\frac{p}{q}\}a_q, \{t + (q-1)\frac{p}{q}\}a_1 \right). \end{aligned}$$

Then

$$\begin{aligned} \mathcal{I}(P) := \Big\{ f \in L^2([0, 1]) : \big(f(t), f\big(\{t + \frac{p}{q}\}\big), \ldots, \\ f\big(\{t + (q-1)\frac{p}{q}\}\big)\big) \in P(t)\mathbb{C}^q \text{ a.e.} \Big\} \end{aligned} \tag{5.22}$$

is a closed T-invariant subspace. Moreover, each T-invariant subspace has the form (5.22) where P has the properties defined above.

Proof By construction $\mathcal{I}(P)$ is a closed T-invariant subspace of $L^2([0, 1])$ (the property of being T-invariant follows from (5.21)).

Now, let $\mathcal{S}$ be a (closed) T-invariant subspace of $L^2([0, 1])$ and then denote by $J(t)$ the (necessarily closed) linear span of

$$\Big\{ \big(e_{m,n}(t), e_{m,n}\big(\{t + \frac{p}{q}\}\big), \ldots, e_{m,n}\big(\{t + (q-1)\frac{p}{q}\}\big)\big) : n \in \mathbb{Z}, m \geq 0 \Big\}, \tag{5.23}$$

where $e_{m,n} = P_{\mathcal{S}} T^m e_n$, $e_n(t) = e^{2i\pi t n}$ and $P_{\mathcal{S}}$ is the orthogonal projection on $\mathcal{S}$. Clearly, $\mathcal{S}$ is contained in the subspace of all $f \in L^2([0, 1])$ such that

$$\big(f(t), f(\{t + \frac{p}{q}\}), \ldots, f\big(\{t + (q-1)\frac{p}{q}\}\big)\big) \in J(t) \qquad \text{a.e. on} \quad [0, 1/q].$$

Suppose that there exists $F \in L^2([0, 1])$ such that

$$(F(t), F(\{t + p/q\}), \ldots, F(\{t + (q-1)p/q\})) \in J(t) \quad \text{a.e. on} \quad [0, 1/q],$$

and F is orthogonal to $\mathcal{S}$. Then we have

$$\int_0^1 F(t) T^{qN} \overline{e_{m,n}}(t)\, dt = 0$$

for all $m, N \geq 0$, and all $n \in \mathbb{Z}$.

Note that

$$\int_0^1 F(t)T^{qN}\overline{e_{m,n}}(t)dt = \int_0^1 F(t)w^N(t)\overline{e_{m,n}}(t)\,dt$$
$$= \int_0^{1/q}\sum_{r=0}^{q-1} F(\{t+rp/q\})w^N(t)\overline{e_{m,n}}(\{t+rp/q\})\,dt.$$

By Lemma 5.4.3, it follows that

$$\sum_{r=0}^{q-1} F(\{t+rp/q\})\overline{e_{m,n}}(\{t+rp/q\}) = 0$$

for all $m \geq 0$, $n \in \mathbb{Z}$ and a.e. on $[0, 1/q]$. By the definition of $J(t)$, as in (5.23), the function F is 0 almost everywhere. The last assertion of the theorem follows. □

Remark 5.4.6 The subspaces $J(t) = \operatorname{Im} P(t)$ that can occur in Theorem 5.4.5 are the invariant subspaces of the operator $V(t) : \mathbb{C}^q \to \mathbb{C}^q$; for almost all t this has the property that $V(t)^q$ is a non-zero multiple of the identity, and hence $V(t)$ has q independent eigenvectors and distinct eigenvalues. The invariant subspaces of $V(t)$ are easily expressed as those subspaces spanned by a subset of the eigenvectors.

Note that for f to be a cyclic vector, it is necessary and sufficient that $J(t) = \mathbb{C}^q$ almost everywhere, which can be rewritten in terms of a linear independence condition on the columns of the determinant $\Delta(f, p/q)$, given in Definition 5.4.1. We may therefore recover Theorem 5.4.4 (this is not a simpler proof, as we have already used Lemma 5.4.3 en route).

To illustrate the above ideas we consider the case $\alpha = 1/2$. Set $w(t) = t\{t + 1/2\}$. Obviously, for all integers n, we have

$$T_{1/2}^{2n} f(t) = w^n(t)f(t) \text{ and } T_{1/2}^{2n+1} f(t) = tw^n(t)f(\{t + 1/2\}).$$

It follows that if f is 1/2-periodic, then the linear hull of $\left\{T_{1/2}^n f : n \geq 0\right\}$ is equal to $\{(p(w) + tq(w))f : p, q \in \mathbb{C}[X]\}$.

The following observation will enable us to give a different characterization of the cyclic vectors; it is also of some interest in its own right.

Lemma 5.4.7 *For all $f \in L^2([0, 1])$, there exists a unique pair of 1/2-periodic functions $f_1, f_2 \in L^2([0, 1])$ such that $f(t) = f_1(t) + tf_2(t)$.*

Proof For $0 \leq t < 1/2$, set

$$f_1(t) = (1 + 2t)f(t) - 2tf(t + 1/2) \tag{5.24}$$

and

$$f_2(t) = 2(f(t + 1/2) - f(t)). \tag{5.25}$$

For $1/2 \leq t < 1$, set $f_1(t) = f_1(t - 1/2)$ and $f_2(t) = f_2(t - 1/2)$. It is an easy calculation to check that $f(t) = f_1(t) + tf_2(t)$.

The uniqueness follows from the fact that if one has $f(t) = f_1(t) + tf_2(t)$, with $f_1, f_2 \in L^2([0, 1])$ 1/2-periodic, then necessarily $f_1(t)$ and $f_2(t)$ are given by (5.24) and (5.25). □

Inspired by Lemma 5.4.7, we may introduce the following notation. For $f \in L^2([0, 1])$, write $f = (f_1, f_2)$ whenever f_1, f_2 are 1/2-periodic functions in $L^2([0, 1])$, such that $f(t) = f_1(t) + tf_2(t)$.

Note that the determinant of the system of equations

$$h_1(t)f(t) + h_2(t)g(t) = h(t),$$
$$h_1(t)f(t + 1/2) + h_2(t)g(t + 1/2) = h(t + 1/2),$$

which is a special case of (5.19), is equal to

$$f(t)g(t + 1/2) - g(t)f(t + 1/2) = 1/2\left(f_1^2(t) - w(t)f_2^2(t)\right).$$

Thus we may make the following definition, in order to rewrite the characterization of the cyclic vectors of $T_{1/2}$.

Definition 5.4.8 Let $f = (f_1, f_2)$ be a function in $L^2([0, 1])$. The *determinant of f associated with $T_{1/2}$*, denoted by $\Delta(f, 1/2)$, is the $L^1([0, 1])$-function defined by

$$\Delta(f, 1/2)(s) = \frac{1}{2}\left(f_1^2(s) - w(s)f_2^2(s)\right),$$

where $w(s) = s\{s + 1/2\}$.

Thus the following result, which follows immediately from Theorem 5.4.4, enables us to write down immediately a large number of cyclic vectors.

Corollary 5.4.9 *A function $f = (f_1, f_2) \in L^2([0, 1])$ is cyclic for $T_{1/2}$ if and only if $\Delta(f, 1/2)$ is non-zero almost everywhere on* [0, 1].

In particular, it follows that the characteristic function χ_Ω, where Ω is a Borel set of $[0, 1]$, is cyclic for $T_{1/2}$ if and only if the measure of the Borel set $\{t \in \Omega \text{ or } \{t + 1/2\} \in \Omega\}$ is equal to 1.

Exercises

1. Verify the properties of A_ρ, asserted in Proposition 5.1.2.
2. Show that if $(\|x_n\|)_{n\in\mathbb{Z}}$ is dominated by a Beurling sequence, then the series (5.7) defining $G_1(z)$ converges absolutely in $\mathbb{C} \setminus \mathbb{T}$.
3. Let U_1 and U_2 be open subsets of $\mathbb{T}$. Show that if A_ρ is a regular Beurling algebra, then there exist functions $f_1, f_2 \in A_\rho$, with each f_k supported on U_k, such that $\chi_{U_1 \cup U_2} = f_1 + f_2$ (this is normally called a *partition of unity*). Deduce that if $f \in A_\rho$ is supported on $U_1 \cup U_2$, then we may write $f = f_1 + f_2$, where each f_k lies in A_ρ and is supported on U_k.
4. Derive the estimates assumed in the proof of Lemma 5.2.5.
5. Show that, if α is irrational, then the transformation $x \mapsto \{x+\alpha\}$ is ergodic, in the sense that there are no (a.e.) invariant subsets except those of measure 0 or 1.
6. Determine whether the function $s \mapsto s^{1/2}$ is cyclic for a rational Bishop operator $T_{p/q}$.

Comments

Our treatment of Beurling algebras, and in particular Theorem 5.1.7, is based on the versions given by Colojoară and Foiaş [82] and Dales [87]. The theorem was stated [82, p. 153] with the additional condition $\lim_{|n|\to\infty} \rho_n^{1/n} = 1$. Note that this is equivalent to $\lim_{|n|\to\infty} \frac{\log \rho_n}{|n|} = 0$. Now, using Fekete's lemma [106, 166], we know that the sub-additivity of $(\log \rho_n)_n$ implies that $\lim_{|n|\to\infty} \frac{\log \rho_n}{|n|}$ exists and is equal to $\inf_{|n|\to\infty} \frac{\log \rho_n}{|n|}$. Moreover, $\inf_{|n|\to\infty} \frac{\log \rho_n}{|n|} = \delta > 0$ would contradict the Beurling condition (5.3). It follows that the condition $\lim_{|n|\to\infty} \rho_n^{1/n} = 1$ is automatically true in a Banach algebra A_ρ satisfying the Beurling condition.

The material of Section 5.2 is based on Atzmon [20] and Wermer [201].

More on the singularities of the Carleman transform, with particular reference to the formula (5.11), can be found in Domar [93].

The pioneering work on the theory of Bishop operators is by Davie [89], and the proof of Theorem 5.3.3 is based on this work. The notion of ρ-regularity is inspired by the work of Flattot [107]. Further work presenting Davie's functional calculus in the context of a product of Bishop operators can be found in Chalendar and Partington [74].

The material of Section 5.3.2 is based on Chalendar, Partington and Pozzi [75], which also presents a multivariable version. Again, many of the original ideas are present in Davie [89].

6
Applications of a fixed-point theorem

One of the oldest results on invariant subspaces is the theorem of Aronszajn and Smith [18], published in 1954, that every compact operator has a non-trivial invariant subspace. Note that, since every non-zero point of the spectrum of a compact operator is an eigenvalue, this result gives new information only in the quasinilpotent case. In fact, the somewhat simpler result for a compact operator on a Hilbert space had been given earlier in unpublished work of von Neumann. The method used to prove the Aronszajn–Smith result was based on the idea of a metric projection onto a finite-dimensional subspace of a Banach space, that is, the construction of a closest point in the subspace.

A much stronger result, given in 1973 by Lomonosov [145], concerns hyperinvariant subspaces for operators that commute with compact operators, and we shall begin by presenting this. The method of proof involves the use of the Schauder–Tychonoff fixed-point theorem,and this will be a unifying theme for this chapter, since we shall then present more recent generalizations due to Simonič [185], which also apply this fixed-point theorem, and which are concerned with essentially self-adjoint operators (that is, operators T for which $T - T^*$ is compact). These results constitute a significant advance on the Lomonosov result, although at present they mainly apply to real normed spaces.

6.1 Operators commuting with compact operators

An alternative proof of the existence of hyperinvariant subspaces for compact operators, using minimal vectors, is given as Corollary 7.4.5. The main result derived here (Theorem 6.1.2) is slightly stronger.

As a consequence of the Schauder–Tychonoff fixed-point theorem (see Theorem 1.1.22) we obtain the following lemma.

Lemma 6.1.1 *Let $\mathcal{X}$ be a Banach space and let $\mathcal{A}$ be a sub-algebra of $\mathcal{L}(\mathcal{X})$ such that*

$$\bigcap_{A\in\mathcal{A}} \operatorname{Lat} A = \{\{0\}, \mathcal{X}\}.$$

Then, for any compact operator $K \in \mathcal{L}(\mathcal{X}) \setminus \{0\}$, there exist $A \in \mathcal{A}$ and $x_0 \in \mathcal{X} \setminus \{0\}$ such that $KAx_0 = x_0$.

Proof For $y \in \mathcal{X}$, denote by $B_1(y)$ the open ball centred at y and of unit norm. Since K is not identically zero, there exists $y \in \mathcal{X}$ such that the closure C of $KB_1(y)$ does not contain 0 (consider $z \in \mathcal{X}$ such that $\|Kz\| = \delta > 0$ and take $y = tz$ for any $t > 0$ with $t\delta > \|K\|$). Note that C is a compact and convex set. Since $\bigcap_{A\in\mathcal{A}} \operatorname{Lat} A$ is trivial, $\overline{\mathcal{A}x}$, the closure of $\mathcal{A}x$, is equal to $\mathcal{X}$ for all $x \in \mathcal{X} \setminus \{0\}$. Now $0 \notin C$, so it follows that $\overline{\mathcal{A}x} = \mathcal{X}$, and thus $\mathcal{A}x \cap B_1(y) \neq \emptyset$, for all $x \in C$. Therefore,

$$C \subset \bigcup_{A\in\mathcal{A}} A^{-1}(B_1(y)).$$

Since C is compact, there exists a finite subset of $\mathcal{A}$, say $\{A_1, \ldots, A_n\}$, such that

$$C \subset \bigcup_{i=1}^{n} A_i^{-1}(B_1(y)).$$

Fix any continuous map $r : \mathbb{R}^+ \to \mathbb{R}^+$ such that $r^{-1}(\{0\}) = [1, +\infty)$. Define $f : C \to B_1(y)$ by

$$f(x) = \sum_{i=1}^{n} \frac{r(\|A_i x - y\|)}{\sum_{j=1}^{n} r(\|A_j x - y\|)} A_i x.$$

The function f is well-defined since, for all $x \in C$, there is an integer $i_0 \in \{1, \ldots, n\}$ such that $A_{i_0}x \in B_1(y)$, which implies that $r(\|A_{i_0}x - y\|) > 0$. We now consider the continuous function $F = K \circ f$ from C to C. By Theorem 1.1.22 there exists $x_0 \in C$ such that $F(x_0) = x_0$. Define $A \in \mathcal{A}$ by

$$Au = \sum_{i=1}^{n} \frac{r(\|A_i x_0 - y\|)}{\sum_{j=1}^{n} r(\|A_j x_0 - y\|)} A_i u.$$

We see immediately that $Ax_0 = f(x_0)$ and then $KAx_0 = F(x_0) = x_0$ with $x_0 \neq 0$ since $0 \notin C$. □

Lemma 6.1.1 and elementary linear algebra yield a proof of the existence of hyperinvariant subspaces for operators commuting with a non-trivial compact operator. This is Lomonosov's theorem.

Theorem 6.1.2 (Lomonosov) *Let $T \in \mathcal{L}(\mathcal{X}) \setminus \mathbb{C}\mathrm{I}$. Suppose that there exists a non-zero compact operator K such that $TK = KT$. Then T has a non-trivial hyperinvariant subspace.*

Proof Assume towards a contradiction that the algebra $\mathcal{A} = \{T\}'$, the commutant of T, has no non-trivial invariant subspace. Then, by Lemma 6.1.1, there are $A \in \{T\}'$ and a non-zero $x \in \mathcal{X}$ such that

$$\mathrm{Ker}(\mathrm{I} - KA) \neq \{0\}.$$

Since KA is compact, the dimension of $\mathrm{Ker}(\mathrm{I} - KA)$ is finite. Moreover, since $KA \in \{T\}'$, T has a finite-dimensional invariant subspace and thus T has a non-empty point spectrum. This is impossible since any eigenspace of T is in $\bigcap_{A \in \{T\}'} \mathrm{Lat}\, A$. □

6.2 Essentially self-adjoint operators

Let $\mathcal{H}$ be a real or complex Hilbert space. An operator $T \in \mathcal{L}(\mathcal{H})$ is said to be *essentially self-adjoint* if it can be written $T = S + K$, where $S \in \mathcal{L}(\mathcal{H})$ is self-adjoint and $K \in \mathcal{L}(\mathcal{H})$ is compact.

In this section we shall show, using fixed-point arguments, that every essentially self-adjoint operator on an infinite-dimensional real Hilbert space has a non-trivial closed invariant subspace. It is still unknown whether, in the complex case, a complex invariant subspace can be found. However, a complex Hilbert space is also a real Hilbert space with the inner product

$$\langle x, y \rangle_{\mathbb{R}} = \mathrm{Re}\langle x, y \rangle,$$

and a complex self-adjoint operator remains real self-adjoint. Thus it will follow that every essentially self-adjoint operator on an infinite-dimensional complex Hilbert space has a non-trivial closed *real* invariant subspace.

6.2.1 Preliminaries

Now let $\mathcal{H}$ be a real Hilbert space, and fix a unit vector $u \in \mathcal{H}$. For a fixed $r \in (0, 1)$ we write $\mathcal{B} = \{x \in \mathcal{H} : \|x - u\| \leq r\}$.

We shall require the following result in order to construct mappings from $\mathcal{B}$ into itself to which a fixed-point theorem can be applied.

Lemma 6.2.1 *Let $\mathcal{W}$ be a subset of $\mathcal{B}$ and let $T \in \mathcal{L}(\mathcal{H})$. Suppose that*

$$\langle Tx, u - x \rangle \geq \delta > 0 \qquad \textit{for all} \quad x \in \mathcal{W}.$$

Then, provided that $\varepsilon > 0$ *is sufficiently small, we have*

$$\|(\mathrm{I} + \varepsilon T)x - u\| < \|x - u\| \qquad \textit{for all} \quad x \in \mathcal{W}.$$

Proof This is a straightforward verification, expanding

$$\|(\mathrm{I} + \varepsilon T)x - u\|^2 = \|x - u\|^2 - 2\varepsilon\langle Tx, u - x\rangle + \varepsilon^2\|Tx\|^2,$$

and noting that, if $\varepsilon > 0$ is sufficiently small, we have $\varepsilon\|Tx\|^2 < 2\delta$ for all x in the bounded set $\mathcal{B}$. □

We define the *polar hyperplane* $\mathcal{P}_{\mathcal{B}}$ by

$$\mathcal{P}_{\mathcal{B}} = \{x \in \mathcal{H} : \langle x, u\rangle = 1 - r^2\},$$

that is, the set of all vectors $x = (1 - r^2)u + x'$, where $x' \perp u$. It plays a role in the analysis of the following non-linear mapping of $\mathcal{B}$ to itself.

Lemma 6.2.2 *The function* $\Lambda_0 : \mathcal{B} \to \mathcal{B}$, *given by*

$$\Lambda_0(x) = \frac{1}{r^2 + \langle x, u\rangle}x,$$

is a weakly continuous mapping of $\mathcal{B}$ *to itself, and its set of fixed points is* $\mathcal{P}_{\mathcal{B}} \cap \mathcal{B}$.

Proof It is easy to see that Λ_0 is well-defined and weakly continuous on $\mathcal{B}$. Moreover, $\Lambda_0 x = x$ if and only if $\langle x, u\rangle = 1 - r^2$, which is to say that $x \in \mathcal{P}_{\mathcal{B}}$.

It remains to show that if $\|x - u\| \le r$ then $\|\Lambda_0(x) - u\| \le r$. To see this, we begin with the case that $\|x - u\| = r$, and adopt an obvious coordinate system in the plane of u and x, namely $u = (1, 0)$ and $x = (1 + r\cos\theta, r\sin\theta)$, where $\theta \in [0, 2\pi]$. Then

$$\Lambda_0(x) - u = \frac{(-r^2, r\sin\theta)}{r^2 + 1 + r\cos\theta},$$

and so

$$\|\Lambda_0(x) - u\|^2 = \frac{r^4 + r^2\sin^2\theta}{(r^2 + 1 + r\cos\theta)^2} \le r^2,$$

since the inequality $r^2 + \sin^2\theta \le (r^2 + 1 + r\cos\theta)^2$ is equivalent to the positivity of

$$(r^2 + 1)\cos^2\theta + 2r(r^2 + 1)\cos\theta + r^2(r^2 + 1),$$

which is just $(r^2 + 1)(\cos\theta + r)^2$.

Now for $\|x-u\| = s < r$ we may write $x = \langle x, u\rangle u + t_0 y$ where y is a unit vector orthogonal to u, and $(\langle x, u\rangle - 1)^2 + t_0^2 = s^2$. Then Λ_0 is linear on the segment $[x_0, x_1]$ where $x_0 = \langle x, u\rangle u - t_1 y$, $x_1 = \langle x, u\rangle u + t_1 y$, and $(\langle x, u\rangle - 1)^2 + t_1^2 = r^2$. Since Λ_0 maps both x_0 and x_1 into $\mathcal{B}$, it does the same for x, which is a convex combination of x_0 and x_1. □

We next wish to make a link with essential norms, as follows.

Lemma 6.2.3 *Let $T \in \mathcal{L}(\mathcal{H})$ and $\delta > 0$. Then there is a weak neighbourhood $\mathcal{W}$ of $0 \in \mathcal{H}$ such that, whenever $x \in \mathcal{W}$ and $\|x\| \le 1$, we have*

$$|\langle Tx, x\rangle| \le \|T\|_e + \delta.$$

Proof For any compact operator K we may find a weak neighbourhood $\mathcal{W}$ of 0 such that $|\langle Kx, x\rangle| < \delta/2$ for $x \in \mathcal{W}$ and $\|x\| \le 1$ (pick a finite-rank operator such that $\|K - F\| < \delta/4$ and choose the weak neighbourhood $\mathcal{W}$ so that $|\langle Fx, x\rangle| < \delta/4$ for $x \in \mathcal{W}$ and $\|x\| \le 1$).

Now, if K is a compact operator such that $\|T - K\| < \|T\|_e + \delta/2$, we have

$$\|T\|_e \ge \|T - K\| - \delta/2 \ge |\langle (T-K)x, x\rangle| - \delta/2 \ge |\langle Tx, x\rangle| - \delta$$

for $x \in \mathcal{W}$ with $\|x\| \le 1$. □

For $T \in \mathcal{L}(\mathcal{H})$ we define a real function $\Delta_T : \mathcal{B} \to \mathbb{R}$ by

$$\Delta_T(x) = \frac{1}{r^2(1-r^2)}\langle Tx, u - x\rangle.$$

We have one more technical lemma to establish before we are ready to use a fixed-point theorem.

Lemma 6.2.4 *Suppose that $x \subset \mathcal{P}_{\mathcal{B}} \cap \mathcal{B}$ and*

$$\Delta_T(x) > \|T\|_e \Delta_I(x)$$

for some $T \in \mathcal{L}(\mathcal{H})$. Then there exists a number $\delta > 0$ and a weak neighbourhood $\mathcal{W}$ of x such that

$$\Delta_T(z) > \|T\|_e |\Delta_I(z)| + \delta$$

for all $z \in \mathcal{W} \cap \mathcal{B}$.

Proof Choose $\delta > 0$ such that $\Delta_T(x) > \|T\|_e \Delta_I(x) + 5\delta$.

We first find a weak neighbourhood $\mathcal{W}_0$ of $\mathcal{P}_{\mathcal{B}}$ such that $\|T\|_e \Delta_I(z) > -\delta$ for all $z \in \mathcal{W}_0 \cap \mathcal{B}$. Suppose that $z \in \mathcal{B}$ and write $z = \lambda u + z'$ where $\lambda = \langle z, u\rangle$ and $z' \perp u$. Then $\|z - u\|^2 = (1-\lambda)^2 + \|z'\|^2 \le r^2$. Also

$$\begin{aligned}\langle z, u - z\rangle &= \langle \lambda u + z', (1-\lambda)u - z'\rangle = \lambda(1-\lambda) - \|z'\|^2 \\ &\geq \lambda(1-\lambda) - r^2 + (1-\lambda)^2 = 1 - \lambda - r^2,\end{aligned}$$

and hence if $z \in \mathcal{B}$ lies in the weak neighbourhood $\mathcal{W}_0$, where

$$\mathcal{W}_0 = \{z \in \mathcal{H} : \langle z, u\rangle < 1 - r^2 + \varepsilon\},$$

and

$$0 < \varepsilon < \frac{\delta r^2(1-r^2)}{1 + \|T\|_e},$$

then

$$\|T\|_e \Delta_I(z) = \|T\|_e \frac{\langle z, u - z\rangle}{r^2(1-r^2)} > -\delta.$$

Next, it follows easily from Lemma 6.2.3 that we may find a weak neighbourhood $\mathcal{W}_1$ of 0 such that

$$\langle Ty, y\rangle < \|T\|_e \|y\|^2 + r^2(1-r^2)\delta$$

for all $y \in \mathcal{W}_1$ with $\|y\| \leq 2$.

We may also find a weak neighbourhood $\mathcal{W}_2$ of 0 such that

$$\langle Ty, u - x\rangle - \langle Tx, y\rangle > \|T\|_e \left(\langle y, u - x\rangle - \langle x, y\rangle\right) - r^2(1-r^2)\delta$$

for all $y \in \mathcal{W}_2$, since both sides of the inequality are weakly continuous in y.

Take $\mathcal{W} = \mathcal{W}_0 \cap (x + \mathcal{W}_1 \cap \mathcal{W}_2)$, so that if $z \in \mathcal{W}$ we have $z = x + y$ where $y \in \mathcal{W}_1 \cap \mathcal{W}_2$ and also $\|T\|_e \Delta_I(z) > -\delta$. Then provided that z also lies in $\mathcal{B}$ we have $\|y\| \leq 2$, and hence

$$\begin{aligned}\Delta_T(z) &= \Delta_T(x + y) \\ &= \Delta_T(x) + \frac{1}{r^2(1-r^2)}(\langle Ty, u - x\rangle - \langle Tx, y\rangle - \langle Ty, y\rangle) \\ &> \|T\|_e \left(\Delta_I(x) + \frac{1}{r^2(1-r^2)}(\langle y, u - x\rangle - \langle x, y\rangle - \langle y, y\rangle)\right) + 3\delta \\ &= \|T\|_e \Delta_I(x + y) + 3\delta \\ &= \|T\|_e \Delta_I(z) + 3\delta > \|T\|_e |\Delta_I(z)| + \delta,\end{aligned}$$

using the fact that $\|T\|_e \Delta_I(z) > -\delta$. This is the required result. □

We now come to the main application of a fixed-point theorem.

Proposition 6.2.5 *Let $\mathcal{H}$ be a real Hilbert space, and $\mathcal{C} \subset \mathcal{L}(\mathcal{H})$ be a convex set of operators. Let $u \in \mathcal{H}$ be a unit vector, and $r \in (0, 1)$. Write*

$c_r = r/\sqrt{1-r^2}$. Suppose that for each $v \perp u$ with $\|v\| \leq 1$ there is an operator $T \in \mathcal{C}$ such that

$$\left\langle T(u+c_r v), u-c_r^{-1}v\right\rangle > \|T\|_{\mathrm{e}}\langle u+c_r v, u-c^{-1}rv\rangle = \|T\|_{\mathrm{e}}(1-\|v\|^2). \quad (6.1)$$

Then $\mathcal{C}$ contains an operator T_0 with an eigenvector in the set

$$\mathcal{B} = \{x \in \mathcal{H} : \|x-u\| \leq r\},$$

for which the eigenvalue λ satisfies $|\lambda| > \|T_0\|_{\mathrm{e}}$.

Proof Every vector $x \in \mathcal{P}_{\mathcal{B}} \cap \mathcal{B}$ can be written as $x = (1-r^2)(u+c_r v)$ with $v \perp u$ and $\|u\| \leq 1$, so that $u - x = r^2\left(u - c_r^{-1}v\right)$. By (6.1), such an x satisfies the hypotheses of Lemma 6.2.4, for some operator $T \in \mathcal{C}$. Hence for every $x \in \mathcal{P}_{\mathcal{B}} \cap \mathcal{B}$ there is an operator $T \in \mathcal{C}$, a basic weak neighbourhood $\mathcal{W}$ of x, and a positive number δ such that for every $y \in \mathcal{W} \cap \mathcal{B}$ one has

$$\Delta_T(y) > \|T\|_{\mathrm{e}}|\Delta_{\mathrm{I}}(y)| + \delta.$$

By Lemma 6.2.1, $\mathrm{I}+\varepsilon T$ maps $\mathcal{W}\cap\mathcal{B}$ into $\mathcal{B}$ whenever $\varepsilon > 0$ is sufficiently small.

By the weak compactness of $\mathcal{P}_{\mathcal{B}}\cap\mathcal{B}$ we can cover the set with finitely many neighbourhoods $\mathcal{W}_1, \ldots, \mathcal{W}_n$ and choose an associated operator $T_k \in \mathcal{C}$ and numbers $\mu_k > 0$ such that, for $\varepsilon \in (0, \mu_k)$, the operator $\mathrm{I}+\varepsilon T_k$ maps $\mathcal{W}_k \cap \mathcal{B}$ into $\mathcal{B}$ and

$$\Delta_{T_k}(y) > \|T_k\|_{\mathrm{e}}|\Delta_{\mathrm{I}}(y)| + \delta \qquad (y \in \mathcal{W}_k \cap \mathcal{B}). \quad (6.2)$$

Fix $\varepsilon \in (0, \min\{\mu_1, \ldots, \mu_n\})$.

We also let $\mathcal{W}_0$ be the weakly open set

$$\mathcal{W}_0 = \{x \in \mathcal{H} : |\langle x, u\rangle - (1-r^2)| > 0\},$$

and define $\Gamma_0(x) = |\langle x, u\rangle - (1-r^2)|$.

For $\mathcal{W}$ any basic weak neighbourhood of a vector $z \in \mathcal{H}$, defined by a finite set $\{w_1, \ldots, w_n\}$, i.e.,

$$\mathcal{W} = \{x \in \mathcal{H} : |\langle x - z, w_k\rangle| < 1 \quad \text{for} \quad k = 1, \ldots, n\},$$

we may define a weakly continuous non-negative function $\Gamma_{\mathcal{W}} : \mathcal{H} \to [0, 1]$ by

$$\Gamma_{\mathcal{W}}(x) = \prod_{k=1}^{n} \max\{0, 1 - |\langle x - z, w_k\rangle|\},$$

satisfying $\mathcal{W} = \{x \in \mathcal{H} : \Gamma_{\mathcal{W}}(x) > 0\}$.

In particular we define $\Gamma_k : \mathcal{B} \to [0, 1]$ by $\Gamma_k = \Gamma_{\mathcal{W}_k}$ for $k = 1, \ldots, n$. Let

$$\alpha_k(x) = \frac{\Gamma_k(x)}{\sum_{j=0}^{n} \Gamma_j(x)} \qquad (k = 0, \ldots, n).$$

We define a mapping $\Lambda : \mathcal{B} \to \mathcal{B}$ by

$$\Lambda(x) = \frac{\alpha_0(x)}{r^2 + \langle x, u\rangle} x + \sum_{k=1}^{n} \alpha_k(x)(\mathrm{I} + \varepsilon T_k)x.$$

Using Lemma 6.2.2, and since $\sum_{k=0}^{n} \alpha_k(x) = 1$ for every $x \in \mathcal{B}$, we see that $\Lambda(x)$ is a convex combination of elements of $\mathcal{B}$, and Λ maps $\mathcal{B}$ into itself weakly continuously.

Now the Schauder–Tychonoff fixed-point theorem (see Theorem 1.1.22) implies that there is a point $x_1 \in \mathcal{B}$ such that $\Lambda(x_1) = x_1$, and hence

$$\varepsilon \left(\sum_{k=1}^{n} \alpha_k(x_1) T_k \right) x_1 = \alpha_0(x_1) \left(1 - \frac{1}{r^2 + \langle x_1, u\rangle} \right) x_1.$$

Now $x_1 \in \mathcal{W}_\ell$ for at least one $\ell > 0$, since otherwise $x_1 \in \mathcal{W}_0$ and consequently $\Lambda(x_1) = (r^2 + \langle x_1, u\rangle)^{-1} x_1 \neq x_1$.

Thus $\alpha_\ell(x_1) > 0$, and we may define

$$\beta_k = \frac{\alpha_k(x_1)}{1 - \alpha_0(x_1)} \qquad (k = 1, \ldots, n),$$

and take $T_0 = \sum_{k=1}^{n} \beta_k T_k \in \mathcal{C}$. Moreover, $T_0 x_1 = \lambda x_1$, where

$$\lambda = \frac{\alpha_0(x_1)}{\varepsilon(1 - \alpha_0(x_1))} \left(1 - \frac{1}{r^2 + \langle x_1, u\rangle} \right).$$

Further, by (6.2),

$$\Delta_{T_0}(x_1) = \sum_{k=1}^{n} \beta_k \Delta_{T_k}(x_1) > \sum_{k=1}^{n} \beta_k \|T_k\|_{\mathrm{e}} |\Delta_{\mathrm{I}}(x_1)| \geq \|T_0\|_{\mathrm{e}} |\Delta_{\mathrm{I}}(x_1)|.$$

So

$$\begin{aligned} |\lambda|\, |\Delta_I(x_1)| \geq \lambda \Delta_{\mathrm{I}}(x_1) &= \frac{1}{r^2(1 - r^2)} \lambda \langle x_1, x_1 - u\rangle \\ &= \Delta_{T_0}(x_1) > \|T_0\|_{\mathrm{e}} |\Delta_{\mathrm{I}}(x_1)|. \end{aligned}$$

Thus $\Delta_{\mathrm{I}}(x_1) \neq 0$ and so $|\lambda| > \|T_0\|_{\mathrm{e}}$. □

6.2.2 Application to invariant subspaces

Our standing hypotheses will be that T is an essentially self-adjoint operator on a real Hilbert space $\mathcal{H}$, and that $T = S + K$, with $S = S^*$ and $K \in \mathcal{K}(\mathcal{H})$. Without loss of generality we take $\|S\| \leq 1$ and let $E \subset [-1, 1]$ denote the essential spectrum of T.

One obstruction to finding invariant subspaces in the real case is that an operator may satisfy an irreducible polynomial equation. This possibility is eliminated in the next result.

Lemma 6.2.6 *Let $T = S + K$ be an essentially self-adjoint operator on a real Hilbert space, and suppose that S is not a multiple of the identity. Then T cannot satisfy an irreducible polynomial equation.*

Proof The only possibility is that T satisfies an irreducible quadratic equation $(T + a\mathrm{I})^2 + b^2\mathrm{I} = 0$. This gives

$$(S + a\mathrm{I})^2 + K(S + a\mathrm{I}) + (S + a\mathrm{I})K + K^2 + b^2\mathrm{I} = 0,$$

implying that $(S + a\mathrm{I})^2 + b^2\mathrm{I}$ is compact. Since S is self-adjoint, this cannot happen unless $b = 0$ and $S = -a\mathrm{I}$. □

Clearly, the case that S is a multiple of the identity is covered by Theorem 6.1.2.

We shall require two (not necessarily complete) normed algebras: let $\mathcal{P}(E)$ denote the algebra of real polynomials, with norm

$$\|p\|_\infty = \max_{t \in E} |p(t)|,$$

and let $\mathcal{A}$ denote the algebra of all operators $p(T)$, where p is a real polynomial.

Definition 6.2.7 A *state* on the algebra $\mathcal{P}(E)$ is a positive functional $\tau : \mathcal{P}(E) \to \mathbb{R}$ such that $\|\tau\| = \tau(1) = 1$. We write $\mathcal{S}(E)$ for the set of all states on $\mathcal{P}(E)$.

One way to obtain a state is as follows. For $x \in \mathcal{H}$ we write

$$\mathcal{S}_x := \{y \in \mathcal{H} : \tau_{x,y} \in \mathcal{S}(E), \text{ where } \tau_{x,y}(p) = \langle p(T)x, y\rangle\}.$$

Clearly if $x \neq 0$ then $\mathcal{S}_x$ is contained in a hyperplane

$$\mathcal{M}_x := \{y \in \mathcal{H} : \langle x, y\rangle = 1\}.$$

Lemma 6.2.8 *It holds that $\tau_{x,y}$ is a bounded positive functional on $\mathcal{P}(E)$, if and only if the* weak von Neumann inequality *is satisfied; namely,*

$$\langle p(T)x, y\rangle \leq \|p\|_\infty \langle x, y\rangle, \qquad p \in \mathcal{P}(E). \tag{6.3}$$

Proof If $\tau_{x,y}$ is a positive functional, then (6.3) follows immediately from the observation that $q := \|p\|_\infty - p$ is a positive polynomial, and thus

$$\|p\|_\infty \langle x, y\rangle - \langle p(T)x, y\rangle = \tau(q) \geq 0.$$

Conversely, if (6.3) holds, then we may suppose that $\langle x, y\rangle \geq 0$, as otherwise $p \equiv -1$ would violate (6.3). For any positive polynomial p we may write $q = \|p\|_\infty - p$. Then

$$\|p\|_\infty \langle x, y\rangle - \langle p(T)x, y\rangle = \langle q(T)x, y\rangle \leq \|q\|_\infty \langle x, y\rangle \leq \|p\|_\infty \langle x, y\rangle,$$

and so $\langle p(T)x, y\rangle \geq 0$, as required. □

We define the set $\mathcal{D} \subset \mathcal{H}$ to be the set of all non-zero vectors $x \in \mathcal{H}$ such that, for some non-zero $y \in \mathcal{H}$, the weak von Neumann inequality (6.3) holds for all $p \in \mathcal{P}(E)$. We shall show that $\mathcal{D}$ is non-empty, and hence that there are many states of the form $\tau_{x,y}$.

Lemma 6.2.9 *Suppose that the essentially self-adjoint operator T has no non-trivial invariant subspaces. Then the set $\mathcal{D}$ is dense in $\mathcal{H}$, and $\mathcal{S}_x \neq \emptyset$ for all $x \in \mathcal{D}$.*

Proof Consider Proposition 6.2.5, as it applies with $\mathcal{C} = \mathcal{A}$. If for some (necessarily non-constant) polynomial p the operator $p(T)$ has an eigenvector $\lambda > \|p(T)\|_e$, then $\operatorname{Ker} p(T) - \lambda I$ is a non-trivial invariant subspace for T unless $p(T) - \lambda I = 0$. In that case T is algebraic with minimal polynomial m_T, say. By Lemma 6.2.6, m_T cannot be irreducible unless T is a multiple of the identity. Now $\operatorname{Ker} q(T)$ is a non-trivial invariant subspace for T whenever q is a non-trivial divisor of m_T.

Hence, given any unit vector $u \in \mathcal{H}$, there is a vector $v \perp u$ such that condition (6.1) fails for every $p \in \mathcal{P}(E)$. Since T is essentially self-adjoint, we have $\|p(T)\|_e = \|p\|_\infty$, and thus $x := u + c_r v$ and $y := u - c_r^{-1} v$ (which depend on r) satisfy the weak von Neumann inequality (6.3). Thus $x \in \mathcal{D}$, since $y \neq 0$. As $r \to 0$, $u + c_r v \to u$, and this implies that $\mathcal{D}$ is dense.

Now if $x \in \mathcal{D}$ and $y \neq 0$, we cannot have $\tau_{x,y} \equiv 0$, as x is a cyclic vector. Thus, as $\tau_{x,y}$ is a non-trivial positive linear functional on $\mathcal{P}(E)$, we may define a state by taking $\tau = \tau_{x,y}/\tau_{x,y}(1)$; that is, $y/\tau_{x,y}(1) \in \mathcal{S}_x$. □

Suppose that $\mathcal{S}_x$ is non-empty. Then, since it is closed and convex, we can define a unique closest-point mapping

$$P_x : \mathcal{H} \to \mathcal{S}_x$$

such that

$$\|z - P_x z\| = \inf_{y \in \mathcal{S}_x} \|z - y\|.$$

Lemma 6.2.10 *For each $x \neq 0$, either x is a non-cyclic vector for T or $\mathcal{S}_x \neq \mathcal{M}_x$.*

Proof If $\mathcal{S}_x = \mathcal{M}_x$ then there is a vector $y \in \mathcal{S}_x$ and a unit vector $u \perp x$ such that $y + \lambda u \in \mathcal{S}_x$ for all $\lambda \geq 0$.

Thus $|\langle T^k x, y + \lambda u\rangle| \leq 1$ for all $k = 0, 1, 2, \ldots$. Hence, letting $\lambda \to \infty$ we have $\langle T^k x, u\rangle = 0$ for each k. Since $u \neq 0$, x is a non-cyclic vector for T. □

We shall also require the following calculation, which will help us produce non-cyclic vectors.

Lemma 6.2.11 *If $x \in \mathcal{D}$, then*

$$\langle T^k \left((\|P_x z\|^2 - \langle P_x z, z\rangle)x + (\mathrm{I} - P_x)z \right), P_x z\rangle = 0$$

for all $k = 0, 1, 2, \ldots$ and $z \in \mathcal{H}$.

Proof Write $A = T^{*k}$ and take $y \in \mathcal{S}_x$. We may define a function $\Phi : (-1, 1) \to \mathbb{R}^+$ by

$$\Phi(\lambda) = \langle (\mathrm{I} + \lambda A)y, x\rangle^{-1} (\mathrm{I} + \lambda A)y,$$

which is well-defined, since $\langle (\mathrm{I} + \lambda T^k)x, y\rangle \neq 0$ when $y \in \mathcal{S}_x$ and $|\lambda| < 1$. Moreover, $\Phi(\lambda)$ also lies in $\mathcal{S}_x$. Now for $z \in \mathcal{H}$ we set

$$\psi(\lambda) = \|\Phi(\lambda) - z\|^2.$$

This is differentiable and it is an easy calculation (Exercise 4) to see that

$$\psi'(0) = 2\left\langle Ay, y - z - (\|y\|^2 - \langle y, z\rangle)x \right\rangle. \tag{6.4}$$

Since $\Phi(0) = y$, the function ψ has a minimum at $\lambda = 0$ when $y = P_x z$; in that case, $\psi'(0) = 0$. The result now follows. □

Theorem 6.2.12 (Simonič) *Let $\mathcal{H}$ be a real infinite-dimensional Hilbert space, and let $T \in \mathcal{L}(\mathcal{H})$ be an essentially self-adjoint operator. Then T has a non-trivial closed invariant subspace.*

Proof By Lemma 6.2.9 we may suppose without loss of generality that $\mathcal{D}$ is dense in $\mathcal{H}$.

Suppose that $x \in \mathcal{D}$; then we note by Lemma 6.2.11 that for each $z \neq 0$, since $P_x z \neq 0$, it follows that the vector

$$y_z := (\|P_x z\|^2 - \langle P_x z, z\rangle)x + (\mathrm{I} - P_x)z$$

is non-cyclic for T.

If x is itself a cyclic vector for T, then by Lemma 6.2.10 we can find $z \in \mathcal{M}_x \setminus \mathcal{S}_x$. Now, if $y_z = 0$, then $P_x z - z = \lambda x$ for some $\lambda \in \mathbb{R}$, and so $\langle \lambda x, P_x z\rangle x = \lambda x$. Either $P_x z = z$ (which would mean that $z \in \mathcal{S}_x$) or $\langle x, P_x z\rangle = 1$, and hence $x \perp (P_x z - z)$, so again $P_x z = z$.

Therefore, $y_z \neq 0$, and hence T has a non-trivial invariant subspace. □

Exercises

1. Let $\mathcal{H}$ be a Hilbert space with orthonormal basis $(e_n)_{n\in\mathbb{Z}}$, and let the mapping $f : \mathcal{H} \to \mathcal{H}$ be defined by

$$f(x) = \frac{1}{2}(1 - \|x\|)e_0 + Ux,$$

where U is the bilateral shift such that $Ue_n = e_{n+1}$ for each $n \in \mathbb{Z}$. Show that f restricts to a continuous map from the closed unit ball of $\mathcal{H}$ into itself which has no fixed point.

2. Prove that the Volterra operator V, given by

$$(Vf)(t) = \int_0^t f(s)\,ds,$$

is compact and quasinilpotent on $L^2(0, 1)$. Show that for $0 < \alpha < 1$ the subspaces $\mathcal{M}_\alpha := \{f \in L^2(0, 1) : f_{|(0,\alpha)} = 0 \text{ a.e.}\}$ are non-trivial closed invariant subspaces for V.

3. Let $\tau : \mathcal{P}(E) \to \mathbb{R}$ be a positive linear functional. Show that $\|\tau\| = 1$ if and only if $\tau(1) = 1$.

4. Verify formula (6.4) by expanding $\|\Phi(\lambda) - z\|^2$ about 0, calculating the linear term in λ.

5. An operator T is said to be *essentially normal* if $T^*T - TT^*$ is compact. Show that the unilateral shift is essentially normal, but cannot be written as the sum of a normal operator and a compact operator.

Comments

A version of Lomonosov's original theorem [145] was given for real spaces by Hooker [125].

Troitsky [195] showed that Lomonosov's theorem cannot be extended to chains of four operators, in the sense that there are operators A, B and K, none of which is a multiple of the identity, such that Read's operator [170], which has no non-trivial invariant subspaces, commutes with A, A commutes with B, B commutes with K, and K is compact.

Section 6.2 is based on the work of Simonič [185]. Lomonosov [146] first showed that every essentially self-adjoint operator on an infinite-dimensional complex Hilbert space has a non-trivial closed *real* invariant subspace.

Exercise 1 is due to Kakutani [128].

7
Minimal vectors

In this chapter we present a construction proposed in the late 1990s by Ansari and Enflo [11] in a Hilbert space context; this was used to show the existence of hyperinvariant subspaces for various classes of operators, and in particular to give new proofs of their existence in the the cases of compact operators and normal operators. Although it was first presented as a Hilbert space construction, it was soon developed further by other authors and extended to general Banach spaces.

The basic idea is that with an operator, a given vector and a parameter, one associates a sequence of 'minimal' vectors, and, broadly speaking, if this sequence converges one can deduce the existence of hyperinvariant subspaces.

We shall now present the construction of minimal vectors and its application to various classes of operators, including weighted composition operators and weighted shifts, in addition to the more standard operators considered by Ansari and Enflo. En route we shall also consider the constrained-approximation problem that needs to be solved if one is to obtain explicit formulae for minimal vectors, since it can be put into a more general context which has applications in other areas of analysis.

7.1 The basic definitions

Let $\mathcal{X}$ be a real or complex Banach space, and $T \in \mathcal{L}(\mathcal{X})$ be injective with dense range. Given a vector $x_0 \in \mathcal{X} \setminus \{0\}$, an integer $n \in \mathbb{N}$ and a parameter ε such that $0 < \varepsilon < \|x_0\|$, we write

$$K = (T^n)^{-1}(\overline{B(x_0, \varepsilon)}) = \{y \in \mathcal{X} : \|T^n y - x_0\| \le \varepsilon\}.$$

The set K has the following easily verified properties:

- K is a closed convex set, since T^n is continuous and linear.
- The zero vector does not lie in K, by the choice of ε.
- $K \neq \emptyset$, since T (and thus T^n) has dense range.

We set

$$d = \inf\{\|z\| : z \in K\},$$

and note that $d > 0$. For each $\lambda > 1$, there clearly exists $y_n \in K$ such that $\|y_n\| \leq \lambda d$. Such a vector y_n is called a *λ-minimal vector* associated with x_0, ε, T and n, and in general it need not be unique.

In the case $\lambda = 1$, such a vector y_n, if it exists, is called simply a *minimal vector* associated with x_0, ε, T and n.

Minimal vectors are sometimes called *backward minimal vectors*. There is a closely-related concept, that of the *forward minimal vector*, which we shall not require (see Exercise 1).

Under certain circumstances, the existence of minimal vectors is guaranteed.

Proposition 7.1.1 *Let $\mathcal{X}$ be a reflexive Banach space. Then minimal vectors exist. Furthermore, the minimal vectors y_n saturate the constraint, in the following sense:*

$$\left\|T^n y_n - x_0\right\| = \varepsilon. \tag{7.1}$$

Moreover, if $\mathcal{X}$ is strictly convex, then each minimal vector is unique.

Proof If $(z_m)_m$ is a sequence in K with $\|z_m\| \to d$, then $(z_m)_m$ is bounded, and hence has a subsequence converging in the weak topology towards a vector z, say. It is easily verified that $\|z\| \leq d$, and since $z \in K$, we have equality.

To see that constraint is saturated, suppose that $\|T^n y_n - x_0\| < \varepsilon$. Take $w = ty_n$ with $0 < t < 1$ and t sufficiently small to guarantee that $y_n - ty_n \in K$. But since $y_n - ty_n$ has smaller norm than y_n, we obtain a contradiction.

Finally, for uniqueness, suppose that there are two distinct minimal vectors y_n and y_n' with $\|y_n\| = \|y_n'\| = d$. Since K is convex, we have $y := \frac{y_n + y_n'}{2} \in K$, and since $\mathcal{X}$ is strictly convex, $\|y\| < d$, which is a contradiction. □

Remark 7.1.2 Note that if $\mathcal{X}$ is reflexive, then it can be given an equivalent strictly convex norm [92, pp. 42, 289]. Therefore, in the construction of invariant subspaces for operators on reflexive spaces, one may always suppose that the space $\mathcal{X}$ is strictly convex, so that minimal vectors are unique.

7.2 Minimal vectors in Hilbert space

The theory of minimal vectors was first developed for Hilbert space, and in this case simple expressions can be derived, as follows.

Proposition 7.2.1 *Let $T \in \mathcal{L}(\mathcal{H})$ with dense range, $\varepsilon > 0$, n a positive integer and $x_0 \in \mathcal{H}$ with $\|x_0\| > \varepsilon$ Then there is a unique positive constant $\mu_n = \mu_n(x_0, \varepsilon)$ such that minimal vector y_n satisfies*

$$y_n = \mu_n T^{*n}(x_0 - T^n y_n), \tag{7.2}$$

and hence

$$x_0 - T^n y_n = (\mathrm{I} + \mu_n T^n T^{*n})^{-1} x_0. \tag{7.3}$$

Proof We employ a variational argument. If $w \in \mathcal{H}$ and $\lambda \in \mathbb{R}$ such that

$$\left\| T^n(y_n + \lambda w) - x_0 \right\| \le \left\| T^n y_n - x_0 \right\|, \tag{7.4}$$

then the optimality condition implies that

$$\|y_n + \lambda w\| \ge \|y_n\|. \tag{7.5}$$

On expanding the inner products, the first condition reduces to

$$2\,\mathrm{Re}\langle \lambda w, T^{*n}(T^n y_n - x_0)\rangle + \lambda^2 \left\| T^n w \right\|^2 \le 0$$

and the second to

$$2\,\mathrm{Re}\langle \lambda w, y_n\rangle + \lambda^2 \|w\|^2 \ge 0.$$

It now follows that there is a number $\mu_n > 0$ such that

$$y_n = -\mu_n T^{*n}(T^n y_n - x_0),$$

for if two non-zero vectors u, v are not real negative multiples of each other, then there is a vector w with $\mathrm{Re}\langle w, u\rangle < 0$ and $\mathrm{Re}\langle w, v\rangle < 0$, for example, $w = -u/\|u\| - v/\|v\|$. Note that $(u_1, u_2) \mapsto \mathrm{Re}\langle u_1, u_2\rangle$ is a *real* inner product, and that

$$\mathrm{Re}\langle w, u\rangle = -\|u\| - \frac{\mathrm{Re}\langle v, u\rangle}{\|v\|} < 0,$$

as we have a strict Cauchy–Schwarz inequality, and similarly for $\mathrm{Re}\langle w, v\rangle$. The observation that condition (7.4) implies (7.5) is now contradicted on taking $\lambda > 0$ sufficiently small.

The uniqueness of μ_n follows from the uniqueness of y_n, which was established in Proposition 7.1.1, and the final expression for $x_0 - T^n y_n$ follows easily. □

Corollary 7.2.2 *Let $T \in \mathcal{L}(\mathcal{H})$ with dense range, $\varepsilon > 0$, n a positive integer and $x_0 \in \mathcal{H}$ such that $\|x_0\| > \varepsilon$. The minimal vector y_n satisfies*

$$\|T^n y_n\| \leq (\|x_0\|^2 - \varepsilon^2)^{1/2}$$

and, moreover, if $z \in \mathcal{H}$, then $z \perp y_n$ if and only if $T^n z \perp T^n y_n - x_0$.

Proof Since $x_0 = (x_0 - T^n y_n) + T^n y_n$, we have

$$\begin{aligned}\|x_0\|^2 &= \langle (x_0 - T^n y_n) + T^n y_n, (x_0 - T^n y_n) + T^n y_n \rangle \\ &= \|x_0 - T^n y_n\|^2 + \|T^n y_n\|^2 - 2\,\mathrm{Re}\langle T^n y_n - x_0, T^n y_n \rangle .\end{aligned}$$

By (7.1) we have $\|T^n y_n - x_0\|^2 = \varepsilon^2$ and

$$\langle T^n y_n - x_0, T^n y_n \rangle = \langle T^{*n}(T^n y_n - x_0), y_n \rangle = - \|y_n\|^2 / \mu_n < 0.$$

Therefore, $\|x_0\|^2 \geq \|T^n y_n\|^2 + \varepsilon^2$ and thus $\|T^n y_n\| \leq (\|x_0\|^2 - \varepsilon^2)^{1/2}$.

For the second assertion, by (7.2) we have $y_n = \mu_n T^{*n}(x_0 - T^n y_n)$. It follows that

$$\begin{aligned}\langle z, y_n \rangle &\Leftrightarrow \langle z, -\mu_n T^{*n}(T^n y_n - x_0) \rangle = 0 \\ &\Leftrightarrow \mu_n \langle T^n z, T^n y_n - x_0 \rangle = 0 \\ &\Leftrightarrow \langle T^n z, T^n y_n - x_0 \rangle = 0,\end{aligned}$$

which ends the proof. □

7.3 A general extremal problem

The construction of minimal vectors can be seen as a special case of a more general extremal problem. In order to formulate this, let $\mathcal{X}$, $\mathcal{Y}$ and $\mathcal{Z}$ be complex Banach spaces, and let $A : \mathcal{X} \to \mathcal{Y}$ and $B : \mathcal{X} \to \mathcal{Z}$ be bounded linear operators such that A and B are coprime, in the sense that there exists a constant $\eta > 0$ such that

$$\|Ax\| + \|Bx\| \geq \eta \|x\| \qquad \text{for all } x \in \mathcal{X}. \tag{7.6}$$

Let $t_0 \in \mathcal{Z}$ and $t_1 \in \mathcal{Y}$ be fixed vectors. For $\varepsilon > 0$ we consider the closed convex set

$$C_{t_0}^{\varepsilon} = \{x \in \mathcal{X} : \|Bx - t_0\| \leq \varepsilon\},$$

and we suppose that ε is chosen so large that $C^{\varepsilon}_{t_0}$ has non-empty interior. We suppose also that $Ax \neq t_1$ for all $x \in C^{\varepsilon}_{t_0}$ (to eliminate a degenerate case). In this situation the associated extremal problem is to find $x_0 \in C^{\varepsilon}_{t_0}$ such that

$$\|Ax_0 - t_1\| = \inf\left\{\|Ax - t_1\| : x \in C^{\varepsilon}_{t_0}\right\} = \beta, \qquad \text{say.} \tag{7.7}$$

In view of the applications that we shall discuss, we also assume that A has dense range (in the case that $\mathcal{Y}$ is a Hilbert space, this is not a restriction, as we may replace $\mathcal{Y}$ by $\overline{A\mathcal{X}}$ and t_1 by its orthogonal projection onto $\overline{A\mathcal{X}}$).

7.3.1 Approximation in Hilbert spaces

In the case when we consider Hilbert spaces, the general extremal problem has a particularly simple solution, which we now describe in detail.

Let $\mathcal{X}$, $\mathcal{Y}$ and $\mathcal{Z}$ be complex Hilbert spaces and $A : \mathcal{X} \to \mathcal{Y}$ and $B : \mathcal{X} \to \mathcal{Z}$ bounded linear operators such that A and B satisfy (7.6). This is the 'right' condition to impose, as will be seen when we come to solve our extremal problem. It is equivalent to saying that $A^*A + B^*B$ is invertible.

Several of the results about minimal vectors have their counterparts in this more general set-up, as we now show.

Lemma 7.3.1 *Under hypothesis (7.6), the extremal problem given by (7.7) has a unique solution and the extremal vector saturates the constraint, in the sense that* $\|Bx_0 - t_0\| = \varepsilon$.

Proof Existence of a solution follows by a standard weak compactness argument. If (z_n) is a sequence in $C^{\varepsilon}_{t_0}$ such that $\|Az_n - t_1\| \to \beta$, then, since $\|Tz_n - t_0\| \leq \varepsilon$, it follows from (7.6) that (z_n) is bounded, and has a weak limit point in $C^{\varepsilon}_{t_0}$, which is easily seen to be an extremal for the problem.

Moreover, the constraint is saturated, as otherwise if $\|Bx_0 - t_0\| < \varepsilon$ we could use the fact that the range of A is dense to find a vector $w \in \mathcal{X}$ with $\|Aw - (Ax_0 - t_1)\| < \|Ax_0 - t_1\|$. Now

$$\begin{aligned}\|A(x_0 - \lambda w) - t_1\| &= \|(1-\lambda)(Ax_0 - t_1) + \lambda(Ax_0 - t_1 - Aw)\| \\ &< (1-\lambda)\|Ax_0 - t_1\| + \lambda\|Ax_0 - t_1\| = \|Ax_0 - t_1\|\end{aligned}$$

for all $\lambda \in (0, 1)$, contradicting the optimality of x_0 if we take λ sufficiently small to ensure that $x_0 - \lambda w \in C^{\varepsilon}_{t_0}$.

The uniqueness follows because, if x_0, x_0' are distinct points in $C^{\varepsilon}_{t_0}$ with $\|Ax_0 - t_0\| = \beta$ and $\left\|Ax_0' - t_0\right\| = \beta$, then the vector $x = \left(x_0 + x_0'\right)/2$ also lies in $C^{\varepsilon}_{t_0}$ and the strict convexity of the norm implies that $\|Ax - t_0\| < \beta$ unless $Ax_0 - t_0 = Ax_0' - t_0$, i.e., $A\left(x_0 - x_0'\right) = 0$. But in that case $Bx_0 - t_0 \neq Bx_0' - t_0$, and the strict convexity of the norm now implies that $\|Bx - t_0\| < \varepsilon$, which is a contradiction of the saturation condition that we have already established. □

The solution to the constrained extremal problem is given by the following theorem.

Theorem 7.3.2 *The solution to (7.7) is given by*

$$(A^*A + \mu B^*B)x_0 = A^*t_1 + \mu B^*t_0,$$

where $\mu > 0$ *is the unique constant such that* $\|Bx_0 - t_0\| = \varepsilon$.

Proof We write out the variational argument in some detail. If w is a vector and λ a real number such that $\|B(x_0 + \lambda w) - t_0\| \leq \|Bx_0 - t_0\|$, then the optimality condition implies that $\|A(x_0 + \lambda w) - t_1\| \geq \|Ax_0 - t_1\|$. On expanding the inner products, the first condition reduces to

$$2\,\mathrm{Re}\langle \lambda w, B^*(Bx_0 - t_0)\rangle + \lambda^2 \|Bw\|^2 \leq 0, \tag{7.8}$$

the second to

$$2\,\mathrm{Re}\langle \lambda w, A^*(Ax_0 - t_1)\rangle + \lambda^2 \|Aw\|^2 \geq 0. \tag{7.9}$$

It now follows that there is a number $\mu > 0$ such that

$$A^*(Ax_0 - t_1) = -\mu B^*(Bx_0 - t_0), \tag{7.10}$$

by the same geometric argument in $\mathbb{R}^2$ that was used in the proof of Proposition 7.2.1. The observation that condition (7.8) implies (7.9) is now contradicted on taking $\lambda > 0$ sufficiently small. Clearly μ is unique by Lemma 7.3.1, and so the result follows. □

We establish now a further relation between the parameters ε, β and μ.

Proposition 7.3.3 *The constant* μ *satisfies*

$$\mu = -\frac{d\beta}{d\varepsilon}\frac{\beta}{\varepsilon}.$$

Proof By a simple change of variables we have

$$\frac{d(\beta^2)}{d(\varepsilon^2)} = \frac{d\beta}{d\varepsilon}\frac{\beta}{\varepsilon}.$$

We now write

$$f(x_0) = \langle Ax_0 - t_1, Ax_0 - t_1\rangle = \beta^2 \quad \text{and} \quad g(x_0) = \langle Bx_0 - t_0, Bx_0 - t_0\rangle = \varepsilon^2.$$

For each $h \in \mathcal{X}$ we have

$$f(x_0 + h) - f(x_0) = 2\,\mathrm{Re}\langle Ah, Ax_0 - t_1\rangle + \|Ah\|^2,$$

which gives

$$Df_{x_0}h = 2\,\mathrm{Re}\langle Ah, Ax_0 - t_1\rangle = \frac{d\beta^2}{dx_0}(h).$$

Similarly, we have

$$Dg_{x_0}h = 2\,\mathrm{Re}\langle Bh, Bx_0 - t_0\rangle = \frac{d\varepsilon^2}{dx_0}(h).$$

It follows that

$$\mathrm{Re}\langle h, B^*(Bx_0 - t_0)\rangle \frac{d(\beta^2)}{d(\varepsilon^2)} = \mathrm{Re}\langle h, A^*(Ax_0 - t_1)\rangle$$

for each $h \in \mathcal{X}$. Using (7.10), we obtain

$$\frac{d(\beta^2)}{d(\varepsilon^2)} = -\mu,$$

which gives the result. □

7.3.2 Approximation in reflexive Banach spaces

By means of more complicated arguments, we shall now derive a solution to the extremal problem (7.7) in a framework more general that that of Hilbert spaces, in particular one that includes the reflexive L^p and H^p spaces.

Our assumptions for this section are that $\mathcal{X}$, $\mathcal{Y}$ and $\mathcal{Z}$ are Banach spaces, such that $\mathcal{X}$ is reflexive and $\mathcal{Y}$ and $\mathcal{Z}$ are smooth.

Lemma 7.3.4 *Under hypothesis (7.6), the extremal problem given by (7.7) has at least one solution and every extremal vector x_0 saturates the constraint, in the sense that $\|Bx_0 - t_0\| = \varepsilon$. If, in addition, $\mathcal{Y}$ and $\mathcal{Z}$ are strictly convex, then the vector x_0 is unique.*

Proof The fact that the solution exists and saturates the constraint follows from the reflexivity of $\mathcal{X}$ and the assumption that A has dense range, as in the proof of Lemma 7.3.1.

The uniqueness of x_0 in the case where $\mathcal{X}$ is strictly convex is not difficult to check. Indeed, if x_0, x_0' are distinct points in $C_{t_0}^{\varepsilon}$ with $\|Ax_0 - t_1\| = \beta$ and $\|Ax_0' - t_1\| = \beta$, then the vector $x = (x_0 + x_0')/2$ also lies in $C_{t_0}^{\varepsilon}$ and the strict convexity of the norm implies that $\|Ax - t_1\| < \beta$ unless $Ax_0 - t_1 = Ax_0' - t_1$, i.e., $A(x_0 - x_0') = 0$. But in that case $Bx_0 - t_0 \neq Bx_0' - t_0$ by coprimeness, and the strict convexity of the norm now implies that $\|Bx - t_0\| < \varepsilon$, which is a contradiction of the saturation condition. □

The solution to the constrained extremal problem is given by the following theorem.

Theorem 7.3.5 *Let x_0 be a solution to (7.7). Then there exists a positive constant μ such that*

$$B^* \nu_{\mathcal{Z}}(Bx_0 - t_0) = -\mu A^* \nu_{\mathcal{Y}}(Ax_0 - t_1), \tag{7.11}$$

where $\nu_{\mathcal{Y}}$ and $\nu_{\mathcal{Z}}$ are duality maps as in Definition 1.1.20. Conversely, if x_0 satisfies (7.11) with $\mu > 0$, and $\varepsilon = \|Bx_0 - t_0\|$, then x_0 satisfies (7.7).

Proof Suppose that x_0 is a solution of (7.7). Define the following subset of $\mathcal{X}$:

$$\mathcal{R}_{t_1}^{\beta} = \{x \in \mathcal{X} : \|Ax - t_1\| \leq \beta\}.$$

Our assumptions imply that $\operatorname{int} \mathcal{C}_{t_0}^{\varepsilon}$ and $\operatorname{int} \mathcal{R}_{t_1}^{\beta}$ are open, disjoint, non-empty convex sets. Hence by the Hahn–Banach theorem (see Theorem 1.1.4) there are a functional $\Lambda \in \mathcal{X}^* \setminus \{0\}$ and a real number γ, such that

$$\operatorname{Re} \Lambda(x) < \gamma < \operatorname{Re} \Lambda(y) \qquad \text{for all } x \in \operatorname{int} \mathcal{C}_{t_0}^{\varepsilon},\ y \in \operatorname{int} \mathcal{R}_{t_1}^{\beta}.$$

Now $x_0 \in \partial \mathcal{C}_{t_0}^{\varepsilon} \cap \partial \mathcal{R}_{t_1}^{\beta}$, and so $\operatorname{Re} \Lambda(x_0) = \gamma$.

Consider the following three hyperplanes in $\mathcal{X}$:

$$\begin{aligned}
\mathcal{H}_1 &= \{x \in \mathcal{X} : \operatorname{Re} \nu(Bx_0 - t_0)(Bx - t_0) = \varepsilon\},\\
\mathcal{H}_2 &= \{x \in \mathcal{X} : \operatorname{Re} \nu(Ax_0 - t_1)(Ax - t_1) = \beta\},\\
\mathcal{H}_3 &= \{x \in \mathcal{X} : \operatorname{Re} \Lambda(x) = \gamma\}.
\end{aligned}$$

Note that $x_0 \in \mathcal{H}_1 \cap \mathcal{H}_2 \cap \mathcal{H}_3$. We want to prove that these hyperplanes coincide. To do this we prove that $\mathcal{H}_1 = \mathcal{H}_3$; a similar argument shows that $\mathcal{H}_2 = \mathcal{H}_3$.

So suppose that $\mathcal{H}_1 \neq \mathcal{H}_3$. Then there is a point $z \in \mathcal{H}_3 \setminus \mathcal{H}_1$, for otherwise $\mathcal{H}_3 \subset \mathcal{H}_1$, and since both have real codimension 1 they must be equal. We may assume that

$$\operatorname{Re} \nu(Bx_0 - t_0)(Bz - t_0) < \varepsilon \qquad \text{and} \qquad \operatorname{Re} \Lambda(z) = \gamma, \tag{7.12}$$

for if $\operatorname{Re} \nu(Bx_0 - t_0)(Bz - t_0) > \varepsilon$, then we may replace z by $z' = 2x_0 - z \in \mathcal{H}_3$, which satisfies

$$\begin{aligned}
&\operatorname{Re} \nu(Bx_0 - t_0)(Bz' - t_0)\\
&\quad = 2 \operatorname{Re} \nu(Bx_0 - t_0)(Bx_0 - t_0) - \operatorname{Re} \nu(Bx_0 - t_0)(Bz - t_0) < \varepsilon.
\end{aligned}$$

It follows from Proposition 1.1.21 and (7.12) that there is a number $\lambda \in (0, 1)$ such that

$$\|\lambda(Bx_0 - t_0) + (1 - \lambda)(Bz - t_0)\| < \varepsilon,$$

and so

$$\|B(\lambda x_0 + (1 - \lambda)z) - t_0\| < \varepsilon.$$

Let $x_1 = \lambda x_0 + (1 - \lambda)z$; then $x_1 \in \operatorname{int} \mathcal{C}_{t_0}^{\varepsilon}$ and

$$\begin{aligned}\operatorname{Re} \Lambda(x_1) &= \lambda \operatorname{Re} \Lambda(x_0) + (1 - \lambda) \operatorname{Re} \Lambda(z) \\ &= \lambda\gamma + (1 - \lambda)\gamma = \gamma,\end{aligned}$$

which is a contradiction since $\operatorname{Re} \Lambda(x) < \gamma$ for all $x \in \operatorname{int} \mathcal{C}_{t_0}^{\varepsilon}$.

Given that $\mathcal{H}_1 = \mathcal{H}_2 = \mathcal{H}_3$, we have constants ε' and β' such that the equations of $\mathcal{H}_1$ and $\mathcal{H}_2$ are, respectively,

$$\operatorname{Re} B^*\nu(Bx_0 - t_0)x = \varepsilon' \qquad \text{and} \qquad \operatorname{Re} A^*\nu(Ax_0 - t_1)x = \beta'.$$

Hence there is a non-zero constant μ such that

$$\operatorname{Re} B^*\nu(Bx_0 - t_0) = -\mu \operatorname{Re} A^*\nu(Ax_0 - t_1), \tag{7.13}$$

and thus $B^*\nu(Bx_0 - t_0) = -\mu A^*\nu(Ax_0 - t_1)$, since both functionals are complex-linear. It remains to show that $\mu > 0$.

Suppose, to the contrary, that $\mu < 0$. Since A has dense range, there exists a $w_0 \in \mathcal{X}$ such that $\|Aw_0 - t_1\| < \beta$. In particular,

$$\operatorname{Re} \nu(Ax_0 - t_1)(Aw_0 - t_1) < \beta,$$

and so

$$\operatorname{Re} \nu(Ax_0 - t_1)(Ax_0 - Aw_0) > 0,$$

since $x_0 \in \mathcal{H}_2$.

Now since $\mu < 0$ we see from (7.13) that

$$\operatorname{Re} \nu(Bx_0 - t_0)(Bx_0 - Bw_0) > 0,$$

and so

$$\operatorname{Re} \nu(Bx_0 - t_0)(Bw_0 - t_0) < \operatorname{Re} \nu(Bx_0 - t_0)(Bx_0 - t_0).$$

Applying Proposition 1.1.21 once more, we see that there is a number $\eta \in (0, 1)$ such that

$$\|\eta(Bx_0 - t_0) + (1 - \eta)(Bw_0 - t_0)\| < \varepsilon,$$

i.e.,

$$\|B(\eta x_0 + (1 - \eta)w_0) - t_0\| < \varepsilon.$$

Since $\|Aw_0 - t_1\| < \beta$ we have

$$\|A(\eta x_0 + (1 - \eta)w_0) - t_1\| \le \eta\|Ax_0 - t_1\| + (1 - \eta)\|Aw_0 - t_1\| < \beta,$$

which is a contradiction, since $\eta x_0 + (1 - \eta)w_0 \in C^{\varepsilon}_{t_0}$. This ends the proof of the first assertion of the theorem.

Conversely, suppose that x_0 satisfies (7.11) with $\mu > 0$ and $\varepsilon = \|Bx_0 - t_0\|$. Let $x \in \mathcal{X}$ satisfy $\|Bx - t_0\| \le \varepsilon$. We have

$$\operatorname{Re} \nu(Bx_0 - t_0)(Bx - t_0) \le \varepsilon = \operatorname{Re} \nu(Bx_0 - t_0)(Bx_0 - t_0),$$

and then

$$\operatorname{Re} \nu(Bx_0 - t_0)(B(x - x_0)) \le 0$$

or, equivalently,

$$\operatorname{Re} B^* \nu(Bx_0 - t_0)(x - x_0) \le 0.$$

Using (7.11) we obtain

$$\operatorname{Re} A^* \nu(Ax_0 - t_1)(x - x_0) \ge 0,$$

which implies that

$$\operatorname{Re} \nu(Ax_0 - t_1)(Ax - t_1) \ge \operatorname{Re} \nu(Ax_0 - t_1)(Ax_0 - t_1) = \|Ax_0 - t_1\|.$$

Hence $\|Ax - t_1\| \ge \|Ax_0 - t_1\|$, and so the second assertion of the theorem is proved. □

7.4 Application to hyperinvariant subspaces

7.4.1 The main theorem

As in Section 7.1, we take $T \in \mathcal{L}(\mathcal{X})$ injective with dense range, and a non-zero vector $x_0 \in \mathcal{X}$. Given $\lambda > 1$, a constant ε with $0 < \varepsilon < \|x_0\|$ and $n = 1, 2, \ldots$, let y_n be a λ-minimal vector associated with x_0, ε, T and n.

The main result relies on the following lemma, which is a direct application of the Hahn–Banach theorem.

Lemma 7.4.1 *Let n be a positive integer and let y_n be a λ-minimal vector. Then there exists $g_n \in \mathcal{X}^*$, $\|g_n\| = 1$ and $\beta > 0$ such that*

1. $|g_n(x_0)| \geq \beta + \varepsilon$.
2. $\beta \leq |g_n(T^n y_n)| \leq \lambda\beta$.
3. $|T^{*n} g_n(y_n)| \geq \frac{1}{\lambda}\|T^{*n} g_n\| \|y_n\|$.

Proof Since $\|x_0\| > \varepsilon$, it follows that $d_n := \inf\{\|y\| : \|T^n y - x_0\| \leq \varepsilon\}$ is positive. It follows from the definition of d_n that the interiors of $B(x_0, \varepsilon)$ and $T^n B(0, d_n)$ are disjoint. Using the Hahn–Banach theorem, there exist $g_n \in X^*$ with $\|g_n\| = 1$ and $\beta > 0$ such that

$$\begin{cases} \mathrm{Re}(g_n(x)) \geq \beta & \text{for all} \quad x \in B(x_0, \varepsilon), \\ \mathrm{Re}(g_n(T^n v)) \leq \beta & \text{for all} \quad v \in B(0, d_n). \end{cases}$$

Note that in particular we have $\mathrm{Re}(g_n(x_0)) \geq \beta + \varepsilon$, and thus property 1. Moreover, we get $\|T^{*n} g_n\| \leq \frac{\beta}{d_n}$ and $\mathrm{Re}(g_n(T^n y_n)) \geq \beta$. It follows that $|g_n(T^n y_n)| \geq \beta$ and

$$|g_n(T^n y_n)| \leq \|y_n\| \|T^{*n} g_n\| \leq \frac{\beta}{d_n} \|y_n\|.$$

Since $\|y_n\| \leq \lambda d_n$, the second assertion of the lemma follows. In order to check property 3, note that

$$|T^{*n} g_n(y_n)| \geq \beta \geq d_n \|T^{*n} g_n\| \geq \frac{1}{\lambda} \|y_n\| \|T^{*n} g_n\|,$$

which ends the proof of the lemma. □

We now come to the main result.

Theorem 7.4.2 *Let $\mathcal{X}$ be a Banach space. Suppose that there exists a subsequence (y_{n_k}) of λ-minimal vectors and a uniformly bounded sequence $(A_k) \in \{T\}'$ such that*

$$\lim_{k \to +\infty} \frac{\|y_{n_k - 1}\|}{\|y_{n_k}\|} = 0 \tag{7.14}$$

and

$$(A_k T^{n_k - 1} y_{n_k - 1}) \text{ converges in norm to } z \neq 0. \tag{7.15}$$

Then T has a non-trivial hyperinvariant subspace.

Proof Let $D \in \{T\}'$ and take $g_n \in \mathcal{X}^*$ defined as in Lemma 7.4.1. Set $f_n = T^{*n} g_n$. By property 2 of Lemma 7.4.1, we get $|f_n(y_n)| \geq \beta > 0$. Write

$$D A_k y_{n_k - 1} = \alpha_{n_k} y_{n_k} + w_{n_k},$$

with $\alpha_{n_k} \in \mathbb{C}$ and $w_{n_k} \in \operatorname{Ker} f_{n_k}$. Then we get $f_{n_k}(DA_k y_{n_k-1}) = \alpha_{n_k} f_{n_k}(y_{n_k})$, and thus, using property 3 of Lemma 7.4.1, we obtain

$$|\alpha_{n_k}| \leq \lambda \frac{\|D\|\|A_k\|\|f_{n_k}\|\|y_{n_k-1}\|}{\|f_{n_k}\|\|y_{n_k}\|} = \lambda \|D\|\|A_k\| \frac{\|y_{n_k-1}\|}{\|y_{n_k}\|}.$$

Since $(\|A_k\|)$ is uniformly bounded, by (7.14), we get $\lim_{k\to+\infty} \alpha_{n_k} = 0$. Now, since $DT = TD$, it follows that

$$DA_k T^{n_k} y_{n_k-1} = \alpha_{n_k} T^{n_k} y_{n_k} + T^{n_k} w_{n_k},$$

and thus

$$g_{n_k}(DA_k T^{n_k} y_{n_k-1}) = \alpha_{n_k} f_{n_k}(y_{n_k}).$$

Using property 2 of Lemma 7.4.1 and the fact that $\lim_{k\to+\infty} \alpha_{n_k} = 0$, we obtain

$$\lim_{k\to+\infty} g_{n_k}(DA_k T^{n_k} y_{n_k-1}) = 0.$$

By (7.15), we obtain $\lim_{k\to+\infty} g_{n_k}(DTz) = 0$. By passing to a further subsequence we may assume that (g_{n_k}) converges to some $\Phi \in \mathcal{X}^*$ for the weak* topology, and then $\Phi(DTz) = 0$. Using property 1 of Lemma 7.4.1, we have also $|\Phi(x_0)| \geq \beta + \varepsilon$ and thus $\Phi \neq 0$. Now, consider the hyperinvariant subspace $\mathcal{M}$ for T defined by

$$\mathcal{M} = \operatorname{span}\{DTz : DT = TD\}.$$

Since $z \neq 0$ and T is injective, obviously $\mathcal{M} \neq \{0\}$. Since $\Phi \neq 0$ and $\mathcal{M} \subset \operatorname{Ker} \Phi$, it follows that $\mathcal{M} \neq \mathcal{X}$, which ends the proof of this theorem. □

The following corollary will be used to show the existence of invariant subspaces for particular classes of operators.

Corollary 7.4.3 *Suppose that there exists a subsequence (y_{n_k}) of λ-minimal vectors such that*

$$\lim_{k\to+\infty} \frac{\|y_{n_k-1}\|}{\|y_{n_k}\|} = 0,$$

and that one of the following two conditions holds:

1. *$(T^{n_k-1} y_{n_k-1})$ converges in norm.*
2. *$(T^{n_k} y_{n_k-1})$ converges in norm to a non-zero vector.*

Then T has a non-trivial hyperinvariant subspace.

Proof Apply Theorem 7.4.2: in the first case with $A_k = I$, and in the second with $A_k = T$, for each k. Note that in the first case the limit is non-zero since $\|T^{n_k-1} y_{n_k-1}\| \geq \|x_0\| - \varepsilon$. □

7.4.2 Compact operators

Our first application of Corollary 7.4.3 will be to compact operators. In fact, the result is easy except in the case when the operator is quasinilpotent, and we begin with a preliminary lemma about such operators.

Lemma 7.4.4 *Let $T \in \mathcal{L}(\mathcal{X})$ with dense range, $\varepsilon > 0$ and $x_0 \in \mathcal{X}$ such that $\|x_0\| > \varepsilon$. Let $(y_n)_n$ be a sequence of λ-minimal vectors. If $T \in \mathcal{L}(\mathcal{X})$ is quasinilpotent, then there exists a subsequence $(y_{n_k})_{k\geq 1}$ of $(y_n)_{n\geq 1}$ such that*

$$\lim_{k\to\infty} \frac{\|y_{n_k-1}\|}{\|y_{n_k}\|} = 0.$$

Proof Suppose, to the contrary, that there exists $t > 0$ with $\dfrac{\|y_{n-1}\|}{\|y_n\|} > t$ for all $n > 0$, so

$$\|y_1\| > t\|y_2\| > \cdots > t^{n-1}\|y_n\|.$$

Then $\|T^n y_n - x_0\| = \|T(T^{n-1}y_n) - x_0\|$, and, by the λ-minimality of y_1, we have $t\|T^{n-1}y_n\| \geq \|y_1\|/\lambda$. Therefore,

$$\|T^{n-1}\|\|y_n\| \geq \|T^{n-1}y_n\| \geq \|y_1\|/\lambda > t^{n-1}\|y_n\|/\lambda.$$

Hence $\|T^{n-1}\| > t^{n-1}/\lambda$, so that $\|T^{n-1}\|^{1/(n-1)} > t/\lambda^{1/(n-1)}$, which contradicts the fact that T is quasinilpotent. □

Using Corollary 7.4.3, we can now deduce a result on compact operators. Note that a stronger version of this result is given as Theorem 6.1.2.

Corollary 7.4.5 *Let $K \in \mathcal{L}(\mathcal{X})$ be a non-zero compact operator. Then K has non-trivial hyperinvariant subspaces.*

Proof We may assume that K is quasinilpotent and injective, since otherwise it has eigenspaces by Theorem 1.2.13, and these are hyperinvariant subspaces for K.

Without loss of generality, $\|K\| = 1$, and we take $x_0 \neq 0$ and $0 < \varepsilon < \|Kx_0\|$.

Using Lemma 7.4.4 and Corollary 7.4.3 with hypothesis 2, it remains to prove that $(K^{n_k} y_{n_k-1})_k$ has a subsequence converging in norm to a non-zero vector. However, this follows since K is compact and $(K^{n_k-1} y_{n_k-1})_k$ is bounded. Moreover, with the choice of x_0 and ε we have

$$\|K^{n_k} y_{n_k-1} - Kx_0\| \le \|K\| \|K^{n_k-1} y_{n_k-1} - x_0\| < \varepsilon,$$

and so the limit of any convergent subsequence of $(K^{n_k} y_{n_k-1})_k$ is necessarily non-zero. □

7.4.3 Weighted composition operators

Let X be either $[0, 1]$ or the closed unit disc $\overline{\mathbb{D}}$, with normalized Lebesgue measure μ. In this section we consider the case where $T \in \mathcal{L}(L^2(X))$ is of the form

$$Tf(t) = T_{w,\gamma} f(t) = w(t) f(\gamma(t)), \tag{7.16}$$

with $w \in L^\infty(X)$, and $\gamma : X \to X$ an injective mapping, such that γ' exists and is piecewise continuous. In addition, we suppose that $w/(\gamma')^\alpha$ belongs to $L^\infty(X)$, where $\alpha = 1$ for $X = [0, 1]$ and $\alpha = 2$ for $X = \overline{\mathbb{D}}$.

The above conditions are sufficient to guarantee the continuity of T and we can easily check that

$$T^* f(t) = \frac{\overline{w}(\gamma^{-1}(t))}{|\gamma'(\gamma^{-1}(t))|^\alpha} \chi_{\gamma(X)}(t) f(\gamma^{-1}(t)),$$

where χ_Ω denotes the characteristic function of a set Ω.

Remark 7.4.6 If γ fails to be essentially surjective in the sense that $\mu(S) > 0$, where $S = X \setminus \gamma(X)$, then $T\chi_S = 0$, and so T has a non-trivial hyperinvariant subspace, namely its kernel. Nonetheless, the approach in this section can be used to produce *non-standard* hyperinvariant subspaces, not contained in the kernel, for example, by constructing minimal vectors associated with functions f_0 supported on the complement of S.

We consider the sequence $(y_n)_n$ of minimal vectors associated with f_0, ε, T and n, where $y_0 \in L^2(X)$ with $0 < \varepsilon < \|y_0\|$.

In order to determine whether $(T^n y_n)_n$ converges, we need to study the behaviour of $A_n := T^n T^{*n}$. Since

$$T^n f(t) = w(t) w(\gamma(t)) \cdots w(\gamma^{n-1}(t)) f(\gamma^n(t))$$

and

$$T^{*n} f(t)$$
$$= \frac{\overline{w}(\gamma^{-1}(t))}{|\gamma'(\gamma^{-1}(t))|^\alpha} \frac{\overline{w}(\gamma^{-2}(t))}{|\gamma'(\gamma^{-2}(t))|^\alpha} \cdots \frac{\overline{w}(\gamma^{-n}(t))}{|\gamma'(\gamma^{-n}(t))|^\alpha} \chi_{\gamma^n(X)}(t) f(\gamma^{-n}(t)),$$

we get

$$A_n = M_{\Phi_n}, \text{ where } \Phi_n(t) = \frac{|w(t)|^2 |w(\gamma(t))|^2 \cdots |w(\gamma^{n-1}(t))|^2}{|\gamma'(t)\gamma'(\gamma(t)) \cdots \gamma'(\gamma^{n-1}(t))|^\alpha}.$$

Therefore we have $A_n = M_{\Phi_n}$ with $\Phi_n = h \cdot h(\gamma) \cdots h(\gamma^{n-1})$ and h the function in $L^\infty(X)$ defined by $h := |w|^2/|\gamma'|^\alpha$.

We can now give a fairly general context in which there is a sufficient condition implying the convergence of $(T^n y_n)_n$. To do this we define a partial order $\prec$ on X by saying that

$$t \prec r \quad \text{if and only if} \quad \limsup_{n\to\infty} \frac{h(\gamma^n(t))}{h(\gamma^n(r))} < 1. \tag{7.17}$$

We shall say that $\prec$ is *left-regular* if every non-empty $Y \subset X$, such that $r \in Y$ and $t \prec r$ implies $t \in Y$, satisfies the following:

For each $\delta > 0$ there exists $r \in Y$ such that $\mu\{t \in Y : t \prec r\} > \mu(Y) - \delta$.

Likewise $\prec$ is *right-regular* if every non-empty $Y \subset X$, such that $r \in Y$ and $r \prec t$ implies $t \in Y$, satisfies the following:

For each $\delta > 0$ there exists $r \in Y$ such that $\mu\{t \in Y : r \prec t\} > \mu(Y) - \delta$.

Later we shall give some examples of such partial orders $\prec$.

Theorem 7.4.7 *Let T be a weighted composition operator on $L^2(X)$ defined as in (7.16). Suppose that h determines a partial order $\prec$ that is both left- and right-regular, such that for every $t \in X$ one has*

$$\mu\{r \in X : r \prec t \text{ or } t \prec r\} = 1. \tag{7.18}$$

Then $(T^n y_n)_n$ converges in norm.

Proof The calculations above show that

$$(f_0 - T^n y_n)(t) = \frac{f_0(t)}{1 + \mu_n h(t) \dots h(\gamma^{n-1}(t))}.$$

Now set

$$\begin{cases} J_n(t) = \dfrac{1}{1 + \mu_n h(t) \dots h(\gamma^{n-1}(t))} \\ L_1(t) = \limsup\limits_{n\to\infty} J_n(t) \\ L_2(t) = \liminf\limits_{n\to\infty} J_n(t). \end{cases}$$

Obviously we have $0 \le L_2(t) \le L_1(t) \le 1$.

It follows from (7.18) that for a fixed $t \in X$, almost all $r \in X$ are comparable to t in the sense that either $r \prec t$ or $t \prec r$.

Suppose that there exists $r \in X$ such that $L_1(r) > 0$. Then there exists a subsequence $(n_k)_k$ such that $(\mu_{n_k} h(r) \dots h(\gamma^{n_k-1}(r)))_k$ is bounded.

Suppose that $t \prec r$; it follows that the sequence $(\mu_{n_k} h(t) \dots h(\gamma^{n_k-1}(t)))_k$ tends to 0 and thus $L_1(t) = 1$. Hence, if in addition $L_1(r) < 1$, we must have $L_1(t) = 0$ for all t with $r \prec t$.

Thus there exist disjoint subsets $X_0^{(1)}$ and $X_1^{(1)}$ of X with total measure 1, such that

$$L_1(t) = \begin{cases} 0 & \text{if } t \in X_0^{(1)}, \\ 1 & \text{if } t \in X_1^{(1)}. \end{cases}$$

Similarly, suppose that there exists $r \in X$ such that $L_2(r) < 1$. Then there exists a subsequence $(n_k)_k$ such that $(\mu_{n_k} h(r) \dots h(\gamma^{n_k-1}(r)))_k$ is bounded below.

Suppose that $r \prec t$; it follows that the sequence $(\mu_{n_k} h(t) \dots h(\gamma^{n_k-1}(t)))_k$ tends to infinity, and thus $L_2(t) = 0$. Hence, if in addition $L_2(r) > 0$, we must have $L_2(t) = 1$ for all t with $t \prec r$.

Thus there exist disjoint subsets $X_0^{(2)}$ and $X_1^{(2)}$ of X with total measure 1, such that

$$L_2(t) = \begin{cases} 0 & \text{if } t \in X_0^{(2)}, \\ 1 & \text{if } t \in X_1^{(2)}. \end{cases}$$

Clearly, $X_0^{(1)} \subset X_0^{(2)}$ and $X_1^{(2)} \subset X_1^{(1)}$.

Since $\prec$ is left-regular, for each $\delta > 0$ we can choose $r \in X_1^{(1)}$ such that

$$\int_{\{t:t\prec r\}} |f_0(t)|^2 \, d\mu(t) \ge \int_{t\in X_1^{(1)}} |f_0(t)|^2 \, d\mu(t) - \delta.$$

Now we can find a subsequence $(n_k)_k$ such that $J_{n_k}(r) \to 1$, and so $J_{n_k}(t) \to 1$ for all $t \prec r$. Thus, for each $\delta > 0$,

$$\begin{aligned} \varepsilon^2 &= \limsup_{n\to\infty} \|J_n f_0\|_2^2 \\ &\geq \lim_{k\to\infty} \int_{\{t:t\prec r\}} |J_{n_k}(t) f_0(t)|^2 \, d\mu(t) \geq \int_{t\in X_1^{(1)}} |f_0(t)|^2 \, d\mu(t) - \delta, \end{aligned}$$

and hence

$$\varepsilon^2 \geq \int_{t\in X_1^{(1)}} |f_0(t)|^2 \, d\mu(t). \tag{7.19}$$

Similarly, since $\prec$ is right-regular, for each $\delta > 0$ we can choose $s \in X_0^{(2)}$ such that

$$\int_{\{t:s\prec t\}} |f_0(t)|^2 \, d\mu(t) \geq \int_{t\in X_0^{(2)}} |f_0(t)|^2 \, d\mu(t) - \delta,$$

and hence, using (7.18), we have

$$\int_{\{t:t\prec s\}} |f_0(t)|^2 \, d\mu(t) \leq \int_{t\in X_1^{(2)}} |f_0(t)|^2 \, d\mu(t) + \delta.$$

Now we can find a subsequence $(n_k)_k$ such that $J_{n_k}(s) \to 0$, and so $J_{n_k}(t) \to 0$ for all t with $s \prec t$. Thus, for each $\delta > 0$,

$$\begin{aligned} \varepsilon^2 &= \liminf_{n\to\infty} \|J_n f_0\|_2^2 \\ &\leq \limsup_{k\to\infty} \int_{\{t:t\prec s\}} |J_{n_k}(t) f_0(t)|^2 \, d\mu(t) \leq \int_{t\in X_1^{(2)}} |f_0(t)|^2 \, d\mu(t) + \delta, \end{aligned}$$

and hence

$$\varepsilon^2 \leq \int_{t\in X_1^{(2)}} |f_0(t)|^2 \, d\mu(t). \tag{7.20}$$

It follows from (7.19) and (7.20) that $f_0(t) = 0$ a.e. for $t \in X_1^{(1)} \setminus X_1^{(2)}$ and hence, using the dominated-convergence theorem, we see that $(f_0 - T^n y_n)_n$ converges in norm to $f_0 \chi_{X_1^{(1)}}$. □

Corollary 7.4.8 *Let T be a weighted composition operator on $L^2(X)$ as in (7.16). Suppose that there is a point $x_0 \in X$ such that whenever $|t - x_0| < |r - x_0|$ we have*

$$\limsup_{n\to\infty} \frac{h(\gamma^n(t))}{h(\gamma^n(r))} < 1. \tag{7.21}$$

Then $(T^n y_n)_n$ converges in norm.

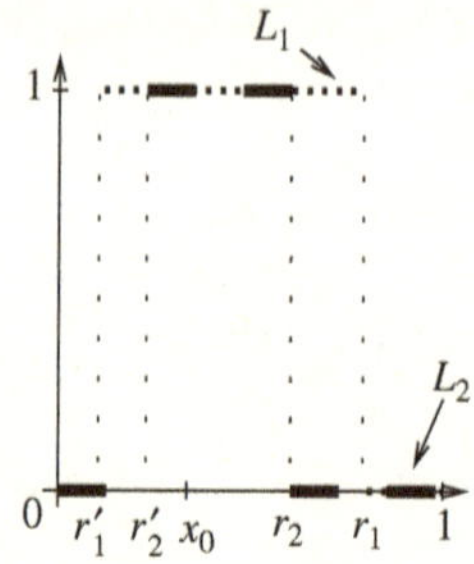

Figure 7.1 The graphs of L_1 and L_2 as in Remark 7.4.9. Dotted curve: L_1; bold curve: L_2.

Proof This result follows from Theorem 7.4.7, since $t \prec r$ means simply that $|t - x_0| < |r - x_0|$. □

Remark 7.4.9 A simple modification of the above arguments shows that for $X = [0, 1]$, if one has (7.21) whenever $x_0 < t < r$ or $r < t < x_0$, then $(T^n y_n)_n$ converges in norm. In this case there exist $r_1, r_1', r_2, r_2' \in [0, 1]$ such that the behaviour of L_1 and L_2 is given by Figure 7.1.

Having ensured the convergence of $(T^n y_n)_n$, we would like to apply Corollary 7.4.3 and Lemma 7.4.4. This requires us to know a condition for T to be quasinilpotent, and we give this in the next result.

Proposition 7.4.10 *Let T be a weighted composition operator on $L^2(X)$ defined as in (7.16). Suppose that γ has a unique attractive fixed point x_0 such that $(\gamma^n(t))_n$ converges to x_0 uniformly on X, and that h is continuous at x_0. Then the spectral radius of T satisfies $\rho(T) = \sqrt{a}$, where $a = h(x_0)$. Therefore T is quasinilpotent if and only if $h(x_0) = 0$.*

Proof For $f \in L^2(X)$ we have

$$\|T^n f\|_2^2 = \int_{s \in \gamma^n(X)} \frac{|w(\gamma^{-n}(s)) \dots w(\gamma^{-1}(s))|^2 |f(s)|^2}{|\gamma'(\gamma^{-n}(s)) \dots \gamma'(\gamma^{-1}(s))|^\alpha} \, d\mu(s).$$

Suppose that $h(x_0) = a$, and take $\varepsilon > 0$. Then there exists a positive integer K such that $k \geq K$ implies $\frac{|w(\gamma^k(t))|^2}{|\gamma'(\gamma^k(t))|^\alpha} < a + \varepsilon$. So, for $n > K$, we have

$$\|T^n f\|^2 \leq \|h\|_\infty^K (a + \varepsilon)^{n-K} \|f\|_2^2 .$$

It follows that $\lim_{n\to\infty} \|T^n\|^{1/n} \leq a + \varepsilon$ and thus $\rho(T) \leq \sqrt{a}$. If $a = 0$, then $\rho(T) = 0$. Suppose now that $a > 0$, and take $\varepsilon < a$. Then there exists

K' such that $k \geq K'$ implies $\frac{|w(\gamma^k(t))|^2}{|\gamma'(\gamma^k(t))|^\alpha} > a - \varepsilon$. Let $a > \delta > 0$ and set $S_\delta = \{t : h(t) \geq \delta\}$. For $n > K'$ and $f = \chi_{\gamma^n(S_\delta)}$, we get

$$\begin{aligned}\|T^n f\|_2^2 &= \int_{s \in \gamma^n(S_\delta)} \frac{|w(\gamma^{-n}(s)) \ldots w(\gamma^{-1}(s))|^2}{|\gamma'(\gamma^{-n}(s)) \ldots \gamma'(\gamma^{-1}(s))|^\alpha} \, d\mu(s) \\ &\geq \delta^{K'} (a - \varepsilon)^{n-K'} \|f\|_2^2 .\end{aligned}$$

It follows that $\lim_{n\to\infty} \|T^n\|^{1/n} \geq \sqrt{a - \varepsilon}$ and thus $\rho(T) \geq \sqrt{a}$. Finally $\rho(T) = \sqrt{a}$, and obviously T is quasinilpotent if and only if $h(x_0) = 0$. □

Using Corollary 7.4.3, Lemma 7.4.4, Theorem 7.4.7 and Proposition 7.4.10, we arrive at the following corollary, which should be interpreted in the light of Remark 7.4.6.

Corollary 7.4.11 *Let T be an operator on $L^2(X)$ defined as in (7.16). Suppose that h determines a partial order $\prec$ that is both left- and right-regular, such that, for every $t \in X$, (7.18) holds. Suppose also that γ has a unique attractive fixed point x_0 such that $(\gamma^n(t))_n$ converges to x_0 uniformly on X, and that h is continuous at x_0 with $h(x_0) = 0$. Then the operator T has non-trivial hyperinvariant subspaces.*

Example 7.4.12 Let $\gamma : [0, 1] \to [0, 1]$ defined by $\gamma(t) = \alpha t + a$ with $0 \leq a \leq 1$ and $-1 < \alpha < 1$. Then γ has a unique fixed point c_0 satisfying $c_0 = \frac{a}{1-\alpha}$. Moreover, $|\gamma(t_1) - \gamma(t_2)| = |\alpha||t_1 - t_2|$ and thus γ is strictly contractive and the sequence $(\gamma^n(t))_n$ converges to c_0 uniformly on $[0, 1]$. Then define w_1 on $[0, 1]$ by

$$w_1(t) = \begin{cases} \dfrac{c_0 - t}{c_0} & \text{if } 0 \leq t \leq c_0, \\[2ex] \dfrac{t - c_0}{1 - c_0} & \text{if } c_0 \leq t \leq 1, \end{cases}$$

and set $w = w_1^{1/2}$. The function w_1 is shown in Figure 7.2.

So, $w \in L^\infty([0, 1])$; $h := \frac{w_1}{|\alpha|} \in L^\infty([0, 1])$ is continuous at c_0 and $h(c_0) = 0$.

We now have to study the quotient $\dfrac{h(\gamma^n(t))}{h(\gamma^n(r))}$ for various $t, r \in [0, 1]$:

$$\frac{h(\gamma^n(t))}{h(\gamma^n(r))} = \begin{cases} \dfrac{c_0 - \gamma^n(t)}{c_0 - \gamma^n(r)} & \text{if } r < t < c_0, \\[2ex] \dfrac{\gamma^n(t) - c_0}{\gamma^n(r) - c_0} & \text{if } c_0 < t < r. \end{cases}$$

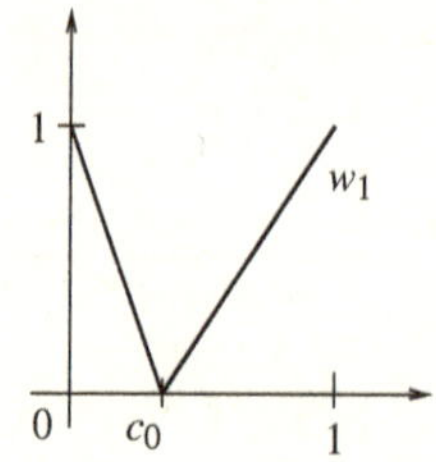

Figure 7.2 The graph of w_1.

But $\gamma^n(t) = \alpha^n t + a \sum_{k=0}^{n-1} \alpha^k = \alpha^n t + c_0(1-\alpha^n)$. So $\gamma^n(t) - c_0 = \alpha^n(t - c_0)$.

Thus,

$$\frac{h(\gamma^n(t))}{h(\gamma^n(r))} = \begin{cases} \dfrac{c_0 - t}{c_0 - r} & \textit{if } r < t < c_0, \\[2mm] \dfrac{t - c_0}{r - c_0} & \textit{if } c_0 < t < r, \end{cases}$$

and in both cases we obtain

$$\limsup_{n\to\infty} \frac{h(\gamma^n(t))}{h(\gamma^n(r))} < 1.$$

We conclude that for these choices of γ and w, T has non-trivial hyperinvariant subspaces.

Now we take the same γ as before and $w = w_2$ defined by

$$w_2(t) = c(t - c_0)^{2m}, \text{ with } m \geq 1, c > 0.$$

Then we have

$$\frac{h(\gamma^n(t))}{h(\gamma^n(r))} = \frac{(\gamma^n(t) - c_0)^{2m}}{(\gamma^n(r) - c_0)^{2m}} = \left(\frac{t - c_0}{r - c_0}\right)^{2m}.$$

If $c_0 < t < r$ or if $r < t < c_0$ we obtain $0 < \dfrac{t - c_0}{r - c_0} < 1$. So, in both cases we have $\limsup_{n\to\infty} \frac{h(\gamma^n(t))}{h(\gamma^n(r))} < 1$ and we conclude that the operator T has non-trivial hyperinvariant subspaces.

For every analytic self-map γ of $\mathbb{D}$ that is not an elliptic automorphism (conjugate to a rotation), there is a unique attractive fixed point $\zeta_0 \in \overline{\mathbb{D}}$, the

Denjoy–Wolff point of f. If $\zeta_0 \in \mathbb{T}$, then γ' exists at ζ_0 in the sense of an *angular derivative*, namely, γ has a non-tangential limit at ζ_0 of modulus 1, and in addition γ' has a non-tangential limit at ζ_0.

Definition 7.4.13 Let γ be an analytic self-map of $\mathbb{D}$ that is not an elliptic automorphism, and let ζ_0 be its Denjoy–Wolff point.

- If $\zeta_0 \in \mathbb{T}$ and the angular derivative satisfies $0 < \gamma'(\zeta_0) < 1$, one says that γ is of *hyperbolic type*.
- If $\zeta_0 \in \mathbb{T}$ and the angular derivative satisfies $\gamma'(\zeta_0) = 1$, one says that γ is of *parabolic type*.
- If $\zeta_0 \in \mathbb{D}$ and $0 < |\gamma'(\zeta_0)| < 1$, one says that γ is of *Schröder type*.

For an arbitrary function $f : \mathbb{D} \to \mathbb{C}$ we define the partial order $\prec$ on $\mathbb{D}$ associated with f by

$$z \prec w \qquad \text{if and only if} \qquad |f(z)| < |f(w)|. \tag{7.22}$$

Let us explore some of the properties of this partial order. First we make the following easy observation (the condition that E is closed enables us to avoid questions of measurability).

Lemma 7.4.14 *Suppose that $f : \mathbb{D} \to \mathbb{C}$ is a non-constant analytic function. Then $\mu(f^{-1}(E)) = 0$ for every closed $E \subset \mathbb{C}$ with $\mu(E) = 0$.*

Proof Let A be a closed set of positive measure, and let a_0 be a point of A with $f'(a_0) \neq 0$ such that a_0 is a Lebesgue density point; i.e., $\mu(A \cap D) \neq 0$ for every open disc D centred at a_0. It is well known that almost all points of A are density points, and that f' can only vanish on a countable set of points. Then, integrating with respect to area measure, we have

$$\int_{D \cap A} |f'(z)|^2 \, dA(z) = \int_{f(D \cap A)} dA(z) \leq \mu(f(A))$$

for every sufficiently small disc D, and hence $f(A)$ has positive measure. □

Corollary 7.4.15 *Let f be a non-constant analytic function on $\mathbb{D}$, and $\prec$ the partial order on $\mathbb{D}$ associated with f by (7.22). Then*

$$\mu(\{z \in \mathbb{D} : z \prec v \quad or \quad v \prec z\}) = 1$$

for every $v \in \mathbb{D}$.

Proof The set of z for which $|f(z)| = |f(v)|$ has measure zero by Lemma 7.4.14, being the inverse image under f of a circle. Hence its complement has full measure, as asserted. □

Proposition 7.4.16 *Let f be a non-constant analytic function on $\mathbb{D}$, and $\prec$ the partial order on $\mathbb{D}$ associated with f, defined by (7.22). Then $\prec$ is regular with respect to μ.*

Proof We begin with a set Y such that whenever $r \in Y$ and $|f(t)| < |f(r)|$ one has $t \in Y$. Now, picking a sequence $(r_n)_n$ in Y with $|f(r_n)| \to \sup_{y \in Y} |f(y)|$ (which may be infinite), we see that the set

$$Y \setminus \bigcup_{n=1}^{\infty} \{t \in Y : |f(t)| < |f(r_n)|\}$$

consists of (at most) all points in Y at which $|f|$ attains its supremum. But this is a set of measure zero, by Lemma 7.4.14. Hence, given $\delta > 0$, we have $\mu\{t \in Y : t \prec r_n\} > \mu(Y) - \delta$ for sufficiently large n, and so $\prec$ is left-regular.

The proof of right-regularity is similar, and is left as an exercise. □

As before, we define the operator $T = T_{w,\gamma}$ by (7.16). We are now ready for the main convergence result.

Theorem 7.4.17 *Let γ be a univalent self-map of $\mathbb{D}$, of hyperbolic or Schröder type, with Denjoy–Wolff point ζ_0. Suppose that w is differentiable at ζ_0, with $w(\zeta_0) = 0$ and $w'(\zeta_0) \neq 0$. Then $(T^n y_n)_n$ converges in norm for every sequence of minimal vectors.*

Proof We claim that the partial order $\prec$ defined by (7.17) can be written as the partial order associated with a non-constant analytic function f by means of the formula (7.22).

Note that we are led to consider $h(\gamma^n(t))/h(\gamma^n(r))$, as in (7.17), where $h = |w|^2/|\gamma'|^2$, and we are given the fact that γ' does not vanish at ζ_0 (interpreting this as an angular derivative if $\zeta_0 \in \mathbb{T}$). Now

$$w(\gamma^n(z)) = w(\zeta_0) + w'(\zeta_0)(\gamma^n(z) - \zeta_0) + O(\gamma^n(z) - \zeta_0)^2,$$

and an easy calculation reduces the claim to consideration of the quantity $\limsup_{n\to\infty} |\gamma^n(t) - \zeta_0|/|\gamma^n(r) - \zeta_0|$.

Let $s \in \mathbb{D}$ be any point such that $\gamma^n(s) \neq \zeta_0$ for all n. Then the limit

$$\lim_{n\to\infty} \frac{\gamma^n(z) - \zeta_0}{\gamma^n(s) - \zeta_0}$$

exists for each $z \in \mathbb{D}$ and defines an analytic function $f(z)$. We omit the details of the arguments used to establish this, which are classical (see the notes) and are left as an exercise. Given this, then $t \prec r$ if and only if $|f(t)| < |f(r)|$, as claimed.

Now by Corollary 7.4.15 and Proposition 7.4.16, the order $\prec$ satisfies the hypotheses of Theorem 7.4.7, and so the result follows. □

Corollary 7.4.18 *Suppose that γ is an analytic self-map of $\mathbb{D}$, satisfying the conditions of Theorem 7.4.17, and in addition that the operator $T = T_{w,\gamma}$ is quasinilpotent. Then $T : L^2(\mathbb{D}) \to L^2(\mathbb{D})$ has a non-trivial hyperinvariant subspace.*

Note that a condition for quasinilpotence was given in Proposition 7.4.10.

7.4.4 Weighted shifts

In this section we apply the method of minimal vectors to certain weighted shifts, beginning with shifts on the Hilbert space $\ell^2(\mathbb{Z})$ and then looking more generally at matricial shifts on vector-valued $\ell^p(\mathbb{Z})$ spaces.

Corollary 7.4.3 may be applied in the following example, in which T is neither quasinilpotent nor compact and it is unclear whether T even commutes with a non-zero compact operator (it is not itself polynomially compact in the sense of Definition 1.2.3).

Example 7.4.19 Let $(\lambda_n)_{n\in\mathbb{Z}}$ be a bounded sequence, and define the weighted shift $T : \ell^2(\mathbb{Z}) \to \ell^2(\mathbb{Z})$ by $Te_n = \lambda_n e_{n+1}$, $n \in \mathbb{Z}$, where $(e_n)_{n\in\mathbb{Z}}$ is the standard orthonormal basis of $\ell^2(\mathbb{Z})$. It follows that T^* satisfies $T^*e_n = \overline{\lambda_{n-1}} e_{n-1}$, $n \in \mathbb{Z}$. Clearly, if $\lambda_n \neq 0$ for all n, then T is injective with dense range. Moreover, we have $\|T^n\| = \sup\{|\lambda_k \cdots \lambda_{k+n-1}| : k \in \mathbb{Z}\}$. Hence T is quasinilpotent if and only if $\lim_{n\to\infty} \sup\{|\lambda_k \cdots \lambda_{k+n-1}|^{1/n} : k \in \mathbb{Z}\} = 0$ and unless $\lambda_n \to 0$ as $|n| \to \infty$, T is not compact.

We begin by taking $x_0 = e_p$ for some $p \in \mathbb{Z}$, and $0 < \varepsilon < 1$. By inspection we see that the backward minimal vector y_n satisfies $T^n y_n = (1-\varepsilon)e_p$, or $y_n = \dfrac{1-\varepsilon}{\lambda_{p-1}\cdots\lambda_{p-n}} e_{p-n}$. Hence $\frac{\|y_{n-1}\|}{\|y_n\|} = |\lambda_{p-n}|$. So, provided that the weights satisfy $\inf\{|\lambda_j| : j < 0\} = 0$, we can find a subsequence $(n_k)_{k\geq 0}$ of $\mathbb{N}$ which satisfies (7.14). Also $(T^{n-1}y_{n-1})_{n\geq 2}$ is a constant sequence and trivially converges to $(1-\varepsilon)e_p$.

By Theorem 7.4.2, a hyperinvariant subspace for T is given by the closed linear hull of the set $\{DTe_p : DT = TD\}$, and it is not equal to $\ell^2(\mathbb{Z})$ itself, since $\langle DTe_p, e_p\rangle = 0$ for all $D \in \{T\}'$. (In Lemma 7.4.1 we may take each g_n to be the functional induced by taking the inner product with e_p.)

In the particular case where $\lambda_k = \begin{cases} -\frac{1}{k} & \text{if } k < 0, \\ 1 & \text{otherwise,} \end{cases}$ we see that T is neither polynomially compact nor quasinilpotent.

Similar arguments apply if we take $x_0 = \sum_{k=-N}^{N} a_k e_k$ for any constants $a_{-N}, \dots, a_N$ not all zero. In this case the expression for y_n is more complicated, and using (7.2), we see that it is given by

$$y_n = \mu_n (\mathrm{I} + \mu_n T^{*n} T^n)^{-1} T^{*n} x_0. \tag{7.23}$$

(Note that $T^{*n}T^n$ is a diagonal operator, which simplifies the calculation.) However, it is clear that $(T^n y_n)_{n \geq 1}$ is a bounded sequence in the finite-dimensional space generated by $e_{-N}, \dots, e_N$, and hence has a norm-convergent subsequence. In this case too a suitable choice of $(\lambda_n)_{n \in \mathbb{Z}}$ guarantees that (7.14) holds.

Remark 7.4.20 In the above case we see that if $DT = TD$ then we have $\langle DTe_p, e_p \rangle = 0$. Hence, for $m \geq 1$, since DT^m also lies in $\{T\}'$, we have $\langle DT^{m+1} e_p, e_p \rangle = 0$ and hence $\langle De_p, e_{p-m-1} \rangle = 0$. This implies that D is *causal*, in the sense that if $\langle w, e_n \rangle = 0$ for all $n < p$, then $\langle Dw, e_n \rangle = 0$ for all $n < p$; this is in contrast with the situation when T is the usual unweighted shift on $\ell^2(\mathbb{Z})$.

We now move on to a more complicated version of the above.

Example 7.4.21 Let $1 \leq p < \infty$ and let $\mathcal{H}$ denote a finite-dimensional Hilbert space. For $k \in \mathbb{Z}$ let $M_k \in \mathcal{L}(\mathcal{H})$ be an invertible linear operator, and suppose that the sequence $(M_k)_{k \in \mathbb{Z}}$ is uniformly norm-bounded. Then we may define a vectorial weighted shift $T : \ell^p(\mathbb{Z}, \mathcal{H}) \to \ell^p(\mathbb{Z}, \mathcal{H})$ by

$$T\left(\sum_{k=-\infty}^{\infty} e_k \otimes v_k\right) = \sum_{k=-\infty}^{\infty} e_{k+1} \otimes M_k v_k,$$

where $v_k \in \mathcal{H}$ for each k (here $e_k \otimes v_k$ denotes the element of $\ell^p(\mathbb{Z}, \mathcal{H})$ whose kth coefficient is v_k, the rest being 0). This is a natural generalization of the scalar weighted shift, and clearly T is an injective operator with dense range. Rather than attempting to analyse a general choice of x_0, we shall fix a vector $v_0 \in \mathcal{H}$, write $x_0 = e_0 \otimes v_0$ and take $0 < \varepsilon < \|x_0\|$. Now consider the minimal vector y_n corresponding to (T, n, x_0, ε).

Theorem 7.4.22 *Let T, n, x_0 and ε be as above. Then the associated minimal vector y_n has the form $y_n = e_{-n} \otimes u_n$, where $u = u_n$ is the unique solution to the problem of minimizing $\|u\|$ subject to the constraint*

$$\|M_{-n}M_{1-n}\dots M_{-1}u - v_0\| \le \varepsilon. \tag{7.24}$$

Suppose that

$$\liminf_{n\to+\infty} \|M_{-n}\|\,\|M_{1-n}\dots M_{-1}\|\,\|(M_{1-n}\dots M_{-1})^{-1}\| = 0. \tag{7.25}$$

Then we may find a subsequence (n_k) such that, simultaneously, $(T^{n_k}y_{n_k-1})$ converges in norm and $\|y_{n_k-1}\|/\|y_{n_k}\| \to 0$; from this one can construct a non-trivial hyperinvariant subspace for T.

Proof Note that the ℓ^p norm is 1-*unconditional*, in the sense that

$$\|(a_n)\|_p \le \|(b_n)\|_p \qquad \text{if} \qquad |a_n| \le |b_n| \quad \text{for all } n.$$

It is clear from this that y_n must be of the form $y_n = e_{-n} \otimes u_n$ where u_n satisfies the conditions of the theorem. Note that the u_n is itself a minimal vector associated with the Hilbert space quadruple $(M_{-n}\dots M_{-1}, 1, v_0, \varepsilon)$, and hence it is unique.

Note that

$$\|y_{n-1}\| = \|u_{n-1}\| \le \|(M_{1-n}\dots M_{-1})^{-1}\|(\|v_0\| + \varepsilon),$$

since $\|M_{1-n}\dots M_{-1}u_{n-1} - v_0\| \le \varepsilon$. Also

$$\|M_{-n}\dots M_{-1}u_n - v_0\| \le \varepsilon,$$

and hence

$$\|v_0\| - \varepsilon \le \|M_{-n}\dots M_{-1}u_n\| \le \|M_{-n}\dots M_{-1}\|\,\|u_n\|,$$

and so

$$\|y_n\| = \|u_n\| \ge \frac{\|v_0\| - \varepsilon}{\|M_{-n}\dots M_{-1}\|}.$$

Therefore,

$$\frac{\|y_{n-1}\|}{\|y_n\|} \le \frac{\|v_0\| + \varepsilon}{\|v_0\| - \varepsilon}\|(M_{1-n}\dots M_{-1})^{-1}\|\,\|M_{1-n}\dots M_{-1}\|\,\|M_{-n}\|.$$

To complete the proof we use a compactness argument in the finite-dimensional space $\mathcal{H}$, and find a subsequence such that the hypotheses of the theorem are satisfied. Finally we apply Corollary 7.4.3. □

Remark 7.4.23 1. In the case when $\dim \mathcal{H} = 1$, condition (7.25) reduces to

$$\liminf_{n\to\infty} \|M_{-n}\| = 0,$$

as in the scalar case.

2. It seems that a full description of the commutant of T is unknown in general, unlike the scalar case described by Gellar [113]. Thus, although Corollary 7.4.3 applied to the weighted shift on ℓ^p appears to give little new in the scalar case, it seems that the consequences in the vectorial case are of independent interest.

7.4.5 Multiplication operators on L^p spaces

We now consider multiplication operators on $L^p(X, \mu)$ for $1 < p < \infty$. This situation includes the well-understood case of normal operators on a Hilbert space, by Theorem 1.2.16. To avoid a conflict of notation, we start with $f_0 \in L^p(X, \mu)$, and $0 < \varepsilon < \|f_0\|$, considering the minimal vectors y_n associated with f_0, ε, T and n. By the results of Section 7.3.2, we know that minimal vectors y_n exist, and can be characterized by the following formula, which is immediate from Theorem 7.3.5, with $A = \mathrm{I}$, $B = T^n$, $t_0 = f_0$, $t_1 = 0$ and $x_0 = y_n$:

$$(T^n)^* \nu(f_0 - T^n y_n) = \mu_n \nu(y_n) \tag{7.26}$$

for some $\mu_n > 0$, where $\nu : L^p(X, \mu) \setminus \{0\} \to L^q(X, \mu)$ is the duality mapping, given by

$$\nu(g) = \frac{\overline{g}|g|^{p-2}}{\|g\|_p^{p-1}}, \tag{7.27}$$

and, as usual, $1/p + 1/q = 1$. We now establish the norm convergence of the sequence $(T^n y_n)$ in the case when $T = M_\phi$, and $\phi \in L^\infty(X, \mu)$.

Theorem 7.4.24 *Let $1 < p < \infty$ and suppose that $T : L^p(X, \mu) \to L^p(X, \mu)$ is the operator M_ϕ of multiplication by $\phi \in L^\infty(X, \mu)$, and that it has dense range. Suppose that $f_0 \in L^p(X, \mu)$. For any ε with $0 < \varepsilon < \|f_0\|$, let (y_n) denote the corresponding sequence of backward minimal vectors. Then the sequence $(T^n y_n)$ converges in norm.*

Proof Let us write $\beta_n = \|y_n\|_p$. By (7.26) and (7.27), we have

$$\phi(z)^n \frac{\overline{(f_0 - T^n y_n)(z)}\, |(f_0 - T^n y_n)(z)|^{p-2}}{\varepsilon^{p-1}} = \mu_n \frac{\overline{y_n(z)}\, |y_n(z)|^{p-2}}{\beta_n^{p-1}},$$

because $T^* = M_\phi$ acting on $L^q(X, \mu)$, since the duality pairing is bilinear rather than sesquilinear. Since $y_n = (T^n y_n)/\phi^n$, it follows that

$$\frac{(f_0 - T^n y_n)(z)}{(T^n y_n)(z)} = \mu_n \frac{\varepsilon^{p-1}}{\beta_n^{p-1}} \frac{1}{|\phi(z)|^{2n}} \frac{|y_n(z)|^{p-2}}{|(f_0 - T^n y_n)(z)|^{p-2}}.$$

Taking absolute values, we obtain

$$\frac{|(f_0 - T^n y_n)(z)|^{p-1}}{|(T^n y_n)(z)|^{p-1}} = \mu_n \frac{\varepsilon^{p-1}}{\beta_n^{p-1}} \frac{1}{|\phi(z)|^{np}}.$$

Write

$$K_n(r) = \mu_n^{1/(p-1)} \frac{\varepsilon}{\beta_n} \frac{1}{r^{np/(p-1)}},$$

so that

$$\frac{(f_0 - T^n y_n)(z)}{(T^n y_n)(z)} = \frac{|(f_0 - T^n y_n)(z)|}{|(T^n y_n)(z)|} = K_n(|\phi(z)|).$$

Thus,

$$(f_0 - T^n y_n)(z) = \frac{K_n(|\phi(z)|)}{1 + K_n(|\phi(z)|)} f_0(z) =: J_n(|\phi(z)|) f_0(z),$$

say, where $0 \le J_n(r) \le 1$ for each $r \ge 0$.

We shall establish pointwise convergence almost everywhere and use the dominated-convergence theorem to deduce norm convergence. The necessary argument is based on the ideas in the proof of Theorem 7.4.7.

Accordingly, let $L_1(r) = \limsup_{n\to\infty} J_n(r)$ and $L_2(r) = \liminf_{n\to\infty} J_n(r)$. Note that $0 \le L_2(r) \le L_1(r) \le 1$ for all r and that L_1 and L_2 are decreasing functions. If $L_1(r) > 0$ for some value of r, then there is a sequence (n_k) such that $(\mu_{n_k} r^{n_k})$ remains bounded, so for $t < r$ we have $\mu_{n_k} t^{n_k} \to 0$, and hence $L_1(t) = 1$ for $t < r$. We conclude immediately that if $L_1(s) < 1$ for some s then $L_1(t) = 0$ for all $t > s$ (if not, we would have $L_1(s) = 1$).

Hence there exist $r_2 \le r_1$ such that

$$L_1(t) = \begin{cases} 1 & \text{for } t < r_1, \\ 0 & \text{for } t > r_1, \end{cases} \tag{7.28}$$

and

$$L_2(t) = \begin{cases} 1 & \text{for } t < r_2, \\ 0 & \text{for } t > r_2. \end{cases} \tag{7.29}$$

Write $g_n = f_0 - T^n y_n$. Then we have

$$\begin{aligned} \varepsilon^p &= \limsup_{n\to\infty} \|g_n\|^p \\ &\ge \int_{0 \le |\phi(z)| < r_1} |f_0(z)|^p \, d\mu(z) + \int_{|\phi(z)| = r_1} L_1(r_1)^p |f_0(z)|^p \, d\mu(z) \end{aligned}$$

and

$$\begin{aligned}\varepsilon^p &= \liminf_{n\to\infty} \|g_n\|^p \\ &\leq \int_{0\leq|\phi(z)|<r_2} |f_0(z)|^p d\mu(z) + \int_{|\phi(z)|=r_2} L_2(r_2)^p |f_0(z)|^p\, d\mu(z).\end{aligned}$$

If $r_1 = r_2$, then either $L_1(r_1) = L_2(r_2)$ or $f_0(z) = 0$ a.e. on the set on which $|\phi(z)| = r_1$, and so the sequence converges almost everywhere. Otherwise,

$$\begin{aligned}&(1 - L_2(r_2)^p) \int_{|\phi(z)|=r_2} |f_0(z)|^p\, d\mu(z) + \int_{r_2<|\phi(z)|<r_1} |f_0(z)|^p\, d\mu(z) \\ &+ L_1(r_1)^2 \int_{|\phi(z)|=r_1} |f_0(z)|^p\, d\mu(z) \leq 0,\end{aligned}$$

and hence each term is zero. It follows that $f_0(z) = 0$ almost everywhere on the set on which $L_1(r) \neq L_2(r)$. Hence (g_n) converges in norm. □

With the aid of the following description of the commutant of a multiplication operator, we can go slightly further. In particular, either T has no cyclic vector (and thus many invariant subspaces) or the minimal vectors provide a hyperinvariant subspace directly.

Lemma 7.4.25 *Let $1 \leq p < \infty$, let $T = M_\phi$ be a multiplication operator on $L^p(X, \mu)$ and suppose that T has a cyclic vector. Then the commutant of T is $\{M_\psi : \psi \in L^\infty(X, \mu)\}$.*

Proof Let f be a cyclic vector for T; clearly $f \neq 0$ a.e. Suppose that $AT = TA$; then $AT^n f = T^n A f$ for all positive integers n, and thus $A(T^n f) = \psi.(T^n f)$, where ψ denotes Af/f. Since A is bounded, and $A = M_\psi$ on a dense subset of $L^p(X, \mu)$, it follows that $\psi \in L^\infty(X, \mu)$, and $A = M_\psi$ everywhere. □

Corollary 7.4.26 *Under the hypotheses of Theorem 7.4.24, suppose that, in addition, $T = M_\phi$ has a cyclic vector. Then $T^n y_n$ converges to a vector h such that the closure of $hL^\infty(X, \mu)$ is a hyperinvariant subspace for T.*

Proof If T is cyclic, then its commutant is $\{M_\psi : \psi \in L^\infty(X, \mu)\}$, by Lemma 7.4.25, and $hL^\infty(X, \mu) \subset \chi_{\{z: |\phi(z)|\geq r_2\}} L^\infty(X, \mu)$. Thus h has the required property. □

Exercises

1. Let $\mathcal{H}$ be a Hilbert space, and let $T \in \mathcal{L}(\mathcal{H})$ be injective with dense range; take $x_0 \in \mathcal{H}$, $0 < \varepsilon < \|x_0\|$ and $n \geq 1$. The *forward minimal vector* associated with x_0, ε, T and n is defined to be a vector v_n such that $\|T^n v_n\|$ is minimized subject to the condition $\|v_n - x_0\| \leq \varepsilon$. Apply Theorem 7.3.2 to find an expression for v_n.
2. Let $\Omega = [-b, b]$ and $J = [-T, T]$ be sub-intervals of $\mathbb{R}$, and $f \in L^2(J)$. For $M > 0$ find an expression for the function $g \in PW(\Omega)$ (the space of functions in $L^2(\mathbb{R})$ whose Fourier transform is supported on Ω) which minimizes $\|g_{|J} - f\|_{L^2(J)}$ subject to the condition $\|g_{|\mathbb{R}\setminus J}\|_{L^2(\mathbb{R}\setminus J)} \leq M$.
3. Fix r with $0 < r < 1$ and let $\phi_1 : \mathbb{D} \to \mathbb{D}$ be defined by $\phi_1(z) := \frac{z+r}{1+rz}$. Determine the Denjoy–Wolff point and type of this mapping. Do the same for ϕ_2, given by $\phi_2(z) := (\phi_1(z))^2$.
4. Complete the proof of Proposition 7.4.16, asserting the regularity of the partial order $\prec$.
5. Write out the explicit details of the proof of Theorem 7.4.17 in the special case when γ is a non-automorphic Möbius transformation of the disc, with two fixed points, of which one lies in $\overline{\mathbb{D}}$ and the other in its complement (possibly at ∞).

Comments

The method of minimal vectors in the case of a Hilbert space was introduced by Ansari and Enflo [11, 97], who presented new proofs of the existence of hyperinvariant subspaces for compact and normal operators. Extensions to super-reflexive spaces by Androulakis [9] and to reflexive spaces by Chalendar, Partington and Smith [76] were given, before Troitsky [196] introduced λ-minimal vectors in general Banach spaces: these are not so explicitly computable, but do have most of the properties needed.

The more general extremal problem in Section 7.3 was solved first in Hilbert spaces by Chalendar and Partington [69] and then extended to reflexive Banach spaces by Chalendar, Partington and Smith [76], providing also an extension of results by Smith [187]. The extremal problem generalizes the problem discussed by Leblond and Partington [140] (with a different notation), where $\mathcal{X}$ is a closed subspace of $\mathcal{Y} \oplus \mathcal{Z}$, and A and B are the restrictions to $\mathcal{X}$ of the orthogonal projections onto $\mathcal{Y}$ and $\mathcal{Z}$, respectively. A simple example of this is the case $\mathcal{X} = H^2$, $\mathcal{Y} = L^2(J)$ and $\mathcal{Z} = L^2(K)$, where J and K are complementary subsets of the unit circle, each with strictly positive measure. Versions of this problems were first studied by Kreĭn and Nudel'man [137], and more recently by Alpay, Baratchart and Leblond [7] and by Baratchart and Leblond [23]. This problem has applications in signal processing and control theory. For further information, see Chalendar, Leblond and Partington [67], for example.

Proposition 7.3.3 generalizes a result of Enflo and Hõim [98] and a result in Alpay, Baratchart and Leblond [7].

Theorem 7.4.2 extends the results of Androulakis [9], which deal with super-reflexive spaces, as well as the main theorem about quasinilpotent operators found in Troitsky [196].

The discussion of weighted composition operators in Section 7.4.3 is based on ideas in Chalendar, Flattot and Partington [64] and Bourdon and Flattot [47]. The diagrams were kindly produced by A. Flattot.

The classical results on convergence of iterates of analytic functions, used in Theorem 7.4.17, are due to Koenigs [135] and Valiron [197]. See also chapter 5 of Shapiro [184].

The discussion of weighted shift operators is based on work by Chalendar and Partington [69, 73]. They also analysed multiplication operators [72], the L^2 case (normal operators) having been done less precisely by Ansari and Enflo [11].

Further results applying techniques of minimal vectors to quasinilpotent operators have been given by Kim [131, 132].

Exercise 1 is based on Ansari and Enflo [11]. Forward minimal vectors provide an alternative approach to backward minimal vectors, from which the same results can be derived, but most authors have chosen to use the backward minimal vectors.

Exercise 2 is a classic problem in signal processing (see Slepian [186]), and is discussed in many places, for example Chalendar, Leblond and Partington [67] and Smith [187]. The notation $PW(\Omega)$ refers to a *Paley–Wiener space*.

8

Universal operators

If the invariant-subspace problem in Hilbert spaces has a negative solution, then there must be a counterexample, but at the time of writing nobody has been able to find one. On the other hand, if the solution is positive, then at first sight it is necessary to prove a theorem that applies to all Hilbert space operators simultaneously. In fact, the situation is somewhat simplified by the existence of universal operators – these have the property that if we could describe their lattice of subspaces precisely enough, then we could solve the invariant-subspace problem.

We therefore begin by presenting an old theorem due to Caradus [55] which gives conditions for an operator to be universal. With this, it is easy to write down various very simple operators, basically various forms of backward shift operator, which have the universality property.

Other operators that are less obviously universal include certain bilateral weighted shifts and composition operators on Hardy spaces; these last have attracted particular interest, since one can use results from function theory to obtain information about their invariant subspaces. We give some of the more interesting results in this area, although of course we cannot yet give a precise description of the lattice of invariant subspaces for any single universal operator.

8.1 Construction of universal models

Definition 8.1.1 Let $\mathcal{X}$ be a Banach space. Then an operator $U \in \mathcal{L}(\mathcal{X})$ is said to be *universal* for $\mathcal{X}$ if for each non-zero $T \in \mathcal{L}(\mathcal{X})$ there is a constant $\lambda \neq 0$ and an invariant subspace $\mathcal{M}$ for U such that $U_{|\mathcal{M}}$ is similar to λT, i.e., $\lambda JT = UJ$, where $J : \mathcal{X} \to \mathcal{M}$ is a linear isomorphism.

This is most useful in the case of a separable Hilbert space, since all its closed infinite-dimensional subspaces are automatically isometric to the space itself (and we can then use the terminology 'universal for Hilbert space'). Let us write down precisely the application of this result to the invariant-subspace problem.

Proposition 8.1.2 *Let $\mathcal{X}$ be an infinite-dimensional Banach space and $U \in \mathcal{L}(\mathcal{X})$ a universal operator. Then the following conditions are equivalent:*

1. *Every non-zero $T \in \mathcal{L}(\mathcal{X})$ has a non-trivial closed invariant subspace.*
2. *Every invariant subspace $\mathcal{M}$ for U that is isomorphic to $\mathcal{X}$ contains a non-trivial proper closed invariant subspace.*

Proof Clearly the first condition implies the second, since $U_{|\mathcal{M}}$ must have a non-trivial closed invariant subspace because $\mathcal{M}$ and $\mathcal{X}$ are isomorphic. But the second implies the first equally easily, since each $T \in \mathcal{L}(\mathcal{X})$ is similar to some $\lambda^{-1}U_{|\mathcal{M}}$. □

In the case when $\mathcal{X}$ is a complex Hilbert space, the second condition of Proposition 8.1.2 may be rewritten in the form 'the minimal non-trivial invariant subspaces for U are one-dimensional', for every infinite-dimensional invariant subspace $\mathcal{M}$ is isomorphic to $\mathcal{X}$ and so is non-minimal; and, by elementary linear algebra, each non-trivial finite-dimensional invariant subspace contains an eigenspace, and so it is non-minimal unless it is one-dimensional.

We now present a theorem that will enable us, in the case of Hilbert space, to show the existence of many 'natural' operators that are universal. We refer to the Exercises for a Banach-space version, which is less useful in general.

Theorem 8.1.3 *Let $\mathcal{H}$ be a separable infinite-dimensional Hilbert space and $U \in \mathcal{L}(\mathcal{H})$. Suppose that*

1. $\mathrm{Ker}(U)$ *is infinite-dimensional; and*
2. $\mathrm{Im}(U)$ *is* $\mathcal{H}$.

Then U is universal for $\mathcal{H}$.

Proof Write $\mathcal{K} = \mathrm{Ker}\, U$. We begin by constructing $V, W \in \mathcal{L}(\mathcal{H})$ such that $UV = \mathrm{I}$, $UW = 0$, $\mathrm{Ker}\, W = \{0\}$, $\mathrm{Im}\, W = \mathcal{K}$ and $\mathrm{Im}\, V = \mathcal{K}^{\perp}$.

To do this, let $\widetilde{U} = U_{|\mathcal{K}^{\perp}}$; this is one-one and maps onto $\mathcal{H}$, since we have $\mathcal{H} = \mathcal{K} \oplus \mathcal{K}^{\perp}$. Then let $V = \widetilde{U}^{-1} : \mathcal{H} \to \mathcal{K}^{\perp}$.

Next, define an isometry $W : \mathcal{H} \to \mathcal{K}$, by taking orthonormal bases $(e_n)_n$ and $(e'_n)_n$ for $\mathcal{H}$ and $\mathcal{K}$ (which is infinite-dimensional), mapping each e_n onto

e'_n, and extending by linearity and continuity to the whole of $\mathcal{H}$. We clearly have all the properties asserted for V and W.

Now, given $T \in \mathcal{L}(\mathcal{H})$, we choose $\lambda \neq 0$ such that $|\lambda| \, \|T\| \, \|V\| < 1$, and define

$$J = \sum_{k=0}^{\infty} \lambda^k V^k W T^k.$$

The operator J satisfies $J = W + \lambda V J T$ and $UJ = \lambda J T$ since $UW = 0$.

We claim that J is an isomorphism onto a closed subspace $\mathcal{M} = \operatorname{Im} J$, for if $Jx = 0$, then $Wx + \lambda V J T x = 0$; but $\operatorname{Im} W \perp \operatorname{Im} V$, so $Wx = 0$ and then $x = 0$.

Moreover, $\operatorname{Im} J$ is closed, since if we have a sequence $(x_n)_n$ with $Jx_n \to y$, then $Wx_n + \lambda V J T x_n \to y$, and so (applying the projection $P_{\mathcal{K}}$) $Wx_n \to P_{\mathcal{K}} y$, whence, since W is an isometry, we have $x_n \to x$, say, and then $Jx_n \to Jx$ and finally $y \in \operatorname{Im} J$. □

Note that on a Hilbert space, an operator $U = V^*$ will therefore be universal if $\operatorname{Im} V$ is closed and has infinite codimension.

The simplest example of a universal operator is the adjoint of a unilateral shift of infinite multiplicity; this can be regarded as a left shift on $L^2(0, \infty)$ by fixing a number $a > 0$ and defining

$$S_a^* f(t) = f(t + a), \qquad \text{for} \quad t > 0 \quad \text{and} \quad f \in L^2(0, \infty).$$

Alternatively, it may be regarded as an operator S^* in $\mathcal{L}(\ell^2(\mathcal{H}))$, defined by

$$S^*((h_0, h_1, h_2, \ldots)) = (h_1, h_2, h_3, \ldots), \qquad \text{for} \quad (h_0, h_1, h_2, \ldots) \in \ell^2(\mathcal{H}).$$

Remark 8.1.4 Note that the backward shift S^* of multiplicity 1 is not universal, as by Beurling's theorem its invariant subspaces $\mathcal{M}$ all satisfy $\dim \mathcal{M} \ominus S\mathcal{M} = 0$ or 1, which means that it does not have any restriction similar to a shift of infinite multiplicity.

For the shift of infinite multiplicity a stronger result is true, namely that we can replace similarity by unitary equivalence in Definition 8.1.1, to obtain what is sometimes called the *universal model*. Recall that an operator $T \in \mathcal{L}(\mathcal{H})$ is of class $C_{0.}$ when $\|T^n x\| \to 0$ for all $x \in \mathcal{H}$, and this clearly includes all operators of norm strictly less than 1.

Theorem 8.1.5 *Let $T \in \mathcal{L}(\mathcal{H})$ be a contraction operator of class $C_{0.}$. Let $S : \mathcal{E} \to \mathcal{E}$ be a unilateral shift on a Hilbert space, such that $\mathcal{K} := \mathcal{E} \ominus \operatorname{Im} S$ is isomorphic to $\mathcal{H}$. Then there is an invariant subspace $\mathcal{M}$ of S^* such that T is unitarily equivalent to $S^*_{|\mathcal{M}}$.*

Proof As in Section 1.2.2, we have the Wold decomposition,

$$\mathcal{E} = \mathcal{K} \oplus S\mathcal{K} \oplus S^2\mathcal{K} \oplus \cdots .$$

Let $J : \mathcal{H} \to \mathcal{K}$ be a unitary operator giving the isomorphism from $\mathcal{H}$ onto $\mathcal{K}$.

The operator $I - T^*T$ is positive and self-adjoint, and has a positive self-adjoint square root, which we denote by D.

Define $W : \mathcal{H} \to \mathcal{E}$ by

$$Wx = \sum_{n=0}^{\infty} S^n JDT^n x, \qquad \text{for} \quad x \in \mathcal{H},$$

and note that $JDT^n x \in \mathcal{K}$. Hence

$$\begin{aligned}
\|Wx\|^2 &= \sum_{n=0}^{\infty} \|DT^n x\|^2 \\
&= \sum_{n=0}^{\infty} \langle T^{*n}(I - T^*T)T^n x, x\rangle \\
&= \sum_{n=0}^{\infty} \Big(\langle T^{*n}T^n x, x\rangle - \langle T^{*(n+1)}T^{n+1}x, x\rangle \Big) \\
&= \lim_{N\to\infty} \Big(\|x\|^2 - \|T^{N+1}x\|^2 \Big) = \|x\|^2,
\end{aligned}$$

since T is of class $C_{0\cdot}$.

Also, since $S^*J = 0$ we have

$$\begin{aligned}
S^*Wx &= \sum_{n=1}^{\infty} S^*S^n JDT^n x \\
&= \sum_{m=0}^{\infty} S^m JDT^m Tx = WTx,
\end{aligned}$$

and hence $\mathcal{M} := W\mathcal{H}$ is a closed S^*-invariant subspace of $\mathcal{H}$, and since $W : \mathcal{H} \to \mathcal{M}$ is unitary, we have the result. □

Let $\phi \in H^\infty(\mathbb{D})$ be an inner function, and consider the multiplication operator $M_\phi \in \mathcal{L}(H^2)$, which is an isometry. In fact it is a shift, as we now show.

Lemma 8.1.6 *Let ϕ be inner. Then M_ϕ is unitarily equivalent to a unilateral shift of multiplicity equal to the degree of ϕ.*

Proof Let $K_\phi = H^2 \ominus \phi H^2$, be the model space associated with ϕ. We have an orthogonal decomposition,

$$H^2 = K_\phi \oplus \phi K_\phi \oplus \phi^2 K_\phi \oplus \cdots .$$

To see this, note that if $f, g \in K_\phi$, then $\langle \phi^m f, \phi^n g \rangle = \langle f, \phi^{n-m} g \rangle = 0$ for $n > m$. Then an easy induction argument shows that $f \perp K_\phi \oplus \phi K_\phi \oplus \cdots \oplus \phi^n K_\phi$ if and only if $f \in \phi^n H^2$, and hence the orthogonal direct sum does equal H^2.

Now M_ϕ acts as a shift on this space, which in the decomposition given is naturally equivalent to $\ell^2(K_\phi)$. Since the dimension of K_ϕ equals the degree of ϕ, the result follows. □

We have an immediate corollary.

Corollary 8.1.7 *Let ϕ be an irrational inner function. Then M_ϕ^* is universal for Hilbert space.*

Proof This follows from Lemma 8.1.6 and Theorem 8.1.3. □

8.2 Bilateral weighted shifts

In this section we develop some of the elementary theory of certain bilateral weighted shifts on $\ell^2(\mathbb{Z})$: first, because they are a source of universal operators in their own right, and, second, because they will be of use when we discuss composition operators in Section 8.3.

In this section T is the (right) bilateral shift defined by $Te_n = w_n e_{n+1}$, where $\{e_n\}_{n\in\mathbb{Z}}$ is the standard orthonormal basis of $\ell^2(\mathbb{Z})$ and $\{w_n\}_{n\in\mathbb{Z}}$ is a sequence of strictly positive weights; note that a shift with complex weights is unitarily equivalent to a shift with real positive weights, simply by renormalizing the basis vectors. We assume further that $w_n \to a$ as $n \to -\infty$ and $w_n \to b$ as $n \to +\infty$ (the theory is also well-developed for weights not tending to a limit, but we do not need this more general situation).

Provided that $a, b \neq 0$, the operator T^{-1} is also unitarily equivalent to a bilateral right shift, with weights $\widetilde{w}_n = w_{-n}^{-1}$ tending to $1/b$ and $1/a$ as $n \to \pm\infty$.

Note that $T^* e_n = w_{n-1} e_{n-1}$, so that T^* is unitarily equivalent to a right shift, for which the weights tend to b and a at $-\infty$ and $+\infty$, respectively. Thus, by considering T^* instead if necessary, we may suppose that $a \leq b$.

We summarize the spectral properties of T and T^* in the following result.

Theorem 8.2.1 *Let T be the bilateral weighted shift defined above. Then $\sigma(T) = \sigma(T^*) = \{z \in \mathbb{C} : a \le |z| \le b\}$. If $a < b$, then for $a < |\lambda| < b$, the value λ is an eigenvalue of T^*, but $T - \lambda I$ is bounded below (i.e., λ does not lie even in the approximate point spectrum of T).*

Proof We shall assume throughout that $a > 0$: the modifications needed for $a = 0$ are minor, and we leave these to the reader. It is easily verified that

$$\sup_{m\in\mathbb{Z}} (w_m w_{m+1} \cdots w_{m+n-1})^{1/n} \to b$$

as $n \to \infty$, and from this we see that $\|T^n\|^{1/n} \to b$. Given that T^{-1} is also a weighted shift, a similar argument shows that $\|T^{-n}\|^{1/n} \to 1/a$. Hence $\sigma(T) \subset \{z \in \mathbb{C} : a \le |z| \le b\}$. Since $\sigma(T^*) = \overline{\sigma(T)}$, the same applies to T^*.

Now take λ with $a < |\lambda| < b$, and consider the equation $T^*x = \lambda x$, with $x = \sum_{n=-\infty}^{\infty} x_n e_n$. That is,

$$\lambda \sum_{n=-\infty}^{\infty} x_n e_n = \sum_{n=-\infty}^{\infty} x_n w_{n-1} e_{n-1} = \sum_{n=-\infty}^{\infty} x_{n+1} w_n e_n,$$

and thus $\lambda x_n = w_n x_{n+1}$ for all n. Choosing $x_0 = 1$ and then

$$x_n = \begin{cases} \lambda^n/(w_0 w_1 \cdots w_{n-1}) & \text{for } n > 0, \\ \lambda^n w_n w_{n+1} \cdots w_{-1} & \text{for } n < 0, \end{cases}$$

we have a solution which clearly lies in $\ell^2(\mathbb{Z})$ because $a < |\lambda| < b$. Thus $\lambda \in \sigma_p(T^*)$ and we may conclude that $\sigma(T) = \sigma(T^*) = \{z \in \mathbb{C} : a \le |z| \le b\}$.

It remains to show that λ is not an approximate eigenvalue of T. We shall suppose otherwise to obtain a contradiction. Assume that for each $k = 1, 2, 3, \ldots$ there is a unit vector $x(k) = \{x_n(k)\}_{n\in\mathbb{Z}}$ such that

$$\|Tx(k) - \lambda x(k)\| < 1/k.$$

There are two possible cases to discuss: either (i) $\liminf_{k\to\infty} |x_0(k)| = 0$, or (ii) $\liminf_{k\to\infty} |x_0(k)| = d > 0$.

In the first case, we may for each $\varepsilon > 0$ find an index k with $|x_0(k)| < \varepsilon$ and $\|Tx(k) - \lambda x(k)\| < \varepsilon$. Thus replacing $x(k)$ by $x'(k) = x(k) - x_0(k)e_0$, we have

$$\begin{aligned} \|Tx'(k) - \lambda x'(k)\| &\le \|Tx(k) - \lambda x(k)\| + \|Tx_0(k)e_0 - \lambda x_0(k)e_0\| \\ &< \varepsilon + (\|T\| + |\lambda|)\varepsilon, \end{aligned}$$

and thus we may find approximate eigenvectors $x'(k)$ of norm 1 such that $x'_0(k) = 0$ and $\|Tx'(k) - \lambda x'(k)\| \to 0$.

Now let $x'(k) = u(k) + v(k)$, where $u(k)$ is supported on the negative integers and $v(k)$ on the positive integers. Thus $Tx'(k) - \lambda x'(k)$ is the orthogonal sum of $Tu(k) - \lambda u(k)$ and $Tv(k) - \lambda v(k)$, since these are still supported on disjoint subsets of $\mathbb{Z}$. Since at least one of $u(k)$ and $v(k)$ has norm greater than $\frac{1}{2}$, it follows that we may find a sequence of approximate eigenvectors supported entirely on either the positive or the negative integers.

Suppose that x is a normalized vector such that $\|Tx - \lambda x\| < 1/k$. We have

$$T^n x - \lambda^n x = (T^{n-1} + \lambda T^{n-2} + \cdots + \lambda^{n-1} \mathrm{I})(Tx - \lambda x),$$

and so $\|T^n x - \lambda^n x\| < C_n/k$, where C_n depends on λ and the weights but not on k. However, if x is supported on the positive integers, then

$$\|T^n x\| \geq \inf_{k>0}(w_k w_{k+1} \cdots w_{k+n-1}) > |\lambda^n| + 2$$

for n sufficiently large. Now choosing k larger than C_n we obtain a contradiction.

A similar argument using T^{-1} disposes of the case when T (and indeed T^{-1}) has approximate eigenvalues supported on the negative integers, and $|\lambda^{-1}| < a^{-1}$.

In the second case, we note that if we have an approximate eigenvector $x = \{x_n\}_n$ of norm 1 with $\|Tx - \lambda x\| \leq 1/k$, then by an easy inductive argument there exist constants $\{D_n\}_{n \geq 0}$, independent of k and depending only on $\{w_n\}_n$ and λ, such that

$$\left| x_{n+1} - \frac{w_0 w_1 \cdots w_n}{\lambda^{n+1}} x_0 \right| \leq D_n/k, \qquad n = 0, 1, 2, \ldots,$$

and hence if $|x_0| \geq d/2$ we may use the fact that $w_0 w_1 \cdots w_n / |\lambda|^{n+1} \to \infty$ to find n such that $|x_{n+1}| > 2 - D_n/k$ for each such k. But x is a normalized vector, and this is a contradiction if k is chosen larger than D_n. □

One implication for universal operators is immediate. However, we need to work with vector-valued $\ell^2(\mathbb{Z}, \mathcal{H})$ spaces rather than scalar $\ell^2(\mathbb{Z})$ spaces.

Corollary 8.2.2 *Let $\mathcal{H}$ be a separable infinite-dimensional Hilbert space and let $T : \ell^2(\mathbb{Z}, \mathcal{H}) \to \ell^2(\mathbb{Z}, \mathcal{H})$ be the weighted right shift $(Tx)_n = w_{n-1} x_{n-1}$ for $n \in \mathbb{Z}$, where $\{w_n\}_n$ is a sequence of non-zero complex weights such that*

$$|w_n| \to \begin{cases} c & \text{as } n \to -\infty, \\ d & \text{as } n \to +\infty, \end{cases}$$

where $d < c$. Then for any complex number λ with $d < |\lambda| < c$, the operator $T - \lambda I$ is universal for Hilbert space.

Proof As discussed above, the adjoint T^* is unitarily equivalent to a weighted right shift, where the weights tend to d at $-\infty$ and c at $+\infty$. Then, it follows from Theorem 8.2.1 that $(T^* - \overline{\lambda} I)^*$ has an infinite-dimensional kernel; moreover, $T^* - \overline{\lambda} I$ has no eigenvalues or approximate eigenvalues, which implies that $T - \lambda I$ is surjective. That is, $U = T - \lambda I$ satisfies the hypotheses of Theorem 8.1.3, and the result follows. □

We take these ideas further in the next section.

8.3 Composition operators

8.3.1 Universality of composition operators

One rich source of universal operators is the class of composition operators, and we now develop some of the theory associated with these. They can be defined on many spaces of functions, including L^p spaces and H^p spaces (and even the Bergman spaces A^p), but we shall be mainly interested in the case of Hardy spaces.

Definition 8.3.1 Let H^2 be the Hardy space on the unit disc $\mathbb{D}$, and suppose that $\phi : \mathbb{D} \to \mathbb{D}$ is analytic. Then the *composition operator* $C_\phi : H^2 \to H^2$ is defined by $(C_\phi f)(z) = f(\phi(z))$, for $f \in H^2$.

At first sight, it is not clear that C_ϕ is necessarily a bounded operator, nor even that it maps into H^2. The following result settles this question.

Theorem 8.3.2 (Littlewood's subordination theorem) *Suppose that $\phi : \mathbb{D} \to \mathbb{D}$ is analytic. Then the operator C_ϕ is bounded on H^2.*

Proof Suppose first that $\phi(0) = 0$ and that f is a polynomial, of degree n, say. Then $f = f(0) + zS^* f$, where as usual S^* denotes the backward shift.

Thus $C_\phi f = f(0) + \phi C_\phi(S^* f)$, which is an orthogonal decomposition, and so

$$\|C_\phi f\|^2 = |f(0)|^2 + \|\phi C_\phi(S^* f)\|^2 \le |f(0)|^2 + \|C_\phi S^* f\|^2.$$

Iterating, we see that

$$\|C_\phi f\|^2 \le \sum_{k=0}^{n} |(S^{*k} f)(0)|^2 + \|C_\phi S^{*(n+1)} f\|^2 = \|f\|^2,$$

since $S^{*(n+1)} f = 0$, and so C_ϕ is a contraction on the space of polynomials, which is dense in H^2. It has a contractive extension to any $f \in H^2$, which clearly takes the pointwise values $f(\phi(z))$.

Next, let $b_\lambda(z) = \dfrac{\lambda - z}{1 - \overline{\lambda} z}$, for $\lambda \in \mathbb{D}$, and consider the composition operator induced by b_λ. We have

$$\begin{aligned}\int_0^{2\pi} |(f \circ b_\lambda)(e^{i\theta})|^2 \, d\theta &= \int_0^{2\pi} |f(e^{i\alpha})|^2 \left| b_\lambda'(e^{i\alpha}) \right| \, d\alpha \\ &= \int_0^{2\pi} |f(e^{i\alpha})|^2 \frac{1 - |\lambda|^2}{|1 - \overline{\lambda} e^{i\alpha}|^2} \, d\alpha,\end{aligned}$$

where we have made the change of variable $b_\lambda(e^{i\theta}) = e^{i\alpha}$, and indeed $b_\lambda(e^{i\alpha}) = e^{i\theta}$. (The reader who feels squeamish about this change of variables may again verify the formula first on polynomials and then pass to a limit.)

Hence C_{b_λ} is bounded. For a general ϕ we now write $\lambda = \phi(0)$ and then $\phi = b_\lambda \circ (b_\lambda \circ \phi)$, where $(b_\lambda \circ \phi)(0) = 0$, so that $\|C_\phi\| \le \|C_{b_\lambda}\| < \infty$. □

One particular case of interest from our point of view is when ϕ is inner.

Proposition 8.3.3 *The composition operator C_ϕ on H^2 is an isometry if and only if ϕ is an inner function with $\phi(0) = 0$.*

Proof If ϕ is inner, with $\phi(0) = 0$, then the powers of ϕ, namely, $1, \phi, \phi^2, \ldots$, form an orthonormal sequence, since $\langle \phi^m, \phi^n \rangle = \langle 1, \phi^{n-m} \rangle = \delta_{mn}$ for $n \ge m$, where δ_{mn} denotes the Kronecker delta. Hence

$$\left\| \sum_{n=0}^{\infty} a_n \phi^n \right\|^2 = \sum_{n=0}^{\infty} |a_n|^2,$$

and so C_ϕ is an isometry.

Conversely, if C_ϕ is an isometry, then $\|\phi\|_2 = 1$, which can only happen if ϕ is inner. Moreover, $\|a + b\phi\|_2^2 = |a|^2 + |b|^2$ for all $a, b \in \mathbb{C}$, from which it is easily deduced (e.g., by the polarization identity) that ϕ is orthogonal to the constant functions, i.e., $\phi(0) = 0$. □

Theorem 8.3.4 *Suppose that ϕ is an inner function with $\phi(0) = 0$ and that ϕ is not an automorphism. Then C_ϕ^* is a universal operator for Hilbert space.*

Proof We know that C_ϕ is an isometry, and hence C_ϕ^* is surjective.

We next show that $\operatorname{Im} C_\phi$ has infinite codimension in H^2. To do this, note that since ϕ is not an automorphism of $\mathbb{D}$, we can, for each $N \geq 1$, find $2N$ distinct points $w_1, \ldots, w_N$ and $w'_1, \ldots, w'_N$ in $\mathbb{D}$ with $\phi(w_j) = \phi(w'_j)$ for each j (indeed, for any w at which $\phi'(w) \neq 0$ there is another point w' with the same image under ϕ). For $w \in \mathbb{D}$, let k_w denote the reproducing kernel function $z \mapsto 1/(1 - \overline{w}z)$. Then the independent functions $h_j = k_{w_j} - k_{w'_j}$ satisfy $\langle C_\phi f, h_j \rangle = f(\phi(w_j)) - f(\phi(w'_j)) = 0$ for all $f \in H^2$, and so they are orthogonal to $C_\phi H^2$, which therefore has codimension at least N.

Thus C_ϕ^* is a surjective operator with infinite-dimensional kernel, and thus, by Theorem 8.1.3, it is universal for Hilbert space. □

Rather than considering adjoints, we now look at C_ϕ, where ϕ is a hyperbolic automorphism of $\mathbb{D}$ (that is, a conformal bijection of $\overline{\mathbb{D}}$ with two fixed points on $\mathbb{T}$). We shall lose no generality by considering $\phi(z) = \dfrac{z + r}{1 + rz}$, where $0 < r < 1$. We can demonstrate certain universality properties by representing C_ϕ in terms of weighted shift operators. To do this, it is easiest to work with C_ϕ^*, because of the reproducing kernel property $C_\phi^* k_w = k_{\phi(w)}$ for $w \in \mathbb{D}$, which follows easily from the calculation

$$\left\langle f, C_\phi^* k_w \right\rangle = \langle C_\phi f, k_w \rangle = f(\phi(w)) = \langle f, k_{\phi(w)} \rangle,$$

for all $f \in H^2$.

Let $z_n = \phi^n(0)$, the nth iterate, for $n \in \mathbb{Z}$ (note that $\phi^{-1}(z) = \dfrac{z - r}{1 - rz}$, and the negative iterates of ϕ can be defined in the obvious way as the positive iterates of ϕ^{-1}). We now have the following useful facts.

Lemma 8.3.5 *The sequence $(z_n)_{n \in \mathbb{Z}}$ defined above is a Blaschke sequence, and thus there is a Blaschke product B with zeroes $(z_n)_{n \in \mathbb{Z}}$. Defining the model space $K_B = H^2 \ominus BH^2$, we have*

$$H^2 = K_B \oplus BK_B \oplus B^2 K_B \oplus \cdots, \tag{8.1}$$

as an orthogonal direct sum, and each of the spaces $B^j K_B$, for $j \geq 0$, is invariant under C_ϕ^*.

Proof Calculation of z_n can be done simply by noting that if we write $s_n = (1 - z_n)/(1 + z_n)$, then $s_n = \left(\dfrac{1-r}{1+r}\right)^n$ for $n \in \mathbb{Z}$, and so

$$z_n = \frac{1 - s_n}{1 + s_n} = \frac{(1+r)^n - (1-r)^n}{(1+r)^n + (1-r)^n},$$

from which it is easily seen that $(z_n)_{n\in\mathbb{Z}}$ is a Blaschke sequence. Next,

$$\langle B^j f, B^k g\rangle = \langle f, B^{k-j} g\rangle = 0$$

if $f, g \in K_B$ and $k > j$, so the sum in (8.1) is certainly orthogonal. Finally, since $(K_B \oplus \cdots \oplus B^{N-1} K_B)^\perp = B^N H^2$, and $\bigcap_{N=0}^\infty B^N H^2 = \{0\}$, we see that the direct sum of the spaces $B^j K_B$ is dense in H^2 and, being closed, is the whole of H^2.

Now K_B is the closed linear span of the reproducing kernels k_{z_n}, since a function lies in BH^2 if and only if it has zeroes at every z_n. But $C_\phi^* k_{z_n} = k_{z_{n+1}}$ for each n, so that K_B is invariant under C_ϕ^*. Likewise, if $f \in K_B$ and $j, k \geq 0$, we have

$$\left\langle B^j f, C_\phi^*(B^k k_{z_n})\right\rangle = \langle (B \circ \phi)^j (f \circ \phi), B^k k_{z_n}\rangle = 0$$

if $j \neq k$, since $B \circ \phi = B$; whereas, if $j = k$, it is $\langle f, k_{z_{n+1}}\rangle$. That is, C_ϕ^* maps $B^j k_{z_n}$ to $B^j k_{z_{n+1}}$. Thus each space $B^j K_B$ is invariant under C_ϕ^*. □

In fact, understanding the action of C_ϕ^* on $B^j K_B$ is simplified by the following observation.

Lemma 8.3.6 *For each $j = 0, 1, 2, \ldots$, the set of normalized functions*

$$e_{j,n} := (1 - |z_n|^2)^{1/2} B^j k_{z_n}, \qquad n \in \mathbb{Z},$$

forms a Riesz basis for $B^j K_B$. Thus, $C_\phi^*|_{B^j K_B}$ is similar to a weighted bilateral shift, with weights $w_n = (1 - |z_n|^2)^{1/2}/(1 - |z_{n+1}|^2)^{1/2}$.

Proof This follows from the result that a sequence of normalized reproducing kernels $((1 - |z_n|^2)^{1/2} k_{z_n})_n$ is a Riesz basis for its closed span if and only if the sequence $(z_n)_n$ is a Carleson sequence (see Theorem 1.4.28). Since multiplication by B is an isometry, the same applies to the sequence $(e_{j,n})_{n\in\mathbb{Z}}$ for each $j \geq 0$. □

Note that $|z_n| = (1 - a^{|n|})/(1 + a^{|n|})$, where $a = (1 - r)/(1 + r)$, and so $|z_n|$ is asymptotic to $1 - 2a^{|n|}$ for large $|n|$; thus, $(1 - |z_n|^2)^{1/2}$ is asymptotic to $2a^{|n|/2}$. Thus the weights satisfy $w_n \to a^{-1/2}$ as $n \to +\infty$ and $w_n \to a^{1/2}$ as $n \to -\infty$.

We are now ready for the main result.

Theorem 8.3.7 *Let C_ϕ be the composition operator on H^2, induced by a hyperbolic automorphism $\phi(z) = \dfrac{z+r}{1+rz}$, and let $a = (1-r)/(1+r)$. Then, for every complex number λ with $\sqrt{a} < |\lambda| < 1/\sqrt{a}$, the operator $C_\phi - \lambda I$ is universal for Hilbert space.*

Proof From Lemma 8.3.5, we see that C_ϕ^* is unitarily equivalent to an infinite orthogonal direct sum of bilateral weighted shifts, where the weights tend to $d = \sqrt{a}$ at $-\infty$ and to $c = 1/\sqrt{a}$ at $+\infty$, with $d < c$. The same now applies to C_ϕ, with the weights reversed. That is, C_ϕ is unitarily equivalent to a weighted shift on $\ell^2(\mathbb{Z}, \mathcal{H})$, satisfying the conditions of Corollary 8.2.2, from which it follows that $C_\phi - \lambda I$ is universal for Hilbert space (being surjective, with infinite-dimensional kernel). □

Remark 8.3.8 The operator C_ϕ has zero kernel and is not itself universal; however, it has the same invariant subspaces as, for example, $C_\phi - I$, which is universal. Thus, a complete description of the subspace lattice of C_ϕ would provide a solution to the invariant-subspace problem for Hilbert space; indeed, if there is any possible counterexample, then it can be found as a restriction of C_ϕ to a minimal infinite-dimensional invariant subspace.

8.3.2 Minimal subspaces and eigenfunctions

In view of Remark 8.3.8, it is of interest to study the minimal invariant subspaces of C_ϕ, where ϕ is a hyperbolic automorphism of $\mathbb{D}$, with fixed points -1 and 1. Given any operator $T \in \mathcal{L}(\mathcal{H})$, and $x \in \mathcal{H}$, we write $\langle x \rangle$ for the closed subspace spanned by $\{T^n x : n = 0, 1, 2, \ldots\}$. The following result holds generally.

Proposition 8.3.9 *Let $T \in \mathcal{L}(\mathcal{H})$ be an invertible operator and $\langle x \rangle$ a minimal invariant subspace. Then $T^{-1}\langle x \rangle \subset \langle x \rangle$. Hence*

$$\langle x \rangle = \operatorname{span}\{T^n x : n \in \mathbb{Z}\}.$$

Proof Since $\langle x \rangle$ is minimal, we have that $\langle Tx \rangle = \langle x \rangle$. So $\langle x \rangle = T^{-1}\langle Tx \rangle = T^{-1}\langle x \rangle$, as required. □

From now on we take $T = C_\phi$, making the choice

$$\phi(z) = \frac{z + \frac{1}{2}}{1 + \frac{1}{2}z}. \tag{8.2}$$

We shall present various criteria that guarantee the non-minimality of $\langle u \rangle$, when $u \in H^2$. Note that a sufficient condition for non-minimality is that $\langle u \rangle$

contains an eigenvector v such that $u \notin \mathbb{C}v$. In particular, if u is non-constant, but $\langle u \rangle$ contains the constants, then $\langle u \rangle$ is non-minimal.

We use the following notation for the iterates:

$$\phi^{(n)} = \phi \circ \phi \circ \cdots \circ \phi \quad \text{and} \quad \phi^{(-n)} = (\phi^{(n)})^{-1},$$

for $n \geq 1$.

It is easily verified (see Exercise 9) that

$$\phi^{(n)}(z) = \frac{z + a_n}{1 + a_n z} \quad \text{and} \quad \phi^{(-n)}(z) = \frac{z - a_n}{1 - a_n z}, \tag{8.3}$$

for all $z \in \mathbb{D}$, where $a_n = (3^n - 1)/(3^n + 1)$; hence $\phi^{(n)}(z) \to 1$ and $\phi^{(-n)}(z) \to -1$, non-tangentially, as $n \to \infty$.

Theorem 8.3.10 *Let $u \in H^2$, $\alpha \in \{\pm 1\}$, and suppose that u has a non-zero non-tangential limit at α. If u is essentially bounded on an open arc containing α, then $\langle u \rangle$ is minimal if and only if u is constant.*

Proof Without loss of generality, $\alpha = 1$, by virtue of Proposition 8.3.9. We shall show first that the sequence $\left(C_\phi^n u\right)_n$ is norm-bounded. To do this, observe that if $v \in L^1(\mathbb{T})$ then $v \circ \phi \in L^1(\mathbb{T})$, and

$$\frac{1}{2\pi} \int_{-\pi}^{\pi} (v \circ \phi)(e^{i\theta}) P_{r,t}(e^{i\theta})\, d\theta = \frac{1}{2\pi} \int_{-\pi}^{\pi} v(e^{i\theta}) P_{r',t'}(e^{i\theta})\, d\theta,$$

where $\phi(re^{it}) = r'e^{it'}$ and $P_{r,t}$ is the usual Poisson kernel. This identity holds because the first integral involves evaluating the harmonic extension of $v \circ \phi$ at re^{it}, which is the same as evaluating the harmonic extension of v at $\phi(re^{it})$ (i.e., the second integral).

Applying this principle with $\phi^{(n)}$ rather than ϕ, we see that

$$\left\| C_\phi^n u \right\|^2 = \frac{1}{2\pi} \int_{-\pi}^{\pi} |u(e^{i\theta})|^2 P_{a_n,0}(e^{i\theta})\, d\theta,$$

where a_n is as in (8.3). Let $\delta \in (0, \pi)$ and $M > 0$ be such that $|u(e^{i\theta})| \leq M$ a.e. for $|\theta| \leq \delta$. We then have

$$\begin{aligned} \left\| C_\phi^n u \right\|^2 &= \frac{1}{2\pi} \int_{-\delta}^{\delta} |u(e^{i\theta})|^2 P_{a_n,0}(e^{i\theta})\, d\theta + \frac{1}{2\pi} \int_{\delta \leq |\theta| \leq \pi} |u(e^{i\theta})|^2 P_{a_n,0}(e^{i\theta})\, d\theta \\ &\leq M^2 + P_{a_n,0}(e^{i\delta}) \|u\|^2, \end{aligned}$$

and $P_{a_n,0}(e^{i\delta}) \to 0$ as $n \to \infty$, since $a_n \to 1$.

We also see that $C_\phi^n u \to u(1)$ pointwise as $n \to \infty$, since, for each $z \in \mathbb{D}$, $\phi^{(n)}(z) \to 1$ non-tangentially. Therefore $C_\phi^n u$ converges weakly to a non-zero constant, and hence the constant functions are in the weak closure (which is the same as the norm closure) of $\langle u \rangle$. The result follows. □

This has the following immediate consequence.

Corollary 8.3.11 *Let $u \in H^\infty$, $\alpha \in \{\pm 1\}$, and suppose that u has a non-zero non-tangential limit at α. Then $\langle u \rangle$ is minimal if and only if u is constant.*

If $u \in H^\infty$ has a *zero* non-tangential limit at α, and u is not identically zero, then Corollary 8.3.11 immediately implies that $\langle u + c \rangle$ is non-minimal for any constant $c \neq 0$.

Another condition on u that excludes minimality is the presence of certain singular inner factors.

Theorem 8.3.12 *Let $u \in H^2$ and suppose that it has a singular inner factor S corresponding to a measure μ such that $\mu(\{1\}) > 0$ or $\mu(\{-1\}) > 0$. Then $\langle u \rangle$ is non-minimal.*

Proof As before, we may assume without loss of generality that $\mu(\{1\}) > 0$. Let σ denote the singular inner function $z \mapsto \exp(-(1+z)/(1-z))$. Then the hypothesis of the theorem is equivalent to saying that $u = \sigma^a v$ for some $a > 0$ and $v \in H^2$ such that no positive power of σ divides v.

It is easily verified that $\sigma \circ \phi^n = \sigma^{3^n}$, and so σ^a divides every function in $\langle u \rangle$. Hence $\langle u \rangle \subset \sigma^a H^2$.

Suppose now that $\langle u \rangle$ is minimal. Then, by Proposition 8.3.9,

$$u \circ \phi^{(-n)} = (\sigma^a \circ \phi^{(-n)})(v \circ \phi^{(-n)}) \in \langle u \rangle,$$

for all $n \geq 1$. However, no positive power of σ divides $v \circ \phi^{(-n)}$, and so we must have that σ^a divides $\sigma^a \circ \phi^{(-n)} = \sigma^{a/3^n}$ for all $n \geq 1$, which is absurd. □

Note that the only finite-dimensional minimal subspaces of C_ϕ are its one-dimensional eigenspaces, and thus we now look at some of the properties of the eigenfunctions of C_ϕ.

Proposition 8.3.13 *If u is an eigenfunction of C_ϕ with non-tangential limits at both 1 and -1, then u is constant.*

Proof Take $z \in \mathbb{D}$ such that $u(z) \neq 0$ and suppose that $C_\phi u = \lambda u$. Then $u(\phi^{(n)}(z)) = \lambda^n u(z)$, and since $\phi^{(n)}(z) \to 1$ non-tangentially, we have $\lambda^n \to u(1)/u(z)$ as $n \to \infty$. Similarly, $\lambda^{-n} \to u(-1)/u(z)$ as $n \to \infty$. Hence $\lambda = 1$

and $u(\phi^{(n)}(w)) = u(w)$ for all $w \in \mathbb{D}$ and $n \geq 1$. Thus, $u(w) = u(1)$, and u is constant. □

The next result shows that, to find all the eigenfunctions of C_ϕ, we may consider each part of its canonical decomposition separately.

Theorem 8.3.14 *Let $u \in H^2$ factorize canonically as $u = BSv$, where B is a Blaschke product, S is a singular inner function and v is outer. Then u is an eigenfunction of C_ϕ if and only if B, S and v are eigenfunctions of C_ϕ.*

Proof Since $C_\phi u = (C_\phi B)(C_\phi S)(C_\phi v)$, it is clear that the product of eigenfunctions is an eigenfunction.

Conversely, note that since v is outer, it follows that there is a sequence $(p_n)_n$ of polynomials such that $p_n v - 1 \to 0$ in H^2 norm. Then the continuity of C_ϕ implies that $(C_\phi p_n)(C_\phi v) - 1 \to 0$ in H^2 norm, and hence there is a sequence (q_n) of polynomials such that $q_n(C_\phi v) - 1 \to 0$ in H^1 norm (taking each q_n to be a polynomial approximation to $C_\phi p_n$). It follows that $C_\phi v$ is an outer function in H^1, and hence in H^2.

Moreover, for a Blaschke product B it is clear that $C_\phi B$ is also a Blaschke product; similarly, if S is a singular inner function, then by considering the integral representation (1.9), it is straightforward to show that $C_\phi S$ is also a singular inner function. So $C_\phi u$ has two canonical factorizations, namely

$$C_\phi u = \lambda u = \lambda BSv \qquad \text{and} \qquad C_\phi u = (C_\phi B)(C_\phi S)(C_\phi v).$$

By the uniqueness of the factorization (up to a constant of modulus 1), it follows that B, S and v are all eigenfunctions of C_ϕ. □

It is easy to specify the Blaschke products that are eigenfunctions of C_ϕ.

Proposition 8.3.15 *A non-trivial Blaschke product B is an eigenfunction of C_ϕ if and only if, for each zero z_0 of B of multiplicity n_0, the point $\phi(z_0)$ is also a zero of B, of multiplicity exactly n_0. Hence the zero set of B decomposes into an at most countable union of orbits of the form $\{\phi^{(n)}(a) : n \in \mathbb{Z}\}$.*

Proof The Blaschke product B is an eigenfunction of C_ϕ if and only if the Blaschke product $B \circ \phi$ has the same zeroes as B (including multiplicity). But z_0 is a zero of $B \circ \phi$ if and only if $\phi(z_0)$ is a zero of B. □

We now consider the effects of applying C_ϕ to a singular inner function.

Lemma 8.3.16 *Let*

$$S_\mu(z) = \exp\left(-\int_0^{2\pi} \frac{e^{it} + z}{e^{it} - z}\, d\mu(e^{it})\right)$$

be a singular inner function, associated with a positive singular measure μ *on* $\mathbb{T}$*, and let* ϕ *be the automorphism defined in (8.2). Then* $C_\phi S_\mu$ *is also a singular inner function, which can be written as*

$$S_\nu(z) = \gamma \exp\left(-\int_0^{2\pi} \frac{e^{is}+z}{e^{is}-z}\, d\nu(e^{is})\right), \tag{8.4}$$

where γ *is a constant of modulus 1, and* ν *is given by* $d\nu(e^{is}) = \alpha(e^{it})d\mu(e^{it})$*, with* $e^{it} = \phi(e^{is})$ *and* $\alpha(e^{it}) = P_{\frac{1}{2},0}(t)$.

Proof It is easily verified that

$$\frac{e^{it}+\phi(z)}{e^{it}-\phi(z)} = \alpha(e^{it})\frac{\beta(e^{it})+z}{\beta(e^{it})-z} + i\widetilde{\alpha}(e^{it}),$$

where $\beta(e^{it}) = \phi^{-1}(e^{it})$, and α and $\widetilde{\alpha}$ are the real functions satisfying

$$\alpha(e^{it}) + i\widetilde{\alpha}(e^{it}) = \frac{e^{it}+\frac{1}{2}}{e^{it}-\frac{1}{2}}.$$

Letting $e^{is} = \beta(e^{it}) = \phi^{-1}(e^{it})$ and defining ν as above, we have

$$(C_\phi S_\mu)(z) = \gamma \exp\left(-\int_0^{2\pi} \alpha(e^{it})\frac{e^{is}+z}{e^{is}-z}\, d\mu(e^{it})\right),$$

where γ is the unimodular constant given by

$$\gamma = \exp\left(-\int_0^{2\pi} i\widetilde{\alpha}(e^{it})\, d\mu(e^{it})\right).$$

Clearly, this is equivalent to (8.4). □

Theorem 8.3.17 *A singular inner function* S_μ *is an eigenfunction of* C_ϕ *if and only if* $P_{\frac{1}{2},0}(t)d\mu(e^{it}) = d\mu(e^{is})$*, where* $e^{is} = \phi^{-1}(e^{it})$.

Proof This follows directly from Lemma 8.3.16. □

The outer functions which are eigenfunctions are discussed in the Comments below.

Exercises

1. Let $\mathcal{H}$ be a separable Hilbert space, and take $T \in \mathcal{L}(\mathcal{H})$ with spectral radius strictly less than 1. Show that the mapping $J : \mathcal{H} \to \ell^2(\mathcal{H})$ given by $x \mapsto (x, Tx, T^2x, \ldots)$ is an isomorphism onto a closed subspace

of $\ell^2(\mathcal{H})$, and deduce that the backward shift on $\ell^2(\mathcal{H})$ is universal for Hilbert space.

2. Show that a version of Theorem 8.1.3 holds in an arbitrary Banach space $\mathcal{X}$, provided that Ker U is a complemented subspace containing an isomorphic copy of $\mathcal{X}$.

3. Using Exercise 2, and the result of Read [169] that there is an operator on ℓ^1 with no non-trivial invariant subspace, show that the backward shift S^* on $\ell^1(\ell^1)$ (i.e., $S^*(x_0, x_1, x_2, \ldots) = (x_1, x_2, x_3, \ldots)$, where each x_j is in ℓ^1 and $\sum_{j=0}^{\infty} \|x_j\| < \infty$) has at least one minimal invariant subspace of infinite dimension.

4. Fill in the details of Remark 8.1.4.

5. Let $T : \ell^2(\mathbb{Z}) \to \ell^2(\mathbb{Z})$ be the weighted shift with weights

$$w_n = \begin{cases} 1 & \text{for } n < 0, \text{ odd}, \\ 2 & \text{for } n < 0, \text{ even}, \\ 3 & \text{for } n \geq 0, \text{ odd}, \\ 4 & \text{for } n \geq 0, \text{ even}. \end{cases}$$

Calculate the spectrum and approximate point spectrum of T and T^*.

6. Show that, unlike the situation in Theorem 8.3.2, the composition operator C_ϕ on $H^2(\mathbb{C}_+)$, induced by the mapping $\phi(z) = 1/z$, is not bounded.

7. By estimating $\|C_{b_\lambda}\|$ more carefully, as used in the proof of Theorem 8.3.2, show that $\|C_\phi\| \leq \left(\dfrac{1 + |\phi(0)|}{1 - |\phi(0)|}\right)^{1/2}$.

8. Show directly that if $\phi(z) = z^2$, then C_ϕ is a unilateral shift of infinite multiplicity.

9. By adapting the proof of Lemma 8.3.5, prove the identities (8.3), and deduce the non-tangential convergence of the iterates $\phi^{(n)}(z)$ for $z \in \mathbb{D}$.

10. Verify that if ϕ is an automorphism of $\mathbb{D}$, if B is a Blaschke product, and if S is a singular inner function, then $C_\phi B$ is a Blaschke product and $C_\phi S$ is a singular inner function.

11. Take $m \geq 2$, and let B_m be the Blaschke product associated with the sequence $\{\phi^{(mn)}(0)\}_{n \in \mathbb{Z}}$, where ϕ is defined in (8.2). Show that B_m is non-minimal.

12. Let u and v be inner functions. Show that the composition $u \circ v$ is also inner. (Hint: do this first for u a finite Blaschke product; then use the ideas of Theorem 1.4.8 to get the result when u is an infinite Blaschke product; finally, use Frostman's Theorem 1.4.18 to deduce the result when u is a singular inner function.)

Comments

Theorem 8.1.3 is due to Caradus [55], whereas Theorem 8.1.5 can be traced to Sz.-Nagy and Foias [191]. Related work can be found in Sz.-Nagy [189], Schreiber [182] and de Branges and Rovnyak [90].

Some results on the invariant subspaces for operators M_ϕ, where ϕ is inner, are given by Lance and Stessin [139]. However, the description given there is not always easy to work with in practice.

The non-trivial part of Theorem 8.2.1 is due to Ridge [174], whose results also apply to the situation where $|w_n|$ does not tend to a limit. We are able to give a technically simpler proof in this case.

Theorem 8.3.2 was given by Littlewood [143].

Proposition 8.3.3 is due to Nordgren [157]. The proof we give was presented in Chalendar and Partington [70]. The proof of Theorem 8.3.4 also uses ideas to be found in this work, some of which may be traced back to apparently unpublished work of M. Jones.

Exercise 1 is based on the work of Rota [176, 177]. Exercise 2 is from Caradus [55].

Although it is a result with a slightly different flavour, we mention that the shift S_A on the Bergman space $A^2(\mathbb{D})$ has a weaker universality property, which implies that to give an affirmative answer to the invariant-subspace problem for Hilbert space, it would be enough to show that for any pair of invariant-subspaces $\mathcal{M} \subset \mathcal{N}$ of S_A with $\dim \mathcal{N} \ominus \mathcal{M} > 1$ there is a non-trivial invariant subspace for $P_{\mathcal{N}\ominus\mathcal{M}} S_A P_{\mathcal{N}\ominus\mathcal{M}}$. This follows from dual-algebra theory [14, 15, 35], and the problem has been solved affirmatively in the case that $\mathcal{N} = A^2(\mathbb{D})$ [120, 194].

The universality properties of C_ϕ were established by Nordgren, Rosenthal and Wintrobe [156]. The simpler approach we follow (working with C_ϕ^*) is a further development of an exercise given in Cowen and MacCluer [85, exercise 7.2.4].

The discussion in Section 8.3.2 is based on the papers of Matache [150, 151] and Mortini [153], in particular Theorem 8.3.10 [151] and Theorem 8.3.12 [153]. A criterion for the non-minimality of $\langle u \rangle$, when u has a zero limit at one of the fixed points of ϕ, was given by Chkliar [81]. Gallardo-Gutiérrez and Gorkin [110] proved similar non-minimality results in the case when u does not have a radial limit equal to zero at the attractive fixed point, but $u(\phi^{(n)}(z_0))$ converges to zero for some z_0 in the open unit disc. In a more recent paper [111], they also provide a constructive characterization of eigenvectors for the hyperbolic composition operator on H^p.

Mortini [153] gave a complete description of the eigenfunctions of C_ϕ that are singular inner functions corresponding to purely atomic measures. The eigenfunctions that are general singular inner functions or outer functions were analysed in some detail by Matache [151]; in particular, Theorem 8.3.17 was first demonstrated in that paper, by a different method (our approach owes more to Mortini [153]). Further, Matache exhibited for every $\lambda \in \sigma_p(C_\phi)$ an outer eigenfunction of C_ϕ.

Further examples of universal operators are related to the composition operators C_ϕ on $L^2(0, 1)$ induced by functions $\phi(x) = x^s$, where $0 < s < 1$, as in Partington and Pozzi [162]. Their universality is demonstrated by representing them in terms of more complicated vectorial shift operators.

9

Moment sequences and binomial sums

This chapter treats two fairly self-contained ideas. Moment sequences, which can be defined as the integrals of successive powers of the independent variable with respect to a positive Borel measure on the real line, appear in many mathematical contexts, including probability theory and other branches of analysis. We begin with the idea of a moment sequence that can be associated with powers of an operator, and show how this leads to a proof of the existence of invariant subspaces. Initially, the Banach spaces involved will be either real or complex, but later some of the results we present will be appropriate only for real sequence spaces; we then employ a trick to handle certain operators (in particular, tridiagonal operators) on complex spaces.

Next, we explore a variety of themes concerned with sequences defined in terms of specific formulae involving binomial coefficients. We start with some apparently elementary results about such sequences, linking these to results from complex analysis in the form of various Phragmén–Lindelöf principles. Finally, we arrive at some applications, first to the theory of Banach algebras, and then to the existence of invariant subspaces. This particular subject appears to us to be one where interesting techniques have been introduced but not fully exploited: that is to say, we think that further developments will be possible.

9.1 Moment sequences

Definition 9.1.1 Let $\mathcal{X}$ be a separable Banach space, and $T \in \mathcal{L}(\mathcal{X})$. We say that T admits a *moment sequence* if there is a non-zero vector $x_0 \in \mathcal{X}$, a non-zero functional $\phi_0 \in \mathcal{X}^*$ and a non-negative Borel measure μ on $\mathbb{R}$ such that

$$\phi_0(T^n x_0) = \int_{\mathbb{R}} t^n \, d\mu(t), \qquad \text{for} \quad n = 0, 1, 2, \ldots.$$

The pair (x_0, ϕ_0) is called a *moment pair* for T.

Note that every operator with a non-trivial closed invariant subspace $\mathcal{M} \subset \mathcal{X}$ admits a moment pair in an easy way, for we may take a non-zero $x_0 \in \mathcal{M}$ and (by the Hahn–Banach theorem) choose $\phi_0 \in \mathcal{X}^*$ such that $\phi_0(x) = 0$ for all $x \in \mathcal{M}$. Then $\phi_0(T^n x_0) = 0$ for all n, which is the moment sequence corresponding to the zero measure.

If such a μ exists, then it is supported on the interval $[-r(T), r(T)]$, where $r(T)$ denotes the spectral radius, since if for some $\varepsilon > 0$ we have that $\mu(\mathbb{R} \setminus [-\varepsilon - r(T), r(T) + \varepsilon]) = \delta > 0$, then

$$\int_{\mathbb{R}} t^n \, d\mu(t) \geq \delta (r(T) + \varepsilon)^n$$

when n is even, and this yields a contradiction when we use the fact that $\|T^n\| \leq (r(T) + \varepsilon/2)^n$, for sufficiently large n.

It is possible to test whether (x_0, ϕ_0) is a moment pair for T without knowing the measure μ, as we shall now show. Let I denote the closed interval $[-r(T), r(T)]$.

We begin with a technical lemma, which is a consequence of the Hahn–Banach theorem, but for which we give a self-contained proof. Here $\mathbb{R}[t]$ and $C_0(\mathbb{R})$ will denote respectively the space of real polynomials and the space of real continuous functions on $\mathbb{R}$ tending to zero at $\pm\infty$.

Lemma 9.1.2 *Let* $\Psi : \mathbb{R}[t] \to \mathbb{R}$ *be a linear functional that is* order-preserving *(i.e., maps non-negative functions to non-negative real numbers). Then it can be extended to an order-preserving linear functional on the space* $\mathcal{Z} := \mathbb{R}[t] \oplus C_0(\mathbb{R})$.

Proof Let $\mathcal{M}$ be a linear space (not necessarily closed in any particular topology) such that $\mathbb{R}[t] \subset \mathcal{M} \subset \mathcal{Z}$, suppose that Ψ is defined and order-preserving on $\mathcal{M}$, and take $f \in C_0(\mathbb{R}) \setminus \mathcal{M}$. We wish to extend Ψ from $\mathcal{M}$ to the space spanned by $\mathcal{M}$ and f. To do this, define

$$a = \sup\{\Psi(g) : g \in \mathcal{M}, \ g \leq f\}$$

and

$$b = \inf\{\Psi(g) : g \in \mathcal{M}, \ g \geq f\}$$

(since f is bounded we can always find constant polynomials above and below f, at least, so these are not a sup and inf over an empty set). If $g_1 \leq f \leq g_2$, then $\Psi(g_2 - g_1) \geq 0$, since Ψ is order-preserving on $\mathcal{M}$, and hence $a \leq b$. Let c be any number between a and b, and define $\Psi(f) = c$; extend by linearity to the span of $\mathcal{M}$ and f. Now, if $\alpha f + g \leq 0$, where $\alpha \in \mathbb{R} \setminus \{0\}$ and $g \in \mathcal{M}$, we have two possibilities:

- $\alpha > 0$, in which case $f \geq -g/\alpha$ and so $\Psi(-g/\alpha) \leq a$ and $\Psi(\alpha f + g) \geq \alpha a - a\alpha \geq 0$;
- $\alpha < 0$, in which case $f \leq -g/\alpha$ and so $\Psi(-g/\alpha) \geq b$ and $\Psi(\alpha f + g) \geq \alpha b - b\alpha \geq 0$.

By a standard Zorn's lemma argument we arrive at a maximal extension, which must be defined on the whole of $\mathcal{Z}$. Alternatively, a countable sequence of extensions produces an extension to a dense subspace, which, as far as $C_0(\mathbb{R})$ is concerned, is necessarily continuous in the norm topology, for if $-1 \leq g \leq 1$, then $-\Psi(1) \leq \Psi(g) \leq \Psi(1)$. Thus it can be extended to $C_0(\mathbb{R})$ by continuity. □

Proposition 9.1.3 *Let $T \in \mathcal{L}(\mathcal{X})$, and let $x_0 \in \mathcal{X}$ be a non-zero vector and $\phi_0 \in \mathcal{X}^*$ a non-zero functional. Then the following conditions on T are equivalent:*

1. *(x_0, ϕ_0) is a moment pair for T.*
2. *For every real polynomial p with $p \geq 0$ on $\mathbb{R}$, one has $\phi_0(p(T)x_0) \geq 0$.*
3. *For every real polynomial q, one has $\phi_0(q(T)^2 x_0) \geq 0$.*
4. *For every $N \geq 0$ and for all $q_0, q_1, \ldots, q_N$ in $\mathbb{R}$, one has*

$$\sum_{j=0}^{N}\sum_{k=0}^{N} q_j q_k \phi_0(T^{j+k}x_0) \geq 0.$$

Proof We start with the equivalence of parts 1 and 2. Clearly, if part 1 holds, then

$$\phi_0(p(T)x_0) = \int_{-\infty}^{\infty} p(t)\, d\mu(t) \geq 0$$

for all polynomials p with $p \geq 0$ on $\mathbb{R}$. That is, part 2 holds.

Suppose now that part 2 holds, and consider the order-preserving functional

$$\Psi : p \mapsto \phi_0(p(T)x_0)$$

acting on the space of real polynomials on $\mathbb{R}$. By Lemma 9.1.2, Ψ has an order-preserving extension to the real space $\mathcal{Z} = \mathbb{R}[t] \oplus C_0(\mathbb{R})$. Hence, by the Riesz–Markov theorem (see Theorem 1.1.17), there is a positive Borel measure μ on $\mathbb{R}$ such that, in particular,

$$\Psi(f) = \int_{-\infty}^{\infty} f(t)\, d\mu(t),$$

for all $f \in C_0(\mathbb{R})$. It remains to show that the identity holds on $\mathbb{R}[t]$ as well. Consider an even power $p_{2n}(t) = t^{2n}$, and define a sequence of functions $(g_k)_k \in C_0(\mathbb{R})$ by

$$g_k(t) = \begin{cases} p_{2n}(t), & \text{if } |t| \le k, \\ 0, & \text{if } |t| \ge 2k, \\ \text{piecewise linear}, & \text{otherwise.} \end{cases}$$

Note that $p_{2n}(t) - g_k(t) = t^{2n} - k^{2n-1}(2k - t) \le t^{2n} + k^{2n-1}t$ on $[k, 2k]$, so that

$$0 \le \frac{p_{2n}(t) - g_k(t)}{t^{2n+2}} \le \frac{2}{k^2} \qquad \text{on} \quad \mathbb{R},$$

and hence

$$\int_{-\infty}^{\infty} g_k(t)\, d\mu(t) \le \Psi(p_{2n}) \le \int_{-\infty}^{\infty} g_k(t)\, d\mu(t) + 2\Psi(p_{2n+2})/k^2,$$

and so

$$\int_{-k}^{k} p_{2n}(t)\, d\mu(t) \le \Psi(p_{2n}) \le \int_{-2k}^{2k} p_{2n}(t)\, d\mu(t) + 2\Psi(p_{2n+2})/k^2.$$

Letting $k \to \infty$ we see that $\Psi(p_{2n}) = \int_{-\infty}^{\infty} p_{2n}(t)\, d\mu(t)$. For odd powers, we apply the above approximation argument to the positive function $p_{2n} + p_{2n+1} + p_{2n+2}$, to conclude that

$$\Psi(p_{2n} + p_{2n+1} + p_{2n+2}) = \int_{-\infty}^{\infty} (p_{2n} + p_{2n+1} + p_{2n+2})(t)\, d\mu(t);$$

then, using linearity and the result for even powers, we have the formula for odd powers too. Hence part 1 holds.

For the equivalence of parts 2 and 3, note that any positive real polynomial is the sum of two squares of polynomials: this is easily seen by splitting it into quadratic factors of the form $(t - a)^2 + b^2$ (since complex roots occur in conjugate pairs, and real ones have even multiplicity), and then repeatedly using the identity

$$(u^2 + v^2)(w^2 + x^2) = (uw - vx)^2 + (ux + vw)^2.$$

Thus parts 2 and 3 are equivalent.

Finally, if $q(t) = \sum_{j=0}^{N} q_j t^j$, then one has

$$\phi_0(q(T)^2 x_0) = \sum_{j=0}^{N} \sum_{k=0}^{N} q_j q_k \phi(T^{j+k} x_0),$$

and thus parts 3 and 4 are equivalent. □

In order to apply the theory of moment pairs we begin with a result from the geometry of Banach spaces. The following notion is best expressed in terms of real normed spaces, and it is in this case that we shall apply it. However, a version for complex space can also be formulated, since every complex normed space is also a real normed space.

Definition 9.1.4 Let S be a subset of a real normed space $\mathcal{X}$. Then a point $s_0 \in S$ is a *support point* for S if there is a linear functional $\phi \in \mathcal{X}^*$ such that $\phi(s_0) = \sup\{\phi(s) : s \in S\}$.

Even in the case of a closed convex set, it is not the case that every boundary point is a support point (see Exercise 2 below), but such a set does have plenty of support points, as we now see.

Lemma 9.1.5 (Bishop–Phelps) *Let F be a closed convex subset of a real Banach space $\mathcal{X}$. Then the support points of F are dense in the boundary of F.*

Proof Given $z \in \partial F$ and $\varepsilon > 0$, take $y \in \mathcal{X} \setminus F$ such that $\|y - z\| < \varepsilon/2$. By applying the Hahn–Banach theorem to a small ball centred on y, chosen to be disjoint from F, we may find $\phi \in \mathcal{X}^*$ such that $\|\phi\| = 1$ and $\phi(y) > \sup\{\phi(x) : x \in F\}$.

Let K denote the closed cone $K = \left\{x \in \mathcal{X} : \phi(x) \geq \frac{1}{2}\|x\|\right\}$, which has non-empty interior, since any point x with $\phi(x) > \frac{1}{2}\|x\|$ is clearly an interior point.

We now put a partial order on F by writing $x \succeq y$ if and only if $x - y \in K$, i.e.,

$$\phi(x) - \phi(y) \geq \frac{1}{2}\|x - y\|. \tag{9.1}$$

Note that every increasing chain G in F yields a bounded increasing chain of real numbers $\{\phi(x) : x \in G\}$, which (regarded as a net) converges to its supremum; then the corresponding net of points in G is also Cauchy, by (9.1), and so converges to an upper bound in F. Hence we may apply Zorn's lemma and conclude that the set of elements $\{x \in F : x \succeq z\}$ has a maximal element, w, say. Now $w - z \in K$ and $w \in F$, so that

$$\|w - z\| \leq 2\phi(w - z) \leq 2\phi(y - z) < \varepsilon.$$

Finally, $(w+K) \cap F = \{w\}$, by maximality, so, since K has non-empty interior, the Hahn–Banach theorem now provides a non-zero functional $\psi \in \mathcal{X}^*$ such that $\sup \psi(F) \leq \inf \psi(w + K)$, and these quantities must therefore equal $\psi(w)$; that is, ψ is a support functional to F at w. □

We are now ready to derive the first result of this chapter on invariant subspaces. For a closed subset $F \subset \mathcal{X}$, let $\mathcal{O}_F$ denote the set of operators V such that $V(F) \subset F$.

Theorem 9.1.6 *Let $\mathcal{X}$ be a real Banach space, and $F \subset \mathcal{X}$ a non-trivial closed convex subset. Let $T \in \mathcal{L}(\mathcal{X})$ and suppose that for each $k = 0, 1, 2, \ldots$ there is a differentiable function $f_k : \mathbb{R} \to \mathcal{L}(\mathcal{X})$ such that $f_k(0) = \mathrm{I}$, $f_k'(0) = T^k$ and $f_k(\mathbb{R}) \subset \mathcal{O}_F$. Then T has a non-trivial closed invariant subspace.*

Proof By Lemma 9.1.5, there is a non-zero support point $w \in \partial F$, with a corresponding support functional $\psi \in \mathcal{X}^*$.

For $k = 1, 2, 3, \ldots$ we consider the function $h_k : \mathbb{R} \to \mathbb{R}$ defined by $h_k(z) = \psi(f_k(z)(w))$ for $z \in \mathbb{R}$. Note that $h_k(0) = \psi(w)$, so that h_k attains its maximum value at 0, since $f_k(z)(w) \in F$ for all $z \in \mathbb{R}$. Hence $h_k'(0) = 0$, that is, $\psi(T^k w) = 0$ for each k.

Clearly, we may suppose that T is not the zero operator. Consider the spaces $\mathcal{M}_0 = \operatorname{span}\{T^k w : k \in \mathbb{N}\}$ and $\mathcal{M}_1 = \operatorname{Ker} T$. These are closed invariant subspaces for T. By the above, we see that $\mathcal{M}_0 \neq \mathcal{X}$, and if $\mathcal{M}_0 = \{0\}$, then $w \in \mathcal{M}_1$. The conclusion is that at least one of $\mathcal{M}_0$ and $\mathcal{M}_1$ is a non-trivial closed invariant subspace for T. □

Theorem 9.1.7 *Let $\mathcal{X}$ be a real Banach space, and $T \in \mathcal{L}(\mathcal{X})$ an operator admitting a moment sequence. Then T has a non-trivial closed invariant subspace.*

Proof Let (x_0, ϕ_0) and the measure μ, which will necessarily be supported on $[-r(T), r(T)]$, be as in Definition 9.1.1.

Consider the set F of all $x \in X$ such that there is a positive measure ν on $\mathbb{R}$ such that

$$\phi_0(T^n x) = \int_{\mathbb{R}} t^n \, d\nu(t) \tag{9.2}$$

for $n = 0, 1, 2, \ldots$. It is easily verified that F is a non-trivial closed convex subset of $\mathcal{X}$. We shall apply Theorem 9.1.6 to F and the operator-valued functions $f_k(z) = \mathrm{I} + zT^k + z^2T^{2k}$ defined on $\mathbb{R}$. Note that $f_k(\mathbb{R}) \subset \mathcal{O}_F$, since if (9.2) holds then

$$\phi_0(T^n(f_k(z))(x)) = \int_{\mathbb{R}} t^n(1 + zt^k + z^2t^{2k}) \, d\nu(t)$$

for $n = 0, 1, 2, \ldots$, and indeed $1 + zt^k + z^2t^{2k} \geq 0$ for all $z, t \in \mathbb{R}$. Now, by Theorem 9.1.6, we may conclude that T has a non-trivial closed invariant subspace. □

9.2 Operators on sequence spaces

In this section we shall be working with Banach spaces of real or complex sequences, and accordingly we make the following definition, which includes, for example, the ℓ^p spaces for $1 \le p < \infty$ and c_0, as well as various weighted versions.

Definition 9.2.1 A Banach space $\mathcal{X}$ of real or complex sequences $x = (x_n)_{n\in\mathbb{N}}$ is called a *self-adjoint sequence space*, if it satisfies the following:

1. $\mathcal{X}$ contains c_{00}, the space of all finitely supported sequences.
2. For each $n \in \mathbb{N}$, the coordinate functional $x \mapsto x_n$ is bounded on $\mathcal{X}$.
3. (Only required in the complex case) Whenever a sequence $x = (x_n)_{n\in\mathbb{N}}$ lies in $\mathcal{X}$, then so does the conjugate sequence $\overline{x} = (\overline{x}_n)_{n\in\mathbb{N}}$.

It will be convenient to begin with linear mappings defined on c_{00}; we provide the space c_{00} with the usual ℓ^2 inner product $\langle \cdot, \cdot \rangle$, even when regarding it as a subspace of some other self-adjoint sequence space.

For a linear mapping $L : c_{00} \to c_{00}$ we identify L with its matrix with respect to the basis $(e_n)_{n\ge 1}$. The transpose of L, L^t, is associated with the transpose of the matrix of L, although in general it may not map c_{00} into itself.

Lemma 9.2.2 *Let T be a linear mapping on c_{00}. Suppose that there exist two matrices, S and L, mapping c_{00} into itself, such that S is symmetric, L is invertible on c_{00}, and L^t also maps c_{00} into itself. Suppose further that either $LT = SL$ or $TL = LS$ on c_{00}. Then there exist two non-zero vectors $x, y \in c_{00}$ such that $\langle p(T)x, y\rangle \ge 0$ for all positive polynomials p on $\mathbb{R}$.*

Proof Suppose first that $LT = SL$, so that $LT^n = S^n L$ for all $n \ge 0$, and, more generally, $\langle Lp(T)u, v\rangle = \langle p(S)Lu, v\rangle$ for all polynomials $p \in \mathbb{R}[t]$ and $u, v \in c_{00}$. Taking $v = Lu$, we see that, when p is a positive polynomial,

$$\langle p(T)u, L^t Lv\rangle = \langle Lp(T)u, Lu\rangle = \langle p(S)Lu, Lu\rangle \ge 0$$

for all $u \in c_{00}$, using the symmetry of S and Proposition 9.1.3. We take $u \ne 0$, so that $L^t Lu \ne 0$, and then setting $x = u$, $y = L^t Lu$ gives the required result.

If instead we have $TL = LS$, then, taking $u = L^t v$ for some non-zero $v \in c_{00}$, we have for every positive polynomial p that

$$\langle p(T)Lu, v\rangle = \langle Lp(S)u, v\rangle = \langle p(S)u, L^t v\rangle = \langle p(S)u, u\rangle \ge 0,$$

so that setting $x = LL^t v$ and $y = v$ gives the required result. □

Lemma 9.2.3 *Let $\mathcal{X}$ be a sequence space, and $T \in \mathcal{L}(\mathcal{X})$ preserving c_{00} and mapping the subspace of real sequences to itself. If there exist two non-zero* *real vectors $x, y \in c_{00}$ such that $\langle p(T)x, y\rangle \geq 0$ for all positive polynomials p on $\mathbb{R}$, then T has a finitely supported moment pair, and hence a non-trivial closed invariant subspace.*

Proof Suppose first that $\mathcal{X}$ is a real sequence space. Let ϕ denote the linear functional on c_{00} induced by taking an inner product with y, that is, $\phi(z) = \langle z, y\rangle$ for all $z \in c_{00}$. For all polynomials p on $\mathbb{R}$ we have $\phi(p(T)x) = \langle p(T)x, y\rangle \geq 0$, and so (x, ϕ) is a moment pair for T, by Proposition 9.1.3.

In the case that $\mathcal{X}$ is a complex space, we see that $\mathcal{X} = \mathcal{Y} \oplus i\mathcal{Y}$ as a topological direct sum, where $\mathcal{Y}$ is the subspace of $\mathcal{X}$ consisting of real sequences. Since $x, y \in \mathcal{Y}$, we deduce that T has a non-trivial closed invariant real subspace $\mathcal{M} \subset \mathcal{Y}$. Then $\mathcal{N} := \mathcal{M} \oplus i\mathcal{M}$ is a non-trivial closed invariant complex subspace. □

In general we expect that the ideas above will have many applications, and we present one very elegant example.

Definition 9.2.4 A bounded operator T on a sequence space is said to be *tridiagonal* if it maps c_{00} into itself and for each basis vector e_n the image Te_n is a linear combination of e_{n-1}, e_n and e_{n+1} (where we interpret e_0 as 0). Thus, regarded as an infinite matrix, it has no non-zero entries except on the main diagonal and the two diagonals adjacent to it.

Theorem 9.2.5 *Let $\mathcal{X}$ be a sequence space and T a tridiagonal operator on $\mathcal{X}$, such that the associated matrix $(t_{i,j})_{i,j\geq 1}$ is real and $t_{n,n+1}t_{n+1,n} \geq 0$ for all $n = 1, 2, \ldots$. Then T has a finitely supported moment pair, and hence a non-trivial closed invariant subspace.*

Proof By using the technique presented in the proof of Lemma 9.2.3, we may suppose without loss of generality that $\mathcal{X}$ is a real sequence space. Using the inner-product notation introduced above, after Definition 9.2.1, we shall show that there are two non-zero vectors $x, y \in c_{00}$, such that $\langle p(T)x, y\rangle \geq 0$ for all positive real polynomials p. Indeed, at this stage we may work entirely in the real sequence space ℓ^2, since all our calculations involve sequences with finitely many non-zero terms. Thus we identify a vector in c_{00} with the linear functional it induces using the inner product.

If $t_{n+1,n} = 0$ for some n, then (e_1, e_{n+1}) is a moment pair (note that $\langle T^k e_1, e_{n+1}\rangle = 0$ for all k), and likewise if $t_{n,n+1} = 0$, then (e_{n+1}, e_1) is a moment pair for similar reasons. We therefore assume that $t_{n,n+1}t_{n+1,n} > 0$ for all n.

Define an infinite matrix $L = \operatorname{diag}(\alpha_1, \alpha_2, \ldots)$, where $\alpha_1 = 1$ and in general

$$\alpha_{n+1} = \alpha_n \sqrt{\frac{t_{n,n+1}}{t_{n+1,n}}}, \qquad \text{for} \quad n \geq 1.$$

Note that L defines an invertible linear mapping on c_{00}. The matrix $S = LTL^{-1}$ represents T with respect to the basis $(f_n)_{n\geq 1} := (e_n/\alpha_n)_{n\geq 1}$ of c_{00}. Thus it is still tridiagonal, and indeed

$$Tf_n = t_{n-1,n}(\alpha_{n-1}/\alpha_n) f_{n-1} + t_{n,n} f_n + t_{n+1,n}(\alpha_{n+1}/\alpha_n) f_{n+1}, \qquad \text{and}$$

$$Tf_{n+1} = t_{n,n+1}(\alpha_n/\alpha_{n+1}) f_n + t_{n+1,n+1} f_{n+1} + t_{n+2,n+1}(\alpha_{n+2}/\alpha_{n+1}) f_{n+2},$$

which means that S is symmetric, since $t_{n+1,n}(\alpha_{n+1}/\alpha_n) = t_{n,n+1}(\alpha_n/\alpha_{n+1})$.

We have $L = L^t$ and $LT = SL$. Hence, by Proposition 9.2.2, the pair $(x, L^2 x)$ is a moment pair for T, for any non-zero $x \in c_{00}$. □

9.3 Binomial sums

Let $(a_n)_{n\geq 0}$ be a sequence of complex numbers, and let $r \geq 0$. Clearly, if $a_n = 0$ for all $n > r$, then

$$\sum_{k=0}^{n} \binom{n}{k} a_k = O(n^r) \qquad \text{as } n \to \infty.$$

The converse is false. For example the sequence $a_n = (-1)^n$ satisfies

$$\sum_{k=0}^{n} \binom{n}{k} a_k = (1 + (-1))^n = 0, \qquad \text{for all } n \geq 1,$$

but a_n does not even tend to zero. There is, however, a partial converse, which, in view of the example above, is perhaps a little surprising.

Theorem 9.3.1 *Let $(a_n)_{n\geq 0}$ be a sequence of complex numbers, and let r be a non-negative integer. Assume that*

$$\sum_{\substack{k=0 \\ k \text{ even}}}^{n} \binom{n}{k} a_k = O(n^r) \qquad \text{and} \qquad \sum_{\substack{k=0 \\ k \text{ odd}}}^{n} \binom{n}{k} a_k = O(n^r) \qquad \text{as } n \to \infty. \tag{9.3}$$

Then $a_n = 0$ for all $n > r$.

This theorem is the focal point of the section. There are several possible variants of this result.

- If we replace $O(n^r)$ by $o(n^r)$, then we can conclude that $a_n = 0$ for all $n \geq r$, as mentioned in Corollary 9.3.4.
- Theorem 9.3.1 also works if we assume that, for some $m \geq 3$,

$$\sum_{\substack{k=0 \\ k\equiv q \pmod m}}^{n} \binom{n}{k} a_k = O(n^r), \qquad q = 1, \dots, m.$$

The proof is similar to that for $m = 2$ (see Section 9.3.3).

We present a short proof in Section 9.3.1 using the theory of entire functions of exponential type. Then, in Section 9.3.3, we give an application to Banach algebras, which leads to a theorem on invariant subspaces.

9.3.1 Proof of Theorem 9.3.1

An entire function $f : \mathbb{C} \to \mathbb{C}$ is of *exponential type* if

$$\tau := \limsup_{|z|\to\infty} \frac{\log|f(z)|}{|z|} < \infty,$$

in which case τ is called the *type* of f. If, further, $\tau = 0$, then f is said to be of *minimal exponential type*. We refer to Boas's book [43] for background on entire functions. In particular, we shall have recourse to the following version of the Phragmén–Lindelöf principle [43, theorem 6.2.13].

Theorem 9.3.2 (Phragmén–Lindelöf principle) *Let f be an entire function of minimal exponential type, and let $r \geq 0$. Suppose that, on the real axis, $f(x) = O(|x|^r)$ as $x \to \pm\infty$. Then f is a polynomial of degree at most r.*

We begin with a preliminary lemma on entire functions.

Lemma 9.3.3 *Let a, b and c be entire functions such that*

$$a(z) = e^{-z}b(z) = e^{z}c(z), \qquad z \in \mathbb{C}. \tag{9.4}$$

If b and c are of exponential type at most 1, then a is of minimal exponential type.

Proof We consider the Laplace transforms A, B, C of a, b, c, respectively. Thus, for example,

$$A(\zeta) = \int_0^\infty a(x)e^{-x\zeta}\,dx.$$

Since a, b, c are all of exponential type at most 1, it follows that A, B, C are well-defined and holomorphic in $\{\zeta : \operatorname{Re}\zeta > 1\}$. Moreover, for $\operatorname{Re}\zeta > 1$,

$$A(\zeta) = \int_0^\infty \left(\sum_{n=0}^\infty \frac{a_n}{n!}x^n\right) e^{-x\zeta}\,dx = \sum_{n=0}^\infty \frac{a_n}{n!}\int_0^\infty x^n e^{-x\zeta}\,dx = \sum_{n=0}^\infty \frac{a_n}{\zeta^{n+1}}, \tag{9.5}$$

with analogous expansions for B, C. Using standard properties of Laurent series, it follows that A, B, C extend holomorphically to $\{\zeta : |\zeta| > 1\}$. Now, taking Laplace transforms in (9.4) gives, for $\operatorname{Re}\zeta > 1$,

$$A(\zeta) = B(\zeta + 1) = C(\zeta - 1).$$

Therefore A in fact extends holomorphically to

$$\{\zeta : |\zeta + 1| > 1\} \cup \{\zeta : |\zeta - 1| > 1\},$$

in other words, to $\mathbb{C} \setminus \{0\}$. In conjunction with the Laurent expansion (9.5), this implies that

$$|a_n|^{1/n} \to 0 \qquad \text{as } n \to \infty.$$

It follows that a is of minimal exponential type. □

Proof of Theorem 9.3.1 Consider the expression

$$a(z) = \sum_{n=0}^{\infty} \frac{a_n}{n!} z^n. \tag{9.6}$$

We shall show successively that this defines an entire function of exponential type, that it is of minimal exponential type and, finally, that it is a polynomial of degree at most r, which will yield the desired conclusion.

For $n \geq 0$, set

$$b_n = \sum_{k=0}^{n} \binom{n}{k} a_k \qquad \text{and} \qquad c_n = \sum_{k=0}^{n} \binom{n}{k} (-1)^{n-k} a_k. \tag{9.7}$$

The hypotheses on (a_n) ensure that both $b_n, c_n = O(n^r)$ as $n \to \infty$. Hence, if we set

$$b(z) = \sum_{n=0}^{\infty} \frac{b_n}{n!} z^n \qquad \text{and} \qquad c(z) = \sum_{n=0}^{\infty} \frac{c_n}{n!} z^n, \tag{9.8}$$

then b and c are entire functions of exponential type at most 1. Comparing coefficients of z^n in the identities $e^z(e^{-z}b(z)) = b(z)$ and $e^{-z}(e^z c(z)) = c(z)$, we see that both $e^{-z}b(z)$ and $e^z c(z)$ have the same coefficients as $a(z)$, and therefore (9.4) is satisfied.

In particular, a too is an entire function. Further, (9.4) implies that $a^2 = bc$, and therefore a is also of exponential type at most 1. It follows from Lemma 9.3.3 that a is in fact of minimal exponential type.

Finally, we show that a is a polynomial. For $x \geq 0$, we have

$$|b(x)| \leq \sum_{n=0}^{\infty} \frac{|b_n|}{n!} x^n \leq \sum_{n=0}^{\infty} \frac{K(n+r)\cdots(n+1)}{n!} x^n = K(x^r e^x)^{(r)},$$

where K is a constant independent of x. Therefore, using (9.4) again,

$$|a(x)| = |e^{-x}b(x)| \leq Ke^{-x}(x^r e^x)^{(r)} = O(x^r) \quad \text{as } x \to \infty.$$

A similar calculation with $c(x)$ in place of $b(x)$ shows that $a(x) = O(|x|^r)$ as $x \to -\infty$. As a is of minimal exponential type, we can apply the Phragmén–Lindelöf principle cited earlier to deduce that a is a polynomial of degree at most r. Hence $a_n = 0$ for all $n > r$, and the proof is complete. □

We conclude this section with a 'little o' version of Theorem 9.3.1. It is actually a simple consequence of the 'big O' form.

Corollary 9.3.4 *Let $(a_n)_{n\geq 0}$ be a sequence of complex numbers, and let r be a non-negative integer. Assume that*

$$\sum_{\substack{k=0 \\ k \text{ even}}}^{n} \binom{n}{k} a_k = o(n^r) \quad \text{and} \quad \sum_{\substack{k=0 \\ k \text{ odd}}}^{n} \binom{n}{k} a_k = o(n^r) \quad \text{as } n \to \infty.$$

Then $a_n = 0$ for all $n \geq r$.

Proof From Theorem 9.3.1 we already have $a_n = 0$ for all $n > r$, so it only remains to show that $a_r = 0$. Now, for all $n \geq r$, we have

$$\binom{n}{r} a_r = \sum_{k=0}^{n} \binom{n}{k} a_k - \sum_{k=0}^{r-1} \binom{n}{k} a_k.$$

The first term on the right-hand side is $o(n^r)$ by hypothesis, and the second term is even $O(n^{r-1})$ (or $= 0$ if $r = 0$). Therefore the left-hand side is also $o(n^r)$, and this can only happen if $a_r = 0$. □

9.3.2 A technical refinement

In this section we use more complicated arguments to deduce an improvement of Theorem 9.3.1, which will have further consequences for invariant subspaces in Section 9.3.3.

We shall require a further version of the Phragmén–Lindelöf principle [149, lemma 3.3].

Theorem 9.3.5 *Let f be an entire function of minimal exponential type and $\phi : (0, \infty) \to (0, \infty)$ be an increasing differentiable function such that $x\phi'(x)/\phi(x) \to \rho$ as $x \to \infty$, for some $\rho \in [0, 1)$. If*

$$\limsup_{x\to\pm\infty} \frac{\log|f(x)|}{\phi(|x|)} \leq 1,$$

then

$$\limsup_{|z|\to\infty} \frac{\log|f(z)|}{\phi(|z|)} \le \frac{1}{\cos(\pi\rho/2)}.$$

We continue with two easy technical lemmas.

Lemma 9.3.6 *Let $(a_n)_{n\ge 0}$ be a sequence of complex numbers, and let (b_n), (c_n) be defined by (9.7).*

(i) If

$$\frac{|a_n|}{n!} \le \inf_{r>0} \frac{\psi(r)}{r^n} \tag{9.9}$$

for all $n \ge 0$, where ψ is a positive function on $(0,\infty)$, then

$$|b_n|, |c_n| \le (n+1)\psi(n)$$

for all $n \ge 0$.

(ii) If $n|a_n|^{1/n} \to 0$ as n tends to ∞, then $b_n, c_n = O\left(e^{\varepsilon\sqrt{n}}\right)$ for all $\varepsilon > 0$.

Proof (i) Since $\binom{n}{k} \le \frac{n^k}{k!}$, we have

$$|b_n|, |c_n| \le \sum_{k=0}^{n} n^k \frac{|a_k|}{k!} \le \sum_{k=0}^{n} n^k \frac{\psi(n)}{n^k} = (n+1)\psi(n).$$

(ii) Suppose $\eta > 0$. Using the hypothesis that $n|a_n|^{1/n} \to 0$ and applying Stirling's formula, we conclude that

$$\frac{|a_n|}{n!} = o\left(\frac{(e\eta)^n}{n^{2n}}\right),$$

as n tends to ∞. Consider the function

$$\psi(r) = \exp\left(2\sqrt{e\eta r}/e\right),$$

and observe that $\inf_{r>0} \frac{\psi(r)}{r^n}$ is attained at $r = n^2 e/\eta$ and is equal to $(e\eta)^n/n^{2n}$. Therefore, applying (i), it follows that $|b_n|, |c_n| = O(n\psi(n))$. Since η is arbitrary, the result follows. □

Lemma 9.3.7 *Let $\phi : [0,\infty) \to [0,\infty)$ be an increasing function such that $\phi(x)/x$ is decreasing for sufficiently large x and $\phi(x)/x \to 0$ as x tends to ∞. Then*

$$\sum_{n\ge 0} \frac{e^{-x}x^n}{n!} e^{\phi(n)} \le e^{\phi(x)(1+o(1))} \tag{9.10}$$

as x tends to ∞. In particular, for $\varepsilon > 0$ and p a positive integer, we have

$$\sum_{n\geq 0} \frac{e^{-x}x^n}{n!} e^{\varepsilon\sqrt{n}} \leq e^{\varepsilon\sqrt{x}(1+o(1))} \quad \text{and} \quad \sum_{n\geq 0} \frac{e^{-x}x^n}{n!}(n+1)^p \leq e^{p\log(1+x)(1+o(1))} \tag{9.11}$$

as x tends to ∞.

Proof For $x > 0$, since ϕ is increasing, we have

$$\sum_{0\leq n\leq x} \frac{e^{-x}x^n}{n!} e^{\phi(n)} \leq \sum_{0\leq n\leq x} \frac{e^{-x}x^n}{n!} e^{\phi(x)} \leq e^{\phi(x)}.$$

Moreover, for x sufficiently large, we have also

$$\begin{aligned}\sum_{n>x} \frac{e^{-x}x^n}{n!} e^{\phi(n)} \leq \sum_{n>x} \frac{e^{-x}x^n}{n!} e^{n\phi(x)/x} &\leq e^{-x}\exp(xe^{\phi(x)/x}) \\ &= \exp\left[\phi(x)\left(1 + O\left(\phi(x)/x\right)\right)\right].\end{aligned}$$

Adding these inequalities, we obtain

$$\sum_{n\geq 0} \frac{e^{-x}x^n}{n!} e^{\phi(n)} \leq e^{\phi(x)(1+o(1))}. \qquad \square$$

The next lemma is a Phragmén–Lindelöf principle for entire functions of minimal exponential type, which is a consequence of Theorem 9.3.5.

Lemma 9.3.8 *Let f be an entire function of minimal exponential type and p a positive integer. If*

$$\limsup_{x\to+\infty} \frac{\log|f(x)|}{p\log(1+x)} \leq 1 \qquad \text{and} \qquad \limsup_{x\to-\infty} \frac{\log|f(x)|}{\sqrt{|x|}} \leq 0,$$

then

$$\limsup_{|z|\to\infty} \frac{\log|f(z)|}{p\log(1+|z|)} \leq 1.$$

Proof Note that the hypotheses imply that $\max(\log|f(x)|, 0) = o\left(\sqrt{|x|}\right)$ as x tends to $\pm\infty$. Applying Theorem 9.3.5 with $\phi(x) = \varepsilon\sqrt{x}$ with $\varepsilon > 0$, and letting $\varepsilon \to 0$, it follows that

$$\max(\log|f(z)|, 0) = o\left(\sqrt{|z|}\right)$$

as $|z|$ tends to ∞. In other words

$$\max(\log|f(z^2)|, 0) = o(|z|)$$

as $|z|$ tends to ∞. By hypothesis we have also

$$\limsup_{x\to\pm\infty} \frac{\log|f(x^2)|}{2p\log(1+|x|)} \le 1.$$

The result follows on applying Theorem 9.3.5, with $\phi(x) = 2p\log(1+|x|)$, to $f(z^2)$, which is also of minimal exponential type. Note that the constant ρ which appears in Theorem 9.3.5 is equal to 0. □

We have now all the tools needed to deduce a stronger version of Theorem 9.3.1.

Theorem 9.3.9 *Let $(a_n)_{n\ge 0}$ be a complex sequence and let p be a non-negative integer. Assume that*

$$\sum_{k=0}^{n}\binom{n}{k}a_k = O(n^p) \quad \textit{and} \quad \sum_{k=0}^{n}\binom{n}{k}(-1)^{n-k}a_k = O\left(e^{\varepsilon\sqrt{n}}\right) \quad \textit{as } n\to\infty, \tag{9.12}$$

for all $\varepsilon > 0$. Then $a_n = 0$ for all $n > p$.

Proof We define b_n and c_n by (9.7), and the entire functions a, b and c as in (9.6) and (9.8). The hypotheses on b_n and c_n imply that b and c are of exponential type at most 1. One can easily check that we have the identity (9.4). By Lemma 9.3.3, a is of minimal exponential type.

In order to apply Lemma 9.3.8 to the function a, we need to estimate a on the real axis.

As $x\to+\infty$, there exists a positive constant K such that

$$|a(x)| = |e^{-x}b(x)| \le K\sum_{n\ge 0}\frac{e^{-x}x^n}{n!}(n+1)^p.$$

Using (9.11), we have $|a(x)| \le e^{p\log(1+x)(1+o(1))}$, which implies that

$$\limsup_{x\to+\infty}\frac{\log|a(x)|}{p\log(1+x)} \le 1.$$

Likewise, as $x\to-\infty$, using (9.11), $|a(x)| = |e^x c(x)| \le e^{\varepsilon\sqrt{|x|}(1+o(1))}$, which implies that

$$\limsup_{x\to-\infty}\frac{\log|a(x)|}{\sqrt{|x|}} \le \varepsilon,$$

for all $\varepsilon > 0$. Applying Lemma 9.3.8, we get

$$\limsup_{|z|\to\infty}\frac{\log|a(z)|}{p\log(1+|z|)} \le 1.$$

By Cauchy's integral formula, it follows that, for all $r > 0$,

$$\frac{|a_n|}{n!} \le \max_{|z|=r} \frac{|a(z)|}{|z|^n} = \frac{\exp(1+o(1))(p\log(1+r))}{r^n}.$$

Letting r tend to ∞, we see that $a_n = 0$ for all $n > p$. □

9.3.3 Application to Banach algebras and invariant subspaces

Allan [6, lemma 3] proved the following result of Gelfand–Hille type, which he used to give an improved version of a result of Kalton [129] on sums of idempotents.

Theorem 9.3.10 (Allan) *Let $\mathcal{A}$ be a Banach algebra with identity 1, let $x \in \mathcal{A}$ and suppose that, for some integers $m \ge 0$, $r \ge 0$,*

$$\left\|x^m\left((1+x)^n - (1-x)^n\right)\right\| = O(n^r) \qquad \text{as } n \to \infty. \tag{9.13}$$

Then $x^{m+r+2} = 0$ if r is odd, while $x^{m+r+1} = 0$ if r is even. Moreover, if r is odd and if

$$\left\|x^m\left((1+x)^n - (1-x)^n\right)\right\| = o(n^r) \qquad \text{as } n \to \infty,$$

then $x^{m+r} = 0$.

Theorem 9.3.1 and Corollary 9.3.4 permit us to prove a localized form of this result.

Theorem 9.3.11 *Let $\mathcal{A}$ be a Banach algebra with identity 1, let $x \in \mathcal{A}$ and let $\phi \in \mathcal{A}^*$. Suppose that, for some integer $r \ge 0$,*

$$\phi\left((1+x)^n - (1-x)^n\right) = O(n^r) \qquad \text{as } n \to \infty. \tag{9.14}$$

Then $\phi(x^n) = 0$ for all odd integers $n > r$. If, further,

$$\phi\left((1+x)^n - (1-x)^n\right) = o(n^r) \qquad \text{as } n \to \infty,$$

then $\phi(x^n) = 0$ for all odd integers $n \ge r$.

Proof Observe that

$$\phi\left((1+x)^n - (1-x)^n\right) = 2\sum_{\substack{k=0\\ k \text{ odd}}}^{n} \binom{n}{k}\phi(x^k).$$

The result therefore follows by applying Theorem 9.3.1 and Corollary 9.3.4 to the sequence

$$a_n = \begin{cases} \phi(x^n), & n \text{ odd}, \\ 0, & n \text{ even}. \end{cases}$$

Allan's theorem is an easy consequence of Theorem 9.3.11. Indeed, given x, m, r satisfying (9.13), take $\psi \in \mathcal{A}^*$ and define $\phi \in \mathcal{A}^*$ by

$$\phi(y) = \psi(x^m y), \qquad y \in \mathcal{A}.$$

Then (9.14) holds, so applying Theorem 9.3.11 we deduce that $\psi(x^{m+n}) = 0$ for all odd integers $n > r$. As ψ is arbitrary, it follows that $x^{m+n} = 0$ for all odd integers $n > r$. This gives the $O(n^r)$ part of the result, and the $o(n^r)$ part is proved similarly. □

Perhaps more interestingly, Theorem 9.3.11 also leads to the following invariant-subspace theorem.

Theorem 9.3.12 *Let $\mathcal{X}$ be a (real or complex) infinite-dimensional Banach space, and let T be a bounded linear operator on $\mathcal{X}$. Suppose that there exist a non-zero $\xi_0 \in \mathcal{X}$, a non-zero $\psi_0 \in \mathcal{X}^*$ and an integer $r \geq 0$ such that*

$$\big\langle\big((I+T)^n - (I-T)^n\big)\,\xi_0, \psi_0\big\rangle = O(n^r) \qquad \text{as } n \to \infty.$$

Then T^2 has a non-trivial closed invariant subspace.

Proof The result is obvious if $T = 0$, so we can suppose that $T \neq 0$. Further, if T is not injective, then $\operatorname{Ker}(T)$ is a non-trivial closed T^2-invariant subspace, and we are done. So we may as well assume from the outset that T is injective.

Let $\mathcal{A}$ be the Banach algebra of bounded linear operators on $\mathcal{X}$, let $x = T$ and let $\phi \in \mathcal{A}^*$ be given by

$$\phi(S) = \langle S\xi_0, \psi_0\rangle, \qquad S \in \mathcal{A}.$$

Then the hypothesis on T implies that (9.14) holds, so we can apply Theorem 9.3.11 to conclude that $\langle T^n\xi_0, \psi_0\rangle = 0$ for all odd integers $n > r$. Let $\mathcal{M}$ be the closed linear span of $\{T^n\xi_0 : n \text{ odd}, n > r\}$. Then $\mathcal{M}$ is a closed T^2-invariant subspace of $\mathcal{X}$. Further, $\mathcal{M} \neq 0$ since T is injective, and $\mathcal{M} \neq \mathcal{X}$ since $\psi_0(\mathcal{M}) = 0$. Thus $\mathcal{M}$ is the required subspace. □

There is also a version of Theorem 9.3.11 in which the central '−' is replaced by '+'.

Theorem 9.3.13 *Let $\mathcal{A}$ be a Banach algebra with identity 1, let $x \in \mathcal{A}$ and let $\phi \in \mathcal{A}^*$. Suppose that, for some integer $r \geq 0$,*

$$\phi\big((1+x)^n + (1-x)^n\big) = O(n^r) \qquad \text{as } n \to \infty.$$

Then $\phi(x^n) = 0$ for all even integers $n > r$. If, further,

$$\phi\big((1+x)^n + (1-x)^n\big) = o(n^r) \qquad \text{as } n \to \infty,$$

then $\phi(x^n) = 0$ for all even integers $n \geq r$.

Proof This time,

$$\phi\big((1+x)^n+(1-x)^n\big)=2\sum_{\substack{k=0\\k\text{ even}}}^{n}\binom{n}{k}\phi(x^k),$$

so the result follows by applying Theorem 9.3.1 and Corollary 9.3.1 to the sequence

$$a_n=\begin{cases}\phi(x^n), & n\text{ even},\\ 0, & n\text{ odd}.\end{cases}$$ □

Both Allan's theorem and Theorem 9.3.13 likewise have '+' analogues, which we shall not state here. Finally, putting the '−' and the '+' results together, we obtain an invariant-subspace theorem for T (rather than merely for T^2).

Theorem 9.3.14 *Let $\mathcal{X}$ be a (real or complex) infinite-dimensional Banach space, and let T be a bounded linear operator on $\mathcal{X}$. Suppose that there exist a non-zero $\xi_0\in\mathcal{X}$, a non-zero $\psi_0\in\mathcal{X}^*$ and an integer $r\geq 0$ such that both*

$$\langle(\mathrm{I}+T)^n\xi_0,\psi_0\rangle=O(n^r)\quad\textit{and}\quad\langle(\mathrm{I}-T)^n\xi_0,\psi_0\rangle=O(n^r)\qquad\textit{as } n\to\infty.$$

Then T has a non-trivial closed invariant subspace.

Proof This follows the same lines as the proof of Theorem 9.3.12. As before, we can suppose that T is injective. This time, applying both Theorems 9.3.11 and 9.3.13, we have $\langle T^n\xi_0,\psi_0\rangle=0$ for *all* integers $n>r$. Let $\mathcal{M}$ be the closed linear span of $\{T^n\xi_0 : n>r\}$. Then $\mathcal{M}$ is now T-invariant and, as before, $\mathcal{M}\neq 0$ and $\mathcal{M}\neq\mathcal{X}$. □

The next result is an application of Theorem 9.3.9, which will lead to a further theorem about invariant subspaces.

Theorem 9.3.15 *Let $\mathcal{X}$ be a real or complex Banach space, let $T\in\mathcal{L}(\mathcal{X})$, and suppose that there exist $x\in\mathcal{X}$ and $\ell\in\mathcal{X}^*$ such that*

$$|\langle T^nx,\ell\rangle|=o(n^k)\qquad\textit{and}\qquad n|\langle(T-\mathrm{I})^nx,\ell\rangle|^{1/n}\to 0$$

as n tends to ∞, for some integer $k\geq 1$. Then

$$\langle(T-\mathrm{I})^nx,\ell\rangle=0\qquad\textit{for all}\qquad n\geq k.$$

Proof Set $a_n=\langle(T-\mathrm{I})^nx,\ell\rangle$. By part (ii) of Lemma 9.3.6, the constants $b_n=\sum_{k=0}^n\binom{n}{k}\langle(T-\mathrm{I})^kx,\ell\rangle$ and $c_n=\sum_{k=0}^n\binom{n}{k}(-1)^{n-k}\langle(T-\mathrm{I})^kx,\ell\rangle$ satisfy

$b_n, c_n = O\left(e^{\varepsilon\sqrt{n}}\right)$ as n tends to ∞ and for all $\varepsilon > 0$. Note that $b_n = \langle T^n x, \ell \rangle$, since

$$T^n = (T - \mathrm{I} + \mathrm{I})^n = \sum_{k=0}^{n} \binom{n}{k} (T - \mathrm{I})^k.$$

In other words, $b_n = o(n^k)$ and $c_n = O\left(e^{\varepsilon\sqrt{n}}\right)$ as n tends to ∞. Applying Theorem 9.3.9, we get $a_n = 0$ for all $n \geq k$. □

Corollary 9.3.16 *Let $T \in \mathcal{L}(\mathcal{X})$ and suppose that there exist non-zero $x \in \mathcal{X}$ and $\ell \in \mathcal{X}^*$ such that*

$$|\langle T^n x, \ell \rangle| = o(n^k) \qquad \text{and} \qquad n|\langle (T - \mathrm{I})^n x, \ell \rangle|^{1/n} \to 0$$

as n tends to ∞, for some integer $k \geq 1$. Then T has a non-trivial closed invariant subspace.

Proof We apply Theorem 9.3.15 and get $\langle (T - \mathrm{I})^n x, \ell \rangle = 0$ for all $n \geq k$. We may assume that $(T - \mathrm{I})^k \neq 0$, as otherwise T has non-trivial invariant subspaces. If $(T-\mathrm{I})^k x = 0$, then $\mathrm{Ker}(T-\mathrm{I})^k$ is a non-trivial invariant subspace for T as $x \neq 0$. Finally, if $(T - \mathrm{I})^k x \neq 0$, then the cyclic subspace generated by $(T - \mathrm{I})^k x$ is a non-trivial invariant subspace for T, as $\ell \neq 0$. □

Exercises

1. Show that, if $T \in \mathcal{L}(L^2(0, 1))$ and $(T - r\mathrm{I})^2 f \geq 0$ whenever $r \in \mathbb{R}$ and $f \in L^2(0, 1)$ satisfies $f \geq 0$, then T admits a moment sequence.
2. Let $S \subset c_0$ denote the closed convex hull of the standard basis $(e_n)_{n\geq 1}$. Show that the point $\left(\frac{1}{2}, \frac{1}{4}, \frac{1}{8}, \dots\right) \in \partial S$ is not a support point of S.
3. Every 2×2 real matrix $T = (t_{ij})_{i,j=1}^2$ is tridiagonal in a trivial way. Show that if $t_{12}t_{21} \geq 0$, then T has real eigenvalues, whereas this need not be the case if $t_{12}t_{21} < 0$. Compare this with Theorem 9.2.5.
4. Show that in Theorem 9.2.5 we may even conclude that (e_1, e_1) is a moment pair for T.
5. Theorem 9.3.1 remains true if we replace (9.3) by

$$\sum_{\substack{k=0 \\ k\equiv l \ (\mathrm{mod}\ q)}}^{n} \binom{n}{k} a_k = O(n^r), \qquad l = 1, \dots, q,$$

provided that q is an integer with $q \geq 3$ (note that (9.3) is simply the case $q = 2$). The proof is essentially the same as that in Section 9.3.1, but now using a (slightly easier) variant of the Phragmén–Lindelöf principle, where one assumes polynomial growth of f along the half-lines $\arg(z) = 2\pi l/q$ $(l = 1, \ldots, q)$. Write out the details.

Deduce a sufficient condition for T^q to have a non-trivial closed invariant subspace.

6. Using Jensen's inequality and the convexity of the functions $x \mapsto e^{\varepsilon\sqrt{x}}$ and $x \mapsto (x+1)^p$, prove that the inequalities in (9.11) are equalities.

Comments

Lemma 9.1.5 is due to Bishop and Phelps [41], and is closely related to the Bishop–Phelps–Bollobás theorem [40, 44], which has other applications in operator theory.

The account of moment sequences in Section 9.1 leading up to Theorem 9.1.7 is based on the papers of Atzmon, Godefroy and Kalton [21, 22]. Some simplifications are possible, since, for example, we deal with convex sets in Theorem 9.1.6, and so we do not require the more general variational principle derived by Deville, Godefroy and Zizler [91], which yields further results on invariant subspaces.

The difficult part of Proposition 9.1.3 is due to Riesz. We have based our account on Koosis [136, chapter 5], with certain simplifications.

The application to operators on sequence spaces presented in Section 9.2, leading up to Theorem 9.2.5, is due to Grivaux [115].

There is a link between the existence of moment sequences and invariant subspaces for rational functions of a given operator in Chalendar, Partington and Smith [77, section 3].

Section 9.3.2 is a simplified account of the discussion by Mashreghi and Ransford [149], which contains more sophisticated versions of the Phragmén–Lindelöf principle and more general results linking the growth of the sequences $(a_n)_n$, $(b_n)_n$ and $(c_n)_n$.

It is much less clear whether Theorem 9.3.11 can be deduced from Allan's result. Allan's proof begins by showing that (9.13) implies that x is quasinilpotent, which fact he then exploits via elementary spectral theory and entire function theory to obtain the desired conclusion. However, the weaker hypothesis (9.14) no longer implies that x is quasinilpotent, so a proof of Theorem 9.3.11 has to proceed along different lines. It was the search for a proof independent of spectral theory which eventually led to a result purely about complex numbers, Theorem 9.3.1. It should be added, however, that the last part of the proof of Theorem 9.3.1 (using the Phragmén–Lindelöf principle) owes much to Allan's paper.

Because the proof of Theorem 9.3.11 uses no spectral theory, it is just as valid for real Banach algebras as for complex ones, and hence so is the deduction of Allan's result.

It is unknown whether the hypotheses of Theorem 9.3.12 imply the existence of a non-trivial closed invariant subspace, not just for T^2, but for T itself.

Theorem 9.3.15 is the main result of the paper by Esterle and Zarrabi [102] but the proof uses other techniques.

10

Positive and strictly-singular operators

In this chapter we look at two other important classes of operators that might reasonably be supposed to have invariant subspaces, but for which only partial progress has been made so far.

We have already seen several examples of operators on function spaces that map non-negative functions to non-negative functions; for example, in Chapter 5, we discussed the Bishop operator T_α on $L^2([0, 1])$, defined by

$$T_\alpha f(x) = xf(\{x + \alpha\}),$$

for a fixed $\alpha \in (0, 1)$, where the curly brackets denote fractional part. Similarly, many important operators on sequence spaces map non-negative sequences to non-negative sequences; one obvious example is the bilateral shift U on $\ell^2(\mathbb{Z})$, such that $U(e_n) = e_{n+1}$, for each $n \in \mathbb{Z}$. These ideas are subsumed in the idea of a positive (i.e., order-preserving) operator on an ordered Banach space. In general it is not known whether these have non-trivial invariant subspaces, although some interesting partial results can be given.

We conclude the chapter by discussing the class of strictly singular operators, which may be regarded as a generalization of the notion of a compact operator (and on a Hilbert space these give the same class of operators). Once more, it is not known whether they must have non-trivial invariant subspaces in general, but again we shall give some partial results.

10.1 Ordered spaces and positive operators

Definition 10.1.1 Let $\mathcal{V}$ be a real or complex vector space. A *cone* in $\mathcal{V}$ is a subset $\mathcal{C} \subset \mathcal{V}$ such that

(i) $\mathcal{C} + \mathcal{C} \subset \mathcal{C}$;
(ii) $a\mathcal{C} \subset \mathcal{C}$ for every $a \geq 0$;
(iii) $\mathcal{C} \cap (-\mathcal{C}) = \{0\}$.

An *ordered vector space* is a vector space containing a cone, and we write $y \le x$ (or $x \ge y$) to mean that $x - y \in \mathcal{C}$.

We shall also write $x > 0$ (resp. $x < 0$) to mean that x is non-zero but $x \ge 0$ (resp. $x \le 0$).

Clearly, the definition of a cone is equivalent to saying that the partial order satisfies the following properties:

(i) Whenever $x \ge 0$ and $y \ge 0$, then $x + y \ge 0$.
(ii) Whenever $a \ge 0$ (a scalar) and $x \ge 0$, then $ax \ge 0$.
(iii) If $x \ge 0$ and $x \le 0$, then $x = 0$.

In fact, we usually require more than a partial order, and make the following additional definitions.

Definition 10.1.2 A *Riesz space* is an ordered vector space $\mathcal{V}$ such that every pair of elements $x, y \in \mathcal{V}$ has a *greatest lower bound* $x \wedge y \in \mathcal{V}$ and a *least upper bound* $x \vee y \in \mathcal{V}$. We write $|x| = x \vee (-x)$.

A *Banach lattice* is a Banach space that is also a Riesz space, with the property that if $|x| \le |y|$, then $\|x\| \le \|y\|$.

The examples we have in mind are function spaces, where $f \ge 0$ means that $f(t) \ge 0$ for (almost) all t, and sequence spaces, where $(a_n) \ge 0$ means that $a_n \ge 0$ for all n.

We may sometimes work with a stronger notion than that of an invariant subspace, so we make the following definition.

Definition 10.1.3 An *ideal* $\mathcal{J}$ in a Banach lattice $\mathcal{X}$ is a subspace such that, whenever $x \in \mathcal{X}$, $y \in \mathcal{J}$ and $|x| \le |y|$, one has $x \in \mathcal{J}$. The *principal ideal generated by an element* $x \ge 0$ *in* $\mathcal{X}$ is defined by

$$J[x] = \{y \in \mathcal{X} : |y| \le \lambda x \quad \text{for some} \quad \lambda > 0\}.$$

So far we have not brought in any operator theory, so let us now do so.

Definition 10.1.4 Let $\mathcal{V}$ be an ordered vector space. An operator $T : \mathcal{V} \to \mathcal{V}$ is said to be *positive*, or *order-preserving*, if, whenever $x \in \mathcal{V}$ with $x \ge 0$, one has $Tx \ge 0$.

Sometimes one says that an operator T on a Hilbert space $\mathcal{H}$ is positive (and self-adjoint) if $\langle Tx, x\rangle \ge 0$ for all $x \in \mathcal{H}$. We emphasize that these two notions are unrelated, even for operators on $\mathbb{C}^2$, and we leave the reader to construct some examples (see Exercise 1).

In this chapter we shall consider positivity only in the sense of Definition 10.1.4, and write $T \geq 0$ if T is a positive operator in this sense. In general, we write $S \leq T$ or $T \geq S$, if $T - S \geq 0$.

Example 10.1.5 The classical Banach spaces $L^p(\Omega, \mu)$ (for $1 \leq p \leq \infty$) and $C(K)$ are ordered vector spaces, with the obvious definitions of '$f \geq 0$'). In addition to the Bishop operator described above, we mention the *Volterra operator* $V \in \mathcal{L}(L^p(0,1))$, defined by

$$(Vf)(t) = \int_0^t f(s)\,ds,$$

as another example of a positive operator. Likewise, the sequence spaces ℓ^p (for $1 \leq p \leq \infty$) and c_0 are ordered vector spaces, where we interpret 'positive sequence' in the natural way. As a 'discrete' form of the Volterra operator, we mention here the *Cesàro operator* $A \in \mathcal{L}(\ell^p)$, defined by

$$(Ax)_n = \frac{1}{n+1} \sum_{k=0}^{n} x_k,$$

for $x = \{x_k\}_{k=0}^\infty \in \ell^p$ or c_0.

10.2 Invariant subspaces for positive operators

From now on we assume that T is a positive bounded operator defined on a Banach space $\mathcal{X}$. In general, it is not known whether such operators always have invariant subspaces (the Bishop operators providing examples that have not yet been fully settled). However, some partial results are known, and we now present some of these.

Definition 10.2.1 Let $\mathcal{X}$ be a separable Banach space. Let $\{x_n\}_{n=1}^\infty$ and $\{f_n\}_{n=1}^\infty$ be sequences in $\mathcal{X}$ and $\mathcal{X}^*$, respectively. The system $\{x_n, f_n\}_{n=1}^\infty$ is called a *fundamental, total, biorthogonal system (f.t.b.o.s.)*, or a *Markushevich basis*, if the system is *biorthogonal* in the sense that

$$f_m(x_n) = \delta_{m,n} = \begin{cases} 1 & \text{if } m = n, \\ 0 & \text{otherwise,} \end{cases}$$

fundamental in the sense that the closed linear span of the vectors $\{x_n\}_{n\geq 1}$ is $\mathcal{X}$, and *total* in the sense that if $x \in \mathcal{X}$ and $f_n(x) = 0$ for all n, then $x = 0$.

If $\mathcal{X}$ has a Schauder basis $\{x_n\}_{n\geq 1}$, so that every vector $x \in \mathcal{X}$ can be written uniquely as $x = \sum_{k=1}^\infty a_k x_k$ for some scalars $\{a_k\}_{k=1}^\infty$, then, defining

$f_n\left(\sum_{k=1}^{\infty} a_k x_k\right) = a_n$ makes the system $\{x_n, f_n\}_{n=1}^{\infty}$ into an f.t.b.o.s. Not all separable Banach spaces have bases, but they do have f.t.b.o.s., as we now explain.

Proposition 10.2.2 *Let $\mathcal{X}$ be a separable Banach space. Then it possesses a fundamental, total, biorthogonal system.*

Proof Clearly, we may assume that $\mathcal{X}$ is infinite-dimensional. We begin with a sequence $\{y_n\}_{n=1}^{\infty}$ of non-zero elements in $\mathcal{X}$ with dense linear span, and a sequence $\{g_n\}_{n=1}^{\infty}$ of elements of $\mathcal{X}^*$ such that $\bigcap_{n=1}^{\infty} \operatorname{Ker} g_n = \{0\}$ (these can be constructed using a dense subset in $\mathcal{X}$ and the Hahn–Banach theorem). The idea is to perform a 'double Gram–Schmidt process', constructing a biorthogonal system $\{x_n, f_n\}_{n=1}^{\infty}$.

To start the process, let $x_1 = y_1$ and let $f_1 = g_{k_1}/g_{k_1}(y_1)$, where k_1 is the least index such that $g_{k_1}(y_1) \neq 0$.

Now suppose that we have constructed $\{x_1, \ldots, x_n\}$ and $\{f_1, \ldots, f_n\}$ as a biorthogonal set. Let m be the least integer such that $y_m \notin \operatorname{span}\{x_1, \ldots, x_n\}$, and define

$$x_{n+1} = y_m - \sum_{k=1}^{n} f_k(y_m) x_k.$$

We see that $f_r(x_{n+1}) = 0$ for $1 \leq r \leq n$. Now let p be the least integer such that $g_p(x_{n+1}) \neq 0$, and define

$$f_{n+1} = \frac{1}{g_p(x_{n+1})}\left(g_p - \sum_{k=1}^{n} g_p(x_k) f_k\right).$$

We see that $f_{n+1}(x_{n+1}) = 1$ and $f_{n+1}(x_r) = 0$ for $1 \leq r \leq n$. Thus, $\{x_1, \ldots, x_{n+1}\}$ and $\{f_1, \ldots, f_{n+1}\}$ form a biorthogonal set.

We can inductively construct the entire biorthogonal system $\{x_n, f_n\}_{n=1}^{\infty}$ in this way, and see that the linear span of the set $\{x_n\}_{n=1}^{\infty}$ is dense in $\mathcal{X}$, while the linear span of the set $\{f_n\}_{n=1}^{\infty}$ is the same as the linear span of the set $\{g_n\}_{n=1}^{\infty}$. Thus $\{x_n, f_n\}_{n=1}^{\infty}$ is a fundamental, total, biorthogonal system in $\mathcal{X}$. □

A space with a fundamental, total, biorthogonal sequence $\{x_n, f_n\}_{n=1}^{\infty}$, may be made into an ordered vector space by defining the positive cone to be

$$\mathcal{C} = \{x \in \mathcal{X} : f_n(x) \geq 0 \text{ for all } n\}.$$

(This is of course the natural notion of positivity in a sequence space such as ℓ^p or c_0, with the canonical basis.) *From now on we shall always define our positive cone in this way.*

Under these circumstances, every quasinilpotent positive operator has an invariant subspace, and indeed slightly more is true. We may weaken the notion of quasinilpotence as follows.

Definition 10.2.3 An operator $A \in \mathcal{L}(\mathcal{X})$ is said to be *quasinilpotent at a vector* $z \in \mathcal{X}$ if $\|A^n z\|^{1/n} \to 0$ as $n \to \infty$.

Clearly every quasinilpotent operator is also quasinilpotent at every vector, but the two notions are not equivalent (see Exercise 2).

Theorem 10.2.4 *Let $\mathcal{X}$ be a Banach lattice ordered by a fundamental, total, biorthogonal sequence $\{x_n, f_n\}_{n=1}^{\infty}$. Suppose that $A, T \in \mathcal{L}(\mathcal{X})$ are non-zero positive operators such that $AT = TA$ and A is quasinilpotent at a vector $z > 0$. Then T has a non-trivial closed invariant subspace.*

Proof Since $z > 0$ there is an index j with $f_j(z) > 0$, by rescaling z we may assume that $f_j(z) > 1$. Thus $f_j(z - x_j) > 0$ and $f_k(z - x_j) = f_k(z) \geq 0$ for all $k \neq j$, so in fact $z - x_j \geq 0$.

Suppose that $Ax_j = 0$. Then $\operatorname{Ker} A$ is a non-trivial closed T-invariant subspace of $\mathcal{X}$. So we may assume that $Ax_j > 0$. We define a rank-1 projection $P \in \mathcal{L}(\mathcal{X})$ by $Px = f_j(x)x_j$ for $x \in \mathcal{X}$. Evidently, $0 \leq P \leq \mathrm{I}$.

Let $m \geq 0$ and suppose that a is such that $PT^m Ax_j = ax_j$. Then $a \geq 0$ and

$$0 \leq a^n x_j = (PT^m A)^n x_j \leq (T^m A)^n x_j = T^{mn} A^n x_j \leq T^{mn} A^n z,$$

for $n = 1, 2, \ldots$, and so

$$0 \leq a\|x_j\|^{1/n} \leq \|T^{mn}\|^{1/n} \|A^n z\|^{1/n};$$

but since A is quasinilpotent at z, it follows on letting $n \to \infty$ that $a = 0$.

Now let $\mathcal{M} = \operatorname{span}\{T^m Ax_j : m = 0, 1, 2, \ldots\}$. Now $Ax_j \in \mathcal{M}$ and $Px = 0$ for all $x \in \mathcal{M}$, so $x_j \notin \mathcal{M}$. Thus, $\mathcal{M}$ is the required T-invariant subspace. □

We therefore see that every quasinilpotent positive operator on a Banach space ordered by means of a basis has a non-trivial invariant subspace. But we can extend these ideas further, by considering operators represented by matrices with respect to a basis.

Theorem 10.2.5 *Let $\mathcal{X}$ be a Banach lattice ordered by a Schauder basis $\{x_n\}_{n=1}^{\infty}$ with biorthogonal functionals $\{f_n\}_{n=1}^{\infty}$. Suppose that $A \in \mathcal{L}(\mathcal{X})$ is a positive operator that is quasinilpotent at a vector $z > 0$. Let $B \in \mathcal{L}(\mathcal{X})$ be an operator such that $f_k(Bx_l) = 0$ whenever $f_k(Ax_l) = 0$. Then B has a non-trivial closed invariant subspace.*

Note that, in matrix terms, this means that A, B are represented by matrices $\{a_{jk}\}_{j,k=1}^{\infty}$ and $\{b_{jk}\}_{j,k=1}^{\infty}$, where $b_{jk} = 0$ whenever $a_{jk} = 0$.

Proof We adapt the proof of Theorem 10.2.4, finding (without loss of generality) an index j with $0 \le x_j \le z$, and defining $Px = f_j(x)x_j$. Replacing T by A, we have $f_j(A^m x_j) = 0$ for all $m \ge 1$. Now define

$$\mathcal{M} = \text{span}\{Sx_j : S \in \mathcal{L}(\mathcal{X}),\ 0 \le S \le A^m \text{ for some } m \ge 1\}.$$

Then $\mathcal{M}$ is invariant under every operator R with $0 \le R \le A$, and, since $0 \le f_j(Sx_j) \le f_j(A^m x_j) = 0$, we see that $\mathcal{M} \subset \operatorname{Ker} f_j$. Thus $\mathcal{M}$ is a non-trivial closed subspace, invariant under all operators R with $0 \le R \le A$.

For $k, l \ge 1$, let A_{kl} denote the operator defined by taking just the (k, l) entry of A, i.e., such that $A_{kl}x = f_l(x) f_k(A(x_l))x_k$. Then $0 \le A_{kl} \le A$ and so $\mathcal{M}$ is invariant under A_{kl}. Thus it is invariant under the analogously defined operator B_{kl}, and hence invariant under the operator

$$B_n = \sum_{k=1}^{n} \sum_{l=1}^{n} B_{kl}$$

for each $n \ge 1$. However, $\{B_n\}_n$ converges to B in the strong operator topology, and so $\mathcal{M}$ is also a non-trivial invariant subspace for B. □

When we are working in a space that is not ordered by a basis or biorthogonal sequence, results are harder to prove, in general. Rather than give an exhaustive account here, we pick out one or two of the more elegant results in this field.

Definition 10.2.6 Let A and T be two operators on a Banach lattice $\mathcal{X}$ such that $A \ge 0$. Then A is said to *dominate* T if

$$|T(x)| \le A|x| \qquad \text{for all} \quad x \in \mathcal{X}.$$

Theorem 10.2.7 *Let A be a positive operator on a Banach lattice $\mathcal{X}$, which is quasinilpotent at a vector $z \ge 0$ and dominates a non-zero compact operator K. Then any positive operator B that commutes with A has a non-trivial closed invariant subspace.*

Proof Without loss of generality we may take $\|B\| < 1$, and, by rescaling, assume also that $\|K\| = 1$. Let $T = \sum_{n=0}^{\infty} B^n$. This is also a positive operator, and it commutes with both A and B. For each $x > 0$ we write $\mathcal{J}_x = J[Tx]$, the principal ideal generated by Tx, and note that $x \in \mathcal{J}_x$ since $Tx \ge x$.

Now, $\mathcal{J}_x$ is invariant under B, since if $y \in \mathcal{J}_x$, that is, $|y| \le \lambda Tx$ for some $\lambda > 0$, then

$$|By| \leq B|y| \leq \lambda B(Tx) = \lambda \sum_{n=1}^{\infty} B^n x \leq \lambda Tx.$$

We thus have a collection of non-zero closed B-invariant ideals $\overline{\mathcal{J}_x}$, and the remainder of the proof consists in showing that for some $x > 0$ we have $\overline{\mathcal{J}_x} \neq \mathcal{X}$.

Assume, in order to derive a contradiction, that we have

$$\overline{\mathcal{J}_x} = \mathcal{X} \qquad \text{for all} \quad x > 0. \tag{10.1}$$

We begin by considering the ideal $\mathcal{J}_z$. If $Ky = 0$ for each positive $y \in \mathcal{J}_z$, then $K = 0$ on the whole of $\mathcal{J}_z$, which, given the assumed density of $\mathcal{J}_z$, implies that $K = 0$, which is a contradiction. Thus $Kw \neq 0$ for some $w > 0$ in $\mathcal{J}_z$. Now $0 < w < \lambda Tz$ for some $\lambda > 0$; thus

$$\|A^n w\| \leq |\lambda| \|T\| \|A^n z\|,$$

and hence A is quasinilpotent at w, since it is quasinilpotent at z. If necessary, we may replace z by w, in order to suppose without loss of generality that $Kz \neq 0$. Indeed, we may rescale z so that $\|z\| > 1$ and $\|Kz\| > 1$.

Define the closed ball

$$U = \{x \in \mathcal{X} : \|x - z\| \leq 1\},$$

and observe that

$$0 \notin U \qquad \text{and} \qquad 0 \notin \overline{K(U)}. \tag{10.2}$$

Now, since $\overline{J_{|x|}} = \mathcal{X}$ for each $x \neq 0$, by assumption (10.1), the sequence of positive vectors $\{y \wedge nT|x|\}_n$ converges to y for each $y \geq 0$ in $\mathcal{X}$. To see this, note that, by density, given $\varepsilon > 0$ we can find $w \in \mathcal{X}$ with $\|w - y\| < \varepsilon$ and $|w| \leq nT|x|$ for all sufficiently large n. Replacing w by $w \wedge y$ if necessary, we can take $w \leq y$. Then

$$w = w \wedge nT|x| \leq y \wedge nT|x| \leq y,$$

and it now follows easily that $\|y - (y \wedge nT|x|)\| < \varepsilon$.

So, for each $x \neq 0$ there is an n such that $z - z \wedge nT|x|$ has norm strictly less than 1. We thus have

$$\overline{K(U)} \subset \bigcup_{n=1}^{\infty} \{x \in \mathcal{X} : \|z - z \wedge nT|x|\| < 1\},$$

an increasing union of open sets. But $\overline{K(U)}$ is compact, and so there exists a positive integer N such that $z \wedge NT|x| \in U$ for all $x \in \overline{K(U)}$.

We now define a sequence $\{z_n\}_{n\geq 0}$ in $\mathcal{X}$ by setting $z_0 = z$ and

$$z_{n+1} = z \wedge NT|Kz_n| \qquad \text{for} \quad n \geq 0.$$

Inductively, it follows that $z_n \geq 0$ and $z_n \in U$ for each n. Moreover, we have

$$0 \leq z_n \leq N^n T^n A^n z$$

for each n. This also follows inductively, for the case $n = 0$ is clear, and if $0 \leq z_n \leq N^n T^n A^n z$, then, since A dominates K, we have

$$0 \leq z_{n+1} = z \wedge NT|Kz_n| \leq NT|Kz_n| \leq NT(Az_n) \leq N^{n+1}T^{n+1}A^{n+1}z,$$

as required. Estimating norms, we have $\|z_n\|^{1/n} \leq N\|T\|\|A^n z\|^{1/n}$; recalling that A is quasinilpotent at z we see that $\|z_n\| \to 0$. This is a contradiction, as each z_n lies in U and hence $0 \in \overline{U} = U$, in contradiction of (10.2). The result now follows. □

We now mention briefly one further notion, which seems to be well-adapted to the subject of invariant subspaces for positive operators.

Definition 10.2.8 Let $\mathcal{X}$ be a Banach lattice. A positive operator $B \in \mathcal{L}(\mathcal{X})$ is *compact-friendly* if there is a non-zero operator $T \in \mathcal{L}(\mathcal{X})$ that is dominated simultaneously by a positive operator A in the commutant of B and a non-zero positive compact operator K; that is, $AB = BA$ and $|T(x)| \leq A(|x|)$, $|T(x)| \leq K(|x|)$ for all $x \in \mathcal{X}$.

Thus the class of compact-friendly operators is fairly wide, and includes (i) all positive operators that dominate non-zero compact positive operators, and (ii) all positive operators that are dominated by compact operators.

We conclude with the following result, which is one of the more easily stated in this area.

Theorem 10.2.9 *Let $\mathcal{X}$ be a Banach lattice, and let $B \in \mathcal{L}(\mathcal{X})$ be a compact-friendly operator that is quasinilpotent at some vector $z > 0$. Then B has a non-trivial closed invariant ideal, and if $C \in \mathcal{L}(\mathcal{X})$ is another operator commuting with B, then B and C have a common non-trivial closed invariant ideal.*

We omit the proof, which may be found in Abramovich, Aliprantis and Burkinshaw [3].

10.3 Strictly singular operators

An important generalization of compact operators is the notion of a strictly singular operator.

Definition 10.3.1 An operator $T : \mathcal{X} \to \mathcal{Y}$ is said to be *strictly singular (SS)* if for each $\varepsilon > 0$ and for each infinite-dimensional subspace $\mathcal{Z}$ of $\mathcal{X}$ there is an element $z \in \mathcal{Z}$ with $\|z\| = 1$ and $\|Tz\| < \varepsilon$.

The operator T is said to be *finitely strictly singular (FSS)* if for each $\varepsilon > 0$ there is a positive integer N such that each subspace $\mathcal{Z}$ of dimension greater than N contains an element $z \in \mathcal{Z}$ with $\|z\| = 1$ and $\|Tz\| < \varepsilon$.

Thus T is strictly singular if and only if its restriction to every infinite-dimensional subspace fails to be an isomorphism. Clearly FSS implies SS.

Theorem 10.3.2 *Every compact operator $T : \mathcal{X} \to \mathcal{Y}$ is finitely strictly singular, and hence strictly singular.*

Proof Since T is compact, then the image of the closed unit ball $\mathcal{B}$ of $\mathcal{X}$ is totally bounded. Thus, given $\varepsilon > 0$, there are elements $y_1, \dots, y_N$ in $\mathcal{Y}$ such that for each $x \in \mathcal{B}$ there is an index k such that $\|Tx - y_k\| < \varepsilon/2$.

By the Hahn–Banach theorem, there exist support functionals $f_1, \dots, f_N$ in $\mathcal{Y}^*$ such that $\|f_j\| = 1$ and $f_j(y_j) = \|y_j\|$ for each $j = 1, 2, \dots, N$.

Suppose that $\mathcal{Z} \subset \mathcal{X}$ and $\dim \mathcal{Z} > N$. Then $\mathcal{Z}$ contains a norm-1 vector z such that $f_j(Tz) = 0$ for all $j = 1, \dots, N$. But now if k is an index such that $\|Tz - x_k\| < \varepsilon/2$, we have

$$\|x_k\| = |f_k(x_k)| \leq |f_k(Tz)| + \|f_k\| \, \|Tz - x_k\| < \varepsilon/2,$$

and hence

$$\|Tz\| \leq \|x_k\| + \|Tz - x_k\| < \varepsilon.$$

Thus T is FSS and hence SS. □

Theorem 10.3.3 *Suppose that $1 < p < \infty$. Then for operators $T \in \mathcal{L}(\ell^p)$ the three notions of compactness, FSS and SS are equivalent.*

Proof Clearly it is sufficient to show that all non-compact operators T on ℓ^p fail to be strictly singular.

If T is an operator on ℓ^p that is not compact, then there is an $\varepsilon > 0$ and a sequence $(x_n)_n$ tending weakly to zero, such that $\|Tx_n\| \geq \varepsilon$ for all integers n. For if the image of the unit ball $\mathcal{B}$ is not relatively compact, we may find a sequence $(y_n)_n$ in $\mathcal{B}$ such that $\|Ty_m - Ty_n\| \geq \varepsilon$ for all $m \neq n$. By passing to a subsequence, we may assume that (y_n) converges weakly to a limit. The sequence $x_n = y_n - y_{n+1}$ then has the required properties.

By passing to a subsequence and renormalizing, we may further suppose that $\|x_n\| = 1$ for each n, and $\sum_n \|x_n - z_n\| < 1$, where $(z_n)_n$ is a normalized sequence in ℓ^p supported on disjoint blocks of the standard basis of ℓ^p (note that (z_n) spans a space isometric to ℓ^p again).

Then

$$\left\| \sum_n a_n x_n - \sum_n a_n z_n \right\| < \max |a_n| \leq \left\| \sum a_n z_n \right\|$$

for every sequence $(a_n)_n$ in ℓ^p. Thus the mapping U sending each sum $\sum_n a_n z_n$ to $\sum_n a_n x_n$ is an isomorphism, since $\|I - U\| < 1$.

But we can make the same construction simultaneously with (Tx_n) (which is also weakly null) if we pass to a further subsequence, so that we then have T an isomorphism on an infinite-dimensional space. Hence T cannot be strictly singular. □

Remark 10.3.4 The notions of compactness and strict singularity do not coincide in general, however. For example, the canonical embedding from ℓ^p into ℓ^q for $1 \leq p < q < \infty$ is strictly singular but not compact. Hence one may construct an operator on $\ell^p \oplus \ell^q$ that is strictly singular but not compact. We leave the details as an exercise (one may argue by using block bases, as in the proof of Theorem 10.3.3).

Remark 10.3.5 On general Banach spaces, even finitely strictly singular operators need not have invariant subspaces (see Read [172] and Chalendar *et al.* [65] for a counterexample). Nonetheless, the spectral properties of strictly singular operators are very similar to those of compact operators: the spectrum is countable, it can only accumulate at 0, and every non-zero spectral point is an eigenvalue of finite multiplicity. We omit the proof of this fact, which is surprisingly long and complicated for such an apparently straightforward result, and refer the interested reader to Abramovich and Aliprantis [1] (or to the original source, Kato [130]).

Theorem 10.3.6 *Suppose that $T \in \mathcal{L}(\mathcal{X})$ is not quasinilpotent, but that some power of T is strictly singular. Then T has a non-trivial finite-dimensional hyperinvariant subspace.*

Proof We suppose that T^n is strictly singular. If $\mathcal{X}$ is a complex Banach space, then T^n has non-trivial (finite-dimensional) eigenspaces, by Remark 10.3.5, say $\text{Ker}(T^n - \lambda \text{I})$. Such an eigenspace is a hyperinvariant subspace for T, since if $ST = TS$ then $ST^n = T^n S$, and so if $(T^n - \lambda\text{I})x = 0$ we have $(T^n - \lambda\text{I})Sx = 0$.

If $\mathcal{X}$ is merely a vector space over $\mathbb{R}$, we work with the complexification of $\mathcal{X}$, i.e., $\widetilde{\mathcal{X}} = \mathcal{X} \oplus \mathcal{X}$, which is a Banach space over $\mathbb{C}$ if we define

$$(a+ib)(x,y) = (ax - by, bx + ay), \qquad a, b \in \mathbb{R}, \quad x, y \in \mathcal{X},$$

and equip it with the norm

$$\|(x,y)\| = \sup_{\theta \in [0,2\pi]} \|(\cos\theta)x + (\sin\theta)y\|.$$

We define $\widetilde{T} \in \mathcal{L}(\widetilde{\mathcal{X}})$ by $\widetilde{T}(x,y) = (Tx, Ty)$ and note that $\widetilde{T}$ is still strictly singular. Taking a non-trivial finite-dimensional eigenspace $\mathcal{Y}$ of $\widetilde{T}$, spanned by $\{(x_k, y_k) : k = 1, \ldots, m\}$, say, we write $\mathcal{Z} = \operatorname{span}\{x_1, \ldots, x_m, y_1, \ldots, y_m\}$, a subspace of $\mathcal{X}$. Now if $ST = TS$ then $\widetilde{S}\widetilde{T} = \widetilde{T}\widetilde{S}$, and so $\widetilde{S}(x_k, y_k) \in \mathcal{Y}$ for each k. Thus Sx_k and Sy_k both lie in $\mathcal{Z}$, and so $\mathcal{Z}$ is an invariant subspace for S, and hence a hyperinvariant subspace for T. □

It seems that it is still unknown, without further hypotheses on the Banach space $\mathcal{X}$, whether a quasinilpotent strictly singular operator in $\mathcal{L}(\mathcal{X})$ necessarily has a non-trivial closed invariant subspace, even if we suppose that it is a positive operator on an ordered Banach space.

Exercises

1. Write down operators T_1 and T_2 on the two-dimensional Hilbert space $\mathbb{C}^2$ with standard basis $\{e_1, e_2\}$, such that T_1 and T_2 have real matrix entries; T_1 is a positive operator in the sense of this chapter, but is not even self-adjoint; and T_2 is represented by a positive definite matrix, but $T_2(e_1)$ is not a positive vector.
2. Consider the operator T on ℓ^p (for $1 \le p < \infty$) defined by the matrix

$$\begin{pmatrix} 1 & 0 & 0 & 0 & 0 & \cdots \\ 0 & 0 & 0 & 0 & 0 & \cdots \\ 0 & \frac{1}{2} & 0 & 0 & 0 & \cdots \\ 0 & 0 & \frac{1}{3} & 0 & 0 & \cdots \\ 0 & 0 & 0 & \frac{1}{4} & 0 & \cdots \\ \vdots & \vdots & \vdots & \vdots & \vdots & \ddots \end{pmatrix}$$

with respect to the canonical basis $\{e_n\}_{n=1}^{\infty}$ of ℓ^p. Show that T is not quasinilpotent, but that $\|T^n e_2\|^{1/n} \to 0$; that is, T is quasinilpotent at e_2.
3. Show that $L^2(0,1)$ does not have a fundamental, total, biorthogonal system $\{x_n, f_n\}_{n=1}^{\infty}$ such that the functions x_n are positive in the usual sense.

4. Suppose that the singular values of a Hilbert space operator T do not tend to zero. Show that for some $\varepsilon > 0$ we may construct a sequence $\{x_n\}_{n\geq 1}$ such that $\|x_n\| = 1$, $\|Tx_n\| \geq \varepsilon$, while $x_n \perp x_j$ and $x_n \perp T^*Tx_j$ for all $j < n$.
5. Using Exercise 4, show directly that if T is a strictly singular operator on a Hilbert space, then the singular values of T tend to zero, and hence T is compact.
6. Show directly that every compact operator is strictly singular, without using FSS.
7. Verify the claims in Remark 10.3.4.
8. Show that the space $\widetilde{\mathcal{X}}$, as defined in the proof of Theorem 10.3.6, is indeed a complex Banach space, and that $\widetilde{T}$ is strictly singular when T is.

Comments

The general background to Banach spaces with an order structure is covered by many books with titles containing phrases such as 'Banach lattices', 'ordered Banach spaces', 'positive operators' or 'Riesz spaces'. We mention as examples Aliprantis and Burkinshaw [5], Becker [30], Schaefer [181] and Zaanen [204].

It was shown by Enflo [94] that there are separable Banach spaces with no basis (see Davie [88] for a much simplified treatment). The existence of fundamental, total, biorthogonal systems was shown by Markushevich. In fact for any $\varepsilon > 0$ these systems can be chosen to satisfy $\sup \|x_n\| \, \|f_n\| < 1 + \varepsilon$ (see Ovsepian and Pełczyński [159] and Pełczyński [164]). For an excellent account of bases and biorthogonal systems the reader is referred to the classic book by Lindenstrauss and Tzafriri [142].

Theorem 10.2.4 is due to Ercan and Onal [99], and provides an extension of the results of Abramovich, Aliprantis and Burkinshaw [4], which deal with the case of a Banach space with a basis. Theorem 10.2.5 is taken from Abramovich, Aliprantis and Burkinshaw [4].

Theorem 10.2.7 was given by Abramovich, Aliprantis and Burkinshaw [3], and may also be found in the book by Abramovich and Aliprantis [1]. Its proof is said to be inspired by Hilden's proof of the Lomonosov invariant-subspace theorem, as presented by Michaels [152]. It would be interesting to see a proof using minimal vectors.

Exercise 2 concerning quasinilpotence at a point is a slight simplification of an example from Abramovich, Aliprantis and Burkinshaw [2].

Our discussion of strictly singular operators (including some of the Exercises) draws on ideas to be found in Lindenstrauss and Tzafriri [142]. The first author to show that compact operators and strictly singular operators coincide on Hilbert space seems to have been Kato [130].

Theorem 10.3.6 is given as a lemma in Flores, Tradacete and Troitsky [108]. The ideas of Section 10.2 and 10.3 may be combined in the notion of strictly-singular-friendly operators, which are defined analogously to compact-friendly operators. Results on these may be found in the same paper.

References

[1] Y. A. Abramovich and C. D. Aliprantis. *An Invitation to Operator Theory*. Graduate Studies in Mathematics, 50. American Mathematical Society, Providence, RI, 2002.

[2] Y. A. Abramovich, C. D. Aliprantis, and O. Burkinshaw. Invariant subspaces of operators on l_p-spaces. *J. Funct. Anal.*, **115**(2):418–424, 1993.

[3] Y. A. Abramovich, C. D. Aliprantis, and O. Burkinshaw. Invariant subspace theorems for positive operators. *J. Funct. Anal.*, **124**(1):95–111, 1994.

[4] Y. A. Abramovich, C. D. Aliprantis, and O. Burkinshaw. Invariant subspaces for positive operators acting on a Banach space with basis. *Proc. Amer. Math. Soc.*, **123**(6):1773–1777, 1995.

[5] C. D. Aliprantis and O. Burkinshaw. *Positive Operators*. Springer, Dordrecht, 2006. Reprint of the 1985 original.

[6] G. R. Allan. Sums of idempotents and a lemma of N. J. Kalton. *Studia Math.*, **121**:185–192, 1996.

[7] D. Alpay, L. Baratchart, and J. Leblond. Some extremal problems linked with identification from partial frequency data. In R. F. Curtain, A. Bensoussan, and J.-L. Lions, eds., *Analysis and Optimization of Systems: State and Frequency Domain Approaches for Infinite-Dimensional Systems (Sophia-Antipolis, 1992)*, pages 563–573. Springer, Berlin, 1993.

[8] C. Ambrozie and V. Müller. Invariant subspaces for polynomially bounded operators. *J. Funct. Anal.*, **213**(2):321–345, 2004.

[9] G. Androulakis. A note on the method of minimal vectors. In *Trends in Banach Spaces and Operator Theory (Memphis, TN, 2001)*, pages 29–36. Contemporary Mathematics, 321. American Mathematical Society, Providence, RI, 2003.

[10] G. Androulakis and A. Flattot. Hyperinvariant subspaces for weighted composition operators on $L^p([0, 1]^d)$. *J. Operator Theory*, in press.

[11] S. Ansari and P. Enflo. Extremal vectors and invariant subspaces. *Trans. Amer. Math. Soc.*, **350**(2):539–558, 1998.

[12] C. Apostol. Ultraweakly closed operator algebras. *J. Operator Theory*, **2**:49–61, 1979.

[13] C. Apostol. Functional calculus and invariant subspaces. *J. Operator Theory*, **4**:159–190, 1980.

[14] C. Apostol, H. Bercovici, C. Foias, and C. Pearcy. Invariant subspaces, dilation theory, and the structure of the predual of a dual algebra. I. *J. Funct. Anal.*, **63**(3):369–404, 1985.

[15] C. Apostol, H. Bercovici, C. Foias, and C. Pearcy. Invariant subspaces, dilation theory, and the structure of the predual of a dual algebra. II. *Indiana Univ. Math. J.*, **34**(4):845–855, 1985.

[16] C. Apostol, L. A. Fialkow, D. A. Herrero, and D. Voiculescu. *Approximation of Hilbert Space Operators, Volume II*. Research Notes in Mathematics, 102. Pitman (Advanced Publishing Program), Boston, MA, 1984.

[17] S. A. Argyros and R. G. Haydon. A hereditarily indecomposable L_∞-space that solves the scalar-plus-compact problem. Preprint, 2009 arXiv:0903.3921.

[18] N. Aronszajn and K. T. Smith. Invariant subspaces of completely continuous operators. *Ann. Math. (2)*, **60**:345–350, 1954.

[19] A. Atzmon. An operator without invariant subspaces on a nuclear Fréchet space. *Ann. Math.*, **117**:660–694, 1983.

[20] A. Atzmon. On the existence of hyperinvariant subspaces. *J. Operator Theory*, **11**(1):3–40, 1984.

[21] A. Atzmon and G. Godefroy. An application of the smooth variational principle to the existence of non-trivial invariant subspaces. *C. R. Acad. Sci. Paris Sér. I Math.*, **332**(2):151–156, 2001.

[22] A. Atzmon, G. Godefroy, and N. J. Kalton. Invariant subspaces and the exponential map. *Positivity*, **8**(2):101–107, 2004.

[23] L. Baratchart and J. Leblond. Hardy approximation to L^p functions on subsets of the circle with $1 \leq p < \infty$. *Constr. Approx.*, **14**(1):41–56, 1998.

[24] S. Barclay. Banach spaces of analytic vector-valued functions. PhD thesis, University of Leeds, 2008.

[25] S. Barclay. A solution to the Douglas–Rudin problem for matrix-valued functions. *Proc. London Math. Soc.*, **99**:757–786, 2009.

[26] F. Bayart and É. Matheron. *Dynamics of Linear Operators*. Cambridge Tracts in Mathematics, 179. Cambridge University Press, Cambridge, 2009.

[27] B. Beauzamy. *Introduction to Banach Spaces and Their Geometry*. Notas de Matemática [Mathematical Notes], 86. North-Holland, Amsterdam, 1982.

[28] B. Beauzamy. Un opérateur sans sous-espace invariant: Simplification de l'exemple d'Enflo. *Integr. Equat. Oper. Th.*, **8**(3):314–384, 1985.

[29] B. Beauzamy. *Introduction to Operator Theory and Invariant Subspaces*. North-Holland Mathematical Library, 42. North-Holland, Amsterdam, 1988.

[30] R. Becker. *Ordered Banach Spaces*. Travaux en Cours [Works in Progress], 68. Hermann Éditeurs des Sciences et des Arts, Paris, 2008. With a preface by Gilles Godefroy.

[31] H. Bercovici. Factorization theorems and the structure of operators on Hilbert space. *Ann. Math.*, **128**:399–413, 1988.

[32] H. Bercovici. *Operator Theory and Arithmetic in H^∞*. Mathematical Surveys and Monographs, 26. American Mathematical Society, Providence, RI, 1988.

[33] H. Bercovici. Notes on invariant subspaces. *Bull. Amer. Math. Soc.*, **23**(1):1–36, 1990.

[34] H. Bercovici, C. Foias, and C. Pearcy. Dilation theory and systems of simultaneous equations in the predual of an operator algebra, I. *Michigan Math. J.*, **30**:335–354, 1983.

[35] H. Bercovici, C. Foias, and C. Pearcy. *Dual Algebras With Applications to Invariant Subspaces and Dilation Theory*. CBMS Regional Conference Series in Mathematics, 56. American Mathematical Society, Providence, RI, 1985.

[36] H. Bercovici, C. Foias, and C. Pearcy. Two Banach space methods and dual operator algebras. *J. Funct. Anal.*, **78**:306–345, 1988.

[37] C. A. Berenstein and R. Gay. *Complex Variables*. Graduate Texts in Mathematics, 125. Springer, New York, 1991.

[38] C. A. Berger and J. G. Stampfli. Mapping theorems for the numerical range. *Amer. J. Math.*, **89**:1047–1055, 1967.

[39] A. R. Bernstein and A. Robinson. Solution of an invariant subspace problem of K. T. Smith and P. R. Halmos. *Pac. J. Math.*, **16**:421–431, 1966.

[40] E. Bishop and R. R. Phelps. A proof that every Banach space is subreflexive. *Bull. Amer. Math. Soc.*, **67**:97–98, 1961.

[41] E. Bishop and R. R. Phelps. The support functionals of a convex set. In *Convexity: Proc. Sympos. Pure Math., Vol. VII*, pages 27–35. American Mathematical Society, Providence, RI, 1963.

[42] D. P. Blecher and A. M. Davie. Invariant subspaces for an operator on $L^2(\Pi)$ composed of a multiplication and a translation. *J. Operator Theory*, **23**(1):115–123, 1990.

[43] R. P. Boas. *Entire Functions*. Academic Press, New York, 1954.

[44] B. Bollobás. An extension to the theorem of Bishop and Phelps. *Bull. London Math. Soc.*, **2**:181–182, 1970.

[45] F. F. Bonsall. Decompositions of functions as sums of elementary functions. *Quart. J. Math. Oxford Ser. (2)*, **37**(146):129–136, 1986.

[46] F. F. Bonsall and J. Duncan. *Numerical Ranges, II*. London Mathematical Society Lecture Notes Series, 10. Cambridge University Press, New York, 1973.

[47] P. S. Bourdon and A. Flattot. Images of minimal-vector sequences under weighted composition operators on $L^2(\mathbb{D})$. In J. A. Ball, V. Bolotnikov, J. W. Helton, L. Rodman, and I. M. Spitkovsky, eds., *Topics in Operator Theory, Volume 1: Operators, Matrices and Analytic Functions*, pages 39–52. Operator Theory: Advances and Applications, 202. Birkhäuser, Basel, 2010.

[48] J. Bourgain. A problem of Douglas and Rudin on factorization. *Pac. J. Math.*, **121**(1):47–50, 1986.

[49] L. Brown, A. Shields, and K. Zeller. On absolutely convergent exponential sums. *Trans. Amer. Math. Soc.*, **96**:162–183, 1960.

[50] S. Brown. Some invariant subspaces for subnormal operators. *Integr. Equat. Oper. Th.*, **1**:310–333, 1978.

[51] S. Brown. Contractions with spectral boundary. *Integr. Equat. Oper. Th.*, **11**:49–63, 1988.

[52] S. Brown and B. Chevreau. Toute contraction à calcul fonctionnel isométrique est réflexive. *C. R. Acad. Sci. Paris Sér. I Math.*, **307**:185–188, 1988.

[53] S. Brown, B. Chevreau, and C. Pearcy. Contractions with rich spectrum have invariant subspaces. *J. Operator Theory*, **1**(1):123–136, 1979.

[54] S. Brown, B. Chevreau, and C. Pearcy. On the structure of contraction operators, II. *J. Funct. Anal.*, **76**(1):30–55, 1988.

[55] S. R. Caradus. Universal operators and invariant subspaces. *Proc. Amer. Math. Soc.*, **23**:526–527, 1969.

[56] L. Carleson. An interpolation problem for bounded analytic functions. *Amer. J. Math.*, **80**:921–930, 1958.

[57] G. Cassier and I. Chalendar. The group of the invariants of a finite Blaschke product. *Complex Variables*, **42**:193–206, 2000.

[58] G. Cassier, I. Chalendar, and B. Chevreau. New examples of contractions illustrating membership and non-membership in the classes $\mathbb{A}_{n,m}$. *Acta Sci. Math. (Szeged)*, **64**:707–731, 1998.

[59] G. Cassier, I. Chalendar, and B. Chevreau. Some mapping theorems for the classes $\mathbb{A}_{n,m}$ and boundary sets. *Proc. London Math. Soc.*, **79**(1):222–240, 1999.

[60] G. Cassier and T. Fack. Contractions in von Neumann algebras. *J. Funct. Anal.*, **135**(2):297–338, 1996.

[61] I. Chalendar. The operator-valued Poisson kernel and its application. *Irish Math. Soc. Bull.*, **51**:21–44, 2003.

[62] I. Chalendar and J. Esterle. L^1-factorization for C_{00}-contractions with isometric functional calculus. *J. Funct. Anal.*, **154**:174–194, 1998.

[63] I. Chalendar, A. Flattot, and N. Guillotin-Plantard. On the spectrum of multivariable weighted composition operators. *Arch. Math. (Basel)*, **90**(4) 353–359, 2008.

[64] I. Chalendar, A. Flattot, and J. R. Partington. The method of minimal vectors applied to weighted composition operators. In *The Extended Field of Operator Theory*, pages 89–105. Operator Theory: Advances and Applications, 171. Birkhäuser, Basel, 2007.

[65] I. Chalendar, E. Fricain, A. I. Popov, D. Timotin, and V. G. Troitsky. Finitely strictly singular operators between James spaces. *J. Funct. Anal.*, **256**(4):1258–1268, 2009.

[66] I. Chalendar and F. Jaeck. On the contractions in the classes $\mathbb{A}_{n,m}$. *J. Operator Theory*, **38**(2):265–296, 1997.

[67] I. Chalendar, J. Leblond, and J. R. Partington. Approximation problems in some holomorphic spaces, with applications. In *Systems, Approximation, Singular Integral Operators, and Related Topics (Bordeaux, 2000)*, pages 143–168. Operator Theory: Advances and Applications, 129. Birkhäuser, Basel, 2001.

[68] I. Chalendar and J. R. Partington. L^1 factorizations for some perturbations of the unilateral shift. *C. R. Acad. Sci. Paris Sér. I Math.*, **332**:115–119, 2001.

[69] I. Chalendar and J. R. Partington. Constrained approximation and invariant subspaces. *J. Math. Anal. Appl.*, **280**(1):176–187, 2003.

[70] I. Chalendar and J. R. Partington. On the structure of invariant subspaces for isometric composition operators on $H^2(\mathbb{D})$ and $H^2(\mathbb{C}_+)$. *Arch. Math. (Basel)*, **81**(2):193–207, 2003.

[71] I. Chalendar and J. R. Partington. Spectral density for multiplication operators with applications to factorization of L^1 functions. *J. Operator Theory*, **50**(2):411–422, 2003.

[72] I. Chalendar and J. R. Partington. Convergence properties of minimal vectors for normal operators and weighted shifts. *Proc. Amer. Math. Soc.*, **133**(2):501–510, 2005.

[73] I. Chalendar and J. R. Partington. Variations on Lomonosov's theorem via the technique of minimal vectors. *Acta Sci. Math. (Szeged)*, **71**(3-4):603–617, 2005.

[74] I. Chalendar and J. R. Partington. Invariant subspaces for products of Bishop operators. *Acta Sci. Math. (Szeged)*, **74**(3-4):719–727, 2008.

[75] I. Chalendar, J. R. Partington, and E. Pozzi. Multivariable weighted composition operators: Point spectrum and cyclic vectors. In J. A. Ball, V. Bolotnikov, J. W. Helton, L. Rodman, and I. M. Spitkovsky, eds., *Topics in Operator Theory, Volume 1: Operators, Matrices and Analytic Functions*, pages 63–85. Operator Theory: Advances and Applications, 202. Birkhäuser, Basel, 2010.

[76] I. Chalendar, J. R. Partington, and M. Smith. Approximation in reflexive Banach spaces and applications to the invariant subspace problem. *Proc. Amer. Math. Soc.*, **132**(4):1133–1142, 2004.

[77] I. Chalendar, J. R. Partington, and R. C. Smith. L^1 factorizations, moment problems and invariant subspaces. *Studia Math.*, **167**(2):183–194, 2005.

[78] B. Chevreau. Sur les contractions à calcul fonctionnel isométrique, II. *J. Operator Theory*, **20**:269–293, 1988.

[79] B. Chevreau and C. Pearcy. Growth conditions on the resolvent and membership in the classes $\mathbf{A}$ and $\mathbf{A}_{\aleph_0}$. *J. Operator Theory*, **16**(2):375–385, 1986.

[80] B. Chevreau and C. Pearcy. On the structure of contraction operators, I. *J. Funct. Anal.*, **76**:1–29, 1988.

[81] V. Chkliar. Eigenfunctions of the hyperbolic composition operator. *Integr. Equat. Oper. Th.*, **29**(3):364–367, 1997.

[82] I. Colojoară and C. Foiaş. *Theory of Generalized Spectral Operators*. Mathematics and its Applications, 9. Gordon and Breach Science Publishers, New York, 1968.

[83] J. B. Conway. *The Theory of Subnormal Operators*. Mathematical Surveys and Monographs, 36. American Mathematical Society, Providence, RI, 1991.

[84] J. B. Conway. *A Course in Operator Theory*. Graduate Studies in Mathematics, 21. American Mathematical Society, Providence, RI, 2000.

[85] C. C. Cowen and B. D. MacCluer. *Composition Operators on Spaces of Analytic Functions*. Studies in Advanced Mathematics. CRC Press, Boca Raton, FL, 1995.

[86] R. E. Curto and F. H. Vasilescu. Automorphism invariance of the operator-valued Poisson transform. *Acta Sci. Math. (Szeged)*, **57**:65–78, 1993.

[87] H. G. Dales. *Banach Algebras and Automatic Continuity*. London Mathematical Society Monographs, New Series, 24. Clarendon Press, Oxford University Press, New York, 2000.

[88] A. M. Davie. The approximation problem for Banach spaces. *Bull. London Math. Soc.*, **5**:261–266, 1973.

[89] A. M. Davie. Invariant subspaces for Bishop's operators. *Bull. London Math. Soc.*, **6**:343–348, 1974.

[90] L. de Branges and J. Rovnyak. Canonical models in quantum scattering theory. In *Perturbation Theory and its Applications in Quantum Mechanics: Proceedings of an Advanced Seminar Conducted by The Mathematics Research Center,*

U.S. Army, and the Theoretical Chemistry Institute at the University of Wisconsin, Madison, WI, October 4–6, 1965, pages 295–392. Wiley, New York, 1966.

[91] R. Deville, G. Godefroy, and V. Zizler. A smooth variational principle with applications to Hamilton-Jacobi equations in infinite dimensions. *J. Funct. Anal.*, **111**(1):197–212, 1993.

[92] R. Deville, G. Godefroy, and V. Zizler. *Smoothness and Renormings in Banach Spaces*. Pitman Monographs and Surveys in Pure and Applied Mathematics, 64. Longman Scientific & Technical, Harlow, 1993.

[93] Y. Domar. On the analytic transform of bounded linear functionals on certain Banach algebras. *Studia Math.*, **53**(3):203–224, 1975.

[94] P. Enflo. A counterexample to the approximation problem in Banach spaces. *Acta Math.*, **130**:309–317, 1973.

[95] P. Enflo. On the invariant subspace problem in Banach spaces. In *Séminaire Maurey–Schwartz (1975–1976): Espaces L^p, Applications Radonifiantes et Géométrie des Espaces de Banach, Exp. Nos. 14–15*. Centre de Mathématiques, École de Polytechnique, Palaiseau, 1976.

[96] P. Enflo. On the invariant subspace problem in Banach spaces. *Acta Math.*, **158**:213–313, 1987.

[97] P. Enflo. Extremal vectors for a class of linear operators. In *Functional Analysis and Economic Theory (Samos, 1996)*, pages 61–64. Springer, Berlin, 1998.

[98] P. Enflo and T. Hõim. Some results on extremal vectors and invariant subspaces. *Proc. Amer. Math. Soc.*, **131**(2):379–387, 2003.

[99] Z. Ercan and S. Onal. Invariant subspaces for positive operators acting on a Banach space with Markushevich basis. *Positivity*, **8**(2):123–126, 2004.

[100] J. Eschmeier. Representations of $H^\infty(G)$ and invariant subspaces. *Math. Ann.*, **298**(1):167–186, 1994.

[101] J. Eschmeier. Invariant subspaces for spherical contractions. *Proc. London Math. Soc.*, **75**:157–176, 1997.

[102] J. Esterle and M. Zarrabi. Local properties of powers of operators. *Arch. Math. (Basel)*, **65**:53–60, 1995.

[103] G. Exner and Il Bong Jung. Dual operator algebras and contractions with finite defect indices. *J. Operator Theory*, **36**(1):107–119, 1996.

[104] C. Fefferman. Characterizations of bounded mean oscillation. *Bull. Amer. Math. Soc.*, **77**:587–588, 1971.

[105] C. Fefferman and E. M. Stein. H^p spaces of several variables. *Acta Math.*, **129**(3-4):137–193, 1972.

[106] M. Fekete. Über die verteilung der Wurzeln bei gewissen algebraischen Gleichungen mit ganzzahligen Koeffizienten. *Mathematische Zeitschrift*, **17**:228–249, 1923.

[107] A. Flattot. Hyperinvariant subspaces for Bishop-type operators. *Acta Sci. Math. (Szeged)*, **74**:687–716, 2008.

[108] J. Flores, P. Tradacete, and V. G. Troitsky. Invariant subspaces of positive strictly singular operators on Banach lattices. *J. Math. Anal. Appl.*, **343**(2):743–751, 2008.

[109] C. Foiaş. Sur certains théorèmes de J. von Neumann concernant les ensembles spectraux. *Acta Sci. Math. (Szeged)*, **18**:15–20, 1957.

[110] E. Gallardo-Gutiérrez and P. Gorkin. Minimal invariant subspaces for composition operators. *J. Math. Pure. Appl.*, **95**(3):245–259, 2011.

[111] E. Gallardo-Gutiérrez, P. Gorkin, and D. Suárez. Orbits of non-elliptic disc automorphisms on $\mathcal{H}^p$. Preprint, 2010 arXiv:1002.3833.

[112] J. B. Garnett. *Bounded Analytic Functions.* Graduate Texts in Mathematics, 236. Springer, New York, 2007.

[113] R. Gellar. Operators commuting with a weighted shift. *Proc. Amer. Math. Soc.*, **23**:538–545, 1969.

[114] W. T. Gowers and B. Maurey. Banach spaces with small spaces of operators. *Math. Ann.*, **307**(4):543–568, 1997.

[115] S. Grivaux. Invariant subspaces for tridiagonal operators. *Bull. Sci. Math.*, **126**(8):681–691, 2002.

[116] P. R. Halmos. Spectra and spectral manifolds. *Ann. Soc. Polon. Math.*, **25**:43–49, 1952.

[117] P. R. Halmos. Invariant subspaces of polynomially compact operators. *Pac. J. Math.*, **16**:433–437, 1966.

[118] P. R. Halmos. *A Hilbert Space Problem Book.* Graduate Texts in Mathematics, 19, 2nd edition. Springer, New York, 1982.

[119] G. H. Hardy and E. M. Wright. *An Introduction to the Theory of Numbers*, 5th edition. Clarendon Press, Oxford University Press, New York, 1979.

[120] H. Hedenmalm. Maximal invariant subspaces in the Bergman space. *Ark. Mat.*, **36**(1):97–101, 1998.

[121] E. Heinz. Ein v. Neumannscher Satz über beschränkte Operatoren im Hilbertschen Raum. *Nachr. Akad. Wiss. Göttingen. Math.-Phys. Kl. IIa., Math.-Phys.-Chem. Abt.*, 5–6, 1952.

[122] D. A. Herrero. *Approximation of Hilbert Space Operators, Volume 1.* Pitman Research Notes in Mathematics Series, 224, 2nd edition. Longman Scientific & Technical, Harlow, 1989.

[123] E. Hille and R. S. Phillips. *Functional Analysis and Semi-Groups.* American Mathematical Society Colloquium Publications, 31. American Mathematical Society, Providence, RI, 1957.

[124] J. A. R. Holbrook. On the power-bounded operators of Sz.-Nagy and Foiaş. *Acta Sci. Math. (Szeged)*, **29**:299–310, 1968.

[125] N. D. Hooker. Lomonosov's hyperinvariant subspace theorem for real spaces. *Math. Proc. Cambridge Philos. Soc.*, **89**(1):129–133, 1981.

[126] F. John and L. Nirenberg. On functions of bounded mean oscillation. *Comm. Pure Appl. Math.*, **14**:415–426, 1961.

[127] Il Bong Jung. Dual operator algebras and the classes $\mathbb{A}_{m,n}$, I. *J. Operator Theory*, **27**:309–323, 1992.

[128] S. Kakutani. Topological properties of the unit sphere of a Hilbert space. *Proc. Imp. Acad. Tokyo*, **19**:269–271, 1943.

[129] N. J. Kalton. Sums of idempotents in Banach algebras. *Can. Math. Bull.*, **31**(4): 448–451, 1988.

[130] T. Kato. Perturbation theory for nullity, deficiency and other quantities of linear operators. *J. Anal. Math.*, **6**:261–322, 1958.

[131] H. J. Kim. Hyperinvariant subspace problem for quasinilpotent operators. *Integr. Equat. Oper. Th.*, **61**(1):103–120, 2008.

[132] H. J. Kim. Hyperinvariant subspaces for quasinilpotent operators on Hilbert spaces. *J. Math. Anal. Appl.*, **350**(1):262–270, 2009.

[133] S. V. Kisliakov. Quantitative aspect of correction theorems. *Zap. Nauchn. Sem. LOMI*, **92**:182–191, 1979.

[134] S. V. Kisliakov. A sharp correction theorem. *Studia Math.*, **113**(2):177–196, 1995.

[135] G. Koenigs. Recherches sur les intégrales de certaines équations fonctionnelles. *Ann. Sci. École Norm. Sup. (3)*, **1**:3–41, 1884.

[136] P. Koosis. *The Logarithmic Integral, I.* Cambridge University Press, Cambridge, 1988.

[137] M. G. Kreĭn and P. Ja. Nudel'man. Approximation of functions in $L_2(\omega_1, \omega_2)$ by transmission functions of linear systems with minimal energy. *Problemy Peredači Informacii*, **11**(2):37–60, 1975. English translation.

[138] H.-O. Kreiss. Über die Stabilitätsdefinition für Differenzengleichungen die partielle Differentialgleichungen approximieren. *Nordisk Tidskr. Informations-Behandling*, **2**:153–181, 1962.

[139] T. L. Lance and M. I. Stessin. Multiplication invariant subspaces of Hardy spaces. *Can. J. Math.*, **49**(1):100–118, 1997.

[140] J. Leblond and J. R. Partington. Constrained approximation and interpolation in Hilbert function spaces. *J. Math. Anal. Appl.*, **234**:500–513, 1999.

[141] A. Lebow. A power-bounded operator that is not polynomially bounded. *Michigan Math. J.*, **15**:397–399, 1968.

[142] J. Lindenstrauss and L. Tzafriri. *Classical Banach Spaces, I: Sequence Spaces.* Ergebnisse der Mathematik und ihrer Grenzgebiete, 92. Springer, Berlin, 1977.

[143] J. E. Littlewood. On inequalities in the theory of functions. *Proc. London Math. Soc. (2)*, **23**:481–519, 1925.

[144] M. X. Liu. On hyperinvariant subspaces of contraction operators on a Banach space whose spectrum contains the unit circle. *Acta Math. Sin. (Engl. Ser.)*, **24**(9):1471–1474, 2008.

[145] V. I. Lomonosov. Invariant subspaces for operators commuting with compact operators. *Funct. Anal. Appl.*, **7**:213–214, 1973.

[146] V. I. Lomonosov. On real invariant subspaces of bounded operators with compact imaginary part. *Proc. Amer. Math. Soc.*, **115**(3):775–777, 1992.

[147] G. W. MacDonald. Invariant subspaces for Bishop-type operators. *J. Funct. Anal.*, **91**(2):287–311, 1990.

[148] G. W. MacDonald. Decomposable weighted rotations on the unit circle. *J. Operator Theory*, **35**(2):205–221, 1996.

[149] J. Mashreghi and T. Ransford. Binomial sums and functions of exponential type. *Bull. London Math. Soc.*, **37**:15–24, 2005.

[150] V. Matache. On the minimal invariant subspaces of the hyperbolic composition operator. *Proc. Amer. Math. Soc.*, **119**(3):837–841, 1993.

[151] V. Matache. The eigenfunctions of a certain composition operator. In *Studies on Composition Operators (Laramie, WY, 1996)*, pages 121–136. Contemporary Mathematics, 213. American Mathematical Society, Providence, RI, 1998.

[152] A. J. Michaels. Hilden's simple proof of Lomonosov's invariant subspace theorem. *Adv. Math.*, **25**(1):56–58, 1977.

[153] R. Mortini. Cyclic subspaces and eigenvectors of the hyperbolic composition operator. In *Travaux Mathématiques, Fasc. VII*, pages 69–79. Séminaire de Mathématique de Luxembourg. Centre Universitaire du Luxembourg, Luxembourg, 1995.

[154] N. K. Nikolski. *Operators, Functions, and Systems: An Easy Reading, Volume 1*. Mathematical Surveys and Monographs, 92. American Mathematical Society, Providence, RI, 2002. Translated from the French by A. Hartmann.

[155] N. K. Nikolski. *Operators, Functions, and Systems: An Easy Reading, Volume 2*. Mathematical Surveys and Monographs, 93. American Mathematical Society, Providence, RI, 2002. Translated from the French by A. Hartmann and revised by the author.

[156] E. Nordgren, P. Rosenthal, and F. S. Wintrobe. Invertible composition operators on H^p. *J. Funct. Anal.*, **73**:324–344, 1987.

[157] E. A. Nordgren. Composition operators. *Can. J. Math.*, **20**:442–449, 1968.

[158] A. Octavio and M. Kosiek. Representations of $H^\infty(\mathbf{D}^N)$ and absolute continuity for N-tuples of contractions. *Houston J. Math.*, **23**(3):529–537, 1997.

[159] R. I. Ovsepian and A. Pełczyński. On the existence of a fundamental total and bounded biorthogonal sequence in every separable Banach space, and related constructions of uniformly bounded orthonormal systems in L^2. *Studia Math.*, **54**(2):149–159, 1975.

[160] J. R. Partington. *Interpolation, Identification and Sampling*. London Mathematical Society Monographs, 17. Clarendon Press, Oxford University Press, New York, 1997.

[161] J. R. Partington. *Linear Operators and Linear Systems*. London Mathematical Society Student Texts, 60. Cambridge University Press, Cambridge, 2004.

[162] J. R. Partington and E. Pozzi. Universal shifts and composition operators. *Operators and Matrices*, in press.

[163] V. I. Paulsen. *Completely Bounded Maps and Dilations*. Pitman Research Notes in Mathematics Series, 146. Longman Scientific & Technical, Harlow, 1986.

[164] A. Pełczyński. All separable Banach spaces admit for every $\varepsilon > 0$ fundamental total and bounded by $1 + \varepsilon$ biorthogonal sequences. *Studia Math.*, **55**(3):295–304, 1976.

[165] G. Pisier. A polynomially bounded operator on Hilbert space which is not similar to a contraction. *J. Amer. Math. Soc.*, **10**(2):351–369, 1997.

[166] G. Pólya and G. Szegő. *Problems and Theorems in Analysis, I*. Classics in Mathematics. Springer, Berlin, 1998. Reprint of 1978 English translation from the German by Dorothee Aeppli.

[167] H. Radjavi and P. Rosenthal. *Invariant Subspaces*. Springer, Berlin, 1973.

[168] C. Read. A solution to the invariant subspace problem. *Bull. London Math. Soc.*, **16**:337–401, 1984.

[169] C. Read. A solution to the invariant subspace problem on the space ℓ^1. *Bull. London Math. Soc.*, **17**:305–317, 1985.

[170] C. Read. A short proof concerning the invariant subspace problem. *J. London Math. Soc.*, **34**(2):335–348, 1986.

[171] C. Read. The invariant subspace problem for a class of Banach spaces, II: Hypercyclic operators. *Israel J. Math.*, **63**(1):1–40, 1988.

[172] C. Read. Strictly singular operators and the invariant subspace problem. *Studia Math.*, **132**(3):203–226, 1999.

[173] O. Rejasse. Factorization and reflexivity results for polynomially bounded operators. *J. Operator Theory*, **60**(2):219–238, 2008.

[174] W. C. Ridge. Approximate point spectrum of a weighted shift. *Trans. Amer. Math. Soc.*, **147**:349–356, 1970.

[175] F. Riesz and B. Sz.-Nagy. *Functional Analysis*. Frederick Ungar Publishing, New York, 1955. Translated by Leo F. Boron.

[176] G.-C. Rota. Note on the invariant subspaces of linear operators. *Rend. Circ. Mat. Palermo (2)*, **8**:182–184, 1959.

[177] G.-C. Rota. On models for linear operators. *Comm. Pure Appl. Math.*, **13**:469–472, 1960.

[178] W. Rudin. *Function Theory in the Unit Ball of* $\mathbb{C}^n$. Springer, Berlin, 1980.

[179] W. Rudin. *Real and complex analysis*, 3rd edition. McGraw-Hill, New York, 1987.

[180] W. Rudin. *Functional Analysis*, 2nd edition. International Series in Pure and Applied Mathematics. McGraw-Hill, New York, 1991.

[181] H. H. Schaefer. *Banach Lattices and Positive Operators*. Die Grundlehren der Mathematischen Wissenschaften, 215. Springer, New York, 1974.

[182] M. Schreiber. Unitary dilations of operators. *Duke Math. J.*, **23**:579–594, 1956.

[183] H. S. Shapiro and A. L. Shields. On some interpolation problems for analytic functions. *Amer. J. Math.*, **83**:513–532, 1961.

[184] J. H. Shapiro. *Composition Operators and Classical Function Theory*. Universitext: Tracts in Mathematics. Springer, New York, 1993.

[185] A. Simonič. An extension of Lomonosov's techniques to non-compact operators. *Trans. Amer. Math. Soc.*, **348**(3):975–995, 1996.

[186] D. Slepian. On bandwidth. *Proc. IEEE*, **64**(3):292–300, 1976.

[187] M. Smith. Constrained approximation in Banach spaces. *Constr. Approx.*, **19**(3): 465–476, 2003.

[188] M. N. Spijker, S. Tracogna, and B. D. Welfert. About the sharpness of the stability estimates in the Kreiss matrix theorem. *Math. Comp.*, **72**(242):697–713, 2003.

[189] B. Sz.-Nagy. Sur les contractions de l'espace de Hilbert. *Acta Sci. Math. (Szeged)*, **15**:87–92, 1953.

[190] B. Sz.-Nagy and C. Foiaş. On certain classes of power-bounded operators in Hilbert space. *Acta Sci. Math. (Szeged)*, **27**:17–25, 1966.

[191] B. Sz.-Nagy and C. Foiaş. *Harmonic Analysis of Operators on Hilbert Space*. North-Holland, Amsterdam, 1970. Translated from the French and revised.

[192] F. H. Szafraniec. Some spectral properties of operator-valued representations of function algebras. *Ann. Polon. Math.*, **25**:187–194. 1971/72.

[193] J. E. Thomson. Invariant subspaces for algebras of subnormal operators. *Proc. Amer. Math. Soc.*, **96**(3):462–464, 1986.

[194] T. T. Trent. Maximal invariant subspaces for $A_\alpha^2(D)$. *Proc. Amer. Math. Soc.*, **132**(8):2429–2432, 2004.

[195] V. G. Troitsky. Lomonosov's theorem cannot be extended to chains of four operators. *Proc. Amer. Math. Soc.*, **128**(2):521–525, 2000.

[196] V. G. Troitsky. Minimal vectors in arbitrary Banach spaces. *Proc. Amer. Math. Soc.*, **132**(4):1177–1180, 2004.

[197] G. Valiron. Sur l'iteration des fonctions holomorphes dans un demi-plan. *Bull. Sci. Math. (2)*, **55**:105–128, 1931.

[198] F. H. Vasilescu. An operator-valued Poisson kernel. *J. Funct. Anal.*, **110**(1):47–72, 1992.

[199] F. H. Vasilescu. Operator-valued Poisson kernels and standard models in several variables. In R. E. Curto and P.E.T. Jørgensen, editors, *Algebraic Methods in Operator Theory*, pages 37–46. Birkhäuser, Boston, MA, 1994.

[200] J. L. Walsh. *The Location of Critical Points of Analytic and Harmonic Functions.* American Mathematical Society Colloquium Publications, 34. American Mathematical Society, New York, NY, 1950.

[201] J. Wermer. The existence of invariant subspaces. *Duke Math. J.*, **19**:615–622, 1952.

[202] D. J. Westwood. On C_{00}-contractions with dominating spectrum. *J. Funct. Anal.*, **66**(1):96–104, 1986.

[203] K. Yan. Invariant subspaces for joint subnormal operators. *Chinese Ann. Math. A.*, **9**:561–566, 1988.

[204] A. C. Zaanen. *Introduction to Operator Theory in Riesz Spaces.* Springer, Berlin, 1997.

[205] M. Zarrabi. On polynomially bounded operators acting on a Banach space. *J. Funct. Anal.*, **225**(1):147–166, 2005.

[206] C. Zenger. On convexity properties of the Bauer field of values of a matrix. *Num. Math.*, **12**:96–105, 1968.

Index

For EU product safety concerns, contact us at Calle de José Abascal, 56–1°, 28003 Madrid, Spain or eugpsr@cambridge.org.

www.ingramcontent.com/pod-product-compliance
Lightning Source LLC
LaVergne TN
LVHW102008210926
848483LV00010B/167

* 9 7 8 1 1 0 7 0 1 0 5 1 2 *